I0484564

science for a changing world

Prepared in Cooperation with the Department of Geosciences at the University of Arkansas

# U.S. Geological Survey Karst Interest Group Proceedings, Fayetteville, Arkansas, April 26–29, 2011

Edited By Eve L. Kuniansky

Scientific Investigations Report 2011–5031

U.S. Department of the Interior
U.S. Geological Survey

i

**U.S. Department of the Interior**
KEN SALAZAR, Secretary

**U.S. Geological Survey**
Marcia K. McNutt, Director

U.S. Geological Survey, Reston, Virginia 2011

For product and ordering information:
World Wide Web: http://www.usgs.gov/pubprod
Telephone: 1-888-ASK-USGS

For more information on the USGS—the Federal source for science about the Earth,
its natural and living resources, natural hazards, and the environment:
World Wide Web: http://www.usgs.gov
Telephone: 1-888-ASK-USGS

Suggested citation:
Kuniansky,E.L., 2011, U.S. Geological Survey Karst Interest Group Proceedings, Fayetteville, Arkansas, April 26-29,
2011, U.S. Geological Survey Scientific Investigations Report 2011-5031, 212p.
Online copies of the proceedings area available at:
http://water.usgs.gov/ogw/karst/

# Contents

## INTRODUCTION AND ACKNOWLEDGMENTS

Karst aquifer systems are present throughout parts of the United States and some of its territories and are developed in carbonate rocks (primarily limestone and dolomite) that span the entire geologic time frame. The depositional environments, diagenetic processes, and post-depositional tectonic events that form carbonate rock aquifers are varied and complex, involving both biological and physical processes that can influence the development of permeability. These factors, combined with the diverse climatic regimes under which karst development in these rocks has taken place result in the unique dual or triple porosity nature of karst aquifers. These complex hydrologic systems often present challenges to scientists attempting to study groundwater flow and contaminant transport.

The dissolution of carbonate rocks and the subsequent development of distinct and beautiful landscapes, caverns, and springs have resulted in some karst areas of the United States being designated as National or State parks and commercial caverns. Karst aquifers and landscapes that form in tropical areas, such as the north coast of Puerto Rico, differ greatly from karst areas in more arid climates, such as west-central Texas or western South Dakota. Many of these public and private lands also contain unique flora and fauna associated with the hydrologic systems in these karst areas. As a result, multiple Federal, State, and local agencies have an interest in the study of karst terrains.

Many of the major springs and aquifers in the United States are developed in carbonate rocks and karst areas, such as the Floridan aquifer, the Ozark Plateaus in Missouri and Arkansas, and the Edwards aquifer in west-central Texas. These aquifers and the springs that discharge from them, serve as major water-supply sources and as unique biological habitats. Commonly, there is competition for the water resources of karst aquifers, and urban development in karst areas can impact the ecosystem and water quality of these aquifers.

The concept for developing a Karst Interest Group evolved from the November 1999 National Groundwater Meeting of the U.S. Geological Survey (USGS), Water Resources Division. As a result, the Karst Interest Group was formed in 2000. The Karst Interest Group is a loose-knit grass-roots organization of USGS employees devoted to fostering better communication among scientists working on, or interested in, karst hydrology studies.

The mission of the Karst Interest Group is to encourage and support interdisciplinary collaboration and technology transfer among USGS scientists working in karst areas. Additionally, the Karst Interest Group encourages cooperative studies between the different disciplines of the USGS and other Federal agencies, and university researchers or research institutes.

The first Karst Interest Group workshop was held in St. Petersburg, Florida, February 13-16, 2001, in the vicinity of the large springs and other karst features of the Floridan aquifer system. The proceedings of that first meeting, published in Water-Resources Investigations Report 01-4011, are available online at:

http://water.usgs.gov/ogw/karst/kig/

The second Karst Interest Group workshop was held August 20-22, 2002, in Shepherdstown, West Virginia, in proximity to the carbonate aquifers of the northern Shenandoah Valley. The proceedings of the second workshop were published in Water-Resources Investigations Report 02-4174, which is available online at the previously mentioned website.

The third workshop of the Karst Interest Group was held September 12-15, 2005, in Rapid City, South Dakota, which is in proximity to karst features in the Madison Limestone in the semi-arid Black Hills of South Dakota and Wyoming, including Wind Cave National Park and Jewell Cave National Monument. The proceedings of the third workshop were published in Scientific Investigations Report 2005-5160, which is available online at the previously mentioned website.

The fourth workshop was held at the Hoffman Environmental Research Center and Center for Cave and Karst Studies at Western Kentucky University in Bowling Green, Kentucky, near Mammoth Cave National Park and karst features of the Chester Upland and Pennyroyal Plateau. The proceedings of the fourth workshop were published in Scientific Investigations Report 2008-5023, also available online.

This fifth workshop is a joint workshop of the USGS Karst Interest Group and University of Arkansas HydroDays workshop, sponsored by the USGS, the Department of Geosciences at the University of Arkansas in Fayetteville. Additional sponsors are: the National Cave and Karst Research Institute, the Edwards Aquifer Authority, San Antonio, Texas, and Beaver Water District, northwest Arkansas. The majority of funding for the proceedings preparation and workshop was provided by the USGS Groundwater Resources Program, National Cooperative Mapping Program, and the Regional Executives of the Northeast, Southeast, Midwest, South Central and Rocky Mountain Areas. The University of Arkansas provided the rooms and facilities for the technical and poster presentations of the workshop, vans for the field trips, and sponsored the HydroDays banquet at the Savoy Experimental Watershed on Wednesday after the technical sessions.

The session planning committee for this fifth workshop included Van Brahana, University of Arkansas, and Tom Byl, Allan K. Clark, Daniel Doctor, Brian Katz, James Kaufmann, Eve Kuniansky, Kurt McCoy, Larry Spangler, Chuck Taylor, and Patrick Tucci (retired) of the USGS. The field trip committee included Van Brahana, University of Arkansas, Chuck Bitting, National Park Service, and Mark R. Hudson, Phil Hays, and Patrick Tucci (retired) of the USGS. We sincerely hope that this workshop promotes future collaboration among scientists of varied backgrounds and improves of our understanding of karst systems in the United States and its territories.

The extended abstracts of USGS authors were reviewed and approved for publication by the U.S. Geological Survey. Articles submitted by university researchers and other Federal agencies did not go through the USGS review process, and therefore, may not adhere to our editorial standards or stratigraphic nomenclature. All articles were edited for consistency of appearance in the published proceedings. The use of trade names in any article does not constitute endorsement by the U.S. Government. The USGS, Office of Ground Water, provides financial support for the Karst Interest Group website and public availability of the proceedings from these workshops.

The cover illustration was designed by Ann Tihansky, USGS, St. Petersburg, Florida, for the first Karst Interest Group workshop.


Eve L. Kuniansky

Karst Interest Group Coordinator

# AGENDA U.S. GEOLOGICAL SURVEY KARST INTEREST GROUP AND UNIVERSITY OF ARKANSAS HYDRO DAYS WORKSHOP

## April 26–29, 2011, Fayetteville, Arkansas, University of Arkansas Campus
## TUESDAY, APRIL 26

*Registration*

       Start at 7:45 am-- All day – pick up name tags and proceedings

*Welcome and Introductions*

8:00 – 8:20        Eve Kuniansky, U.S. Geological Survey, Karst Interest Group Coordinator and Dr. Van Brahana, Department of Geosciences, University of Arkansas

*Keynote/Speleogenesis  (Eve Kuniansky moderator)*

8:20 – 9:00        Paleokarst of the USA: A brief overview by Arthur N. Palmer and Margaret V. Palmer

*Karst in Coastal Zones  (Eve Kuniansky moderator)*

9:00 – 9:20        Influence of tides and salinity on the discharge at the coastal Spring Creek Springs group and the connection to discharge at the inland by J. Hal Davis and Richard J.Verdi

9:20 – 9:40        Revised hydrogeologic framework for the Floridan Aquifer System in the northern coastal areas of Georgia and parts of South Carolina by Harold E. Gill and Lester J. Williams

9:40 – 10:00        Horizontal bedding-plane conduit systems in the Floridan Aquifer System and their relation to saltwater intrusion in northeastern Florida and southeastern Georgia by Lester J. Williams and Rick M. Spechler

10:00 – 10:40        **BREAK**

*Geochemisty and Contaminant Transport of Karst Systems (Zelda Bailey, moderator)*

10:40 – 11:00        Water-quality changes and dual response in a karst aquifer to the May 1-2, 2010 flood in middle, Tennessee by Michael W. Bradley and Thomas D. Byl

11:00 – 11:20        Interaction between shallow and deep ground water components in the northern Shenandoah Valley karst by Daniel H. Doctor, Nathan C. Farrar, and  Janet S. Herman

11:20 – 11:40        Aqueous geochemical evidence of volcanogenic karstification: Sistema Zacatón, Mexico by Marcus Gary, Daniel H. Doctor, and John M. Sharp, Jr.

11:40 – 1:00        **LUNCH ON YOUR OWN**

*Geochemisty and Contaminant Transport of Karst Systems continued (Tom Byl, moderator)*

1:00 – 1:20      The influence of land use and occurrence of sinkholes on nitrogen transport in the Ozark Plateaus in Arkansas and Missouri by Timothy M. Kresse, Phillip D. Hays, Mark R. Hudson, and James E. Kaufman

1:20 – 1:40      Seasonal carbon dynamics in a northwestern Arkansas cave: linking climate, nutrients and cave conditions by E. D. Pollock, K.J. Knierim, and P.D. Hays

1:40 – 2:00      Characterization of Urban Impact on Water Quality of Karst Springs in Eureka Springs, Arkansas by Renee Vardy

2:00 – 2:40      **BREAK**

*Geochemisty and Contaminant Transport of Karst Systems continued (Brian G. Katz, moderator)*
2:40 – 3:00      Factors affecting dissolved oxygen concentrations in Barton Springs, Austin, Texas by Barbara Mahler and Renan Bourgeais

3:00 – 3:20      Geochemistry, Water Sources, and Pathways in the Zone of Contribution of a Public-Supply Well in San Antonio, Texas by Lynne Fahlquist, MaryLynn Musgrove, Gregory P. Stanton, and Natalie A. Houston

3:20 – 3:40      Temporal stability of cave sediments by Eric W. Peterson and Kevin Hughes

*Karst Mapping and Geographic Information Systems (Brian G. Katz, moderator)*
3:40 – 4:00      Using a combination of geographic information system techniques and field methods to analyze karst terrain in selected Red River sub-watersheds, Tennessee and Kentucky by David E. Ladd

4:00 – 6:00      **Poster Session at University of Arkansas**

# WEDNESDAY, APRIL 27

*Karst Program Updates (Kurt J. McCoy, moderator)*
8:00 – 8:20      National Cave and Karst Research Institute: Growing capabilities and Federal partnerships by George Veni

8:20 – 8:40      The National karst map: an update on its progress by David J. Weary and Daniel H. Doctor

*Ozarks Plateau Karst  (Kurt J. McCoy, moderator)*
8:40 – 9:00      Karst Hydrogeology of the Ozarks by Van Brahana

9:00 – 9:20      Overview of The Nature Conservancy Ozark Karst Program by Michael E. Slay, Ethan Inlander, and Cory Gallipeau

9:20 – 9:40      Microbial Effects on Ozarks Karst Chemistry -- Medicinal Implications  by John E. Svendsen

9:40 – 10:20     **BREAK**

*Karst Aquifer Systems continued (Allan K. Clark, moderator)*
10:20 – 10:40     An integrated approach to recharge area delineation in northern Arkansas and northeastern Oklahoma by Jonathan A. Gillip, Rheannon M. Hart, Joel M. Galloway

10:40 – 11:00     Groundwater Piracy in Semi-Arid Karst Terrains by Ronald T. Green, F. Paul Bertetti, and Mariano Hernandez

| 11:00 – 11:20 | Sequential spring hydrology, karst valleys, and transition conduits draining Kirby watershed in south-central Indiana by Garre Connor |

*Karst Mapping and Geographic Information Systems (Allan K. Clark, moderator)*

| 11:20 – 11:40 | Identification and classification of karst using a pseudospectral method by James E. Kaufmann |

| 11:40 – 1:00 | **LUNCH ON YOUR OWN** |

*Karst Modeling (Daniel H. Doctor, moderator)*

| 1:00 – 1:20 | Modifications to the Conduit Flow Process Mode2 for MODFLOW-2005 by Thomas Reimann, Steffen Birk, Christoph Rehrl, and W. Barclay Shoemaker |

| 1:20 – 1:40 | Comparison of three model approaches for spring simulation, Woodville Karst Plain, Florida by Eve L. Kuniansky, Josue J. Gallegos, and J. Hal Davis |

| 1:40 – 2:00 | Synthesis of multiple scale modeling in the faulted and folded karst of the Shenandoah Valley, Virginia and West Virginia by Kurt J. McCoy, Mark D. Kozar, Richard M. Yager, George E. Harlow, and David L. Nelms |

| 2:00 – 2:20 | A Hydrograph Recession Technique for Karst Springs with Quickflow Components That Do Not Exhibit Simple, Zero-Order Decay by Darrell Pennington and Van Brahana |

| 2:20 – 3:00 | **BREAK** |

*Geophysical Methods in Karst (Pat Tucci, moderator)*

| 3:00 – 3:20 | New tools for detecting vertical fractures and voids in karst by Jack H. Cole and Phillip West |

| 3:20 – 3:40 | Mapping epikarst using helicopter electromagnetic methods by David V. Smith, Bruce D. Smith, Maryla Deszcz-Pan, and Charles D. Blome |

*Karst Hydraulics (Pat Tucci, moderator)*

| 3:40 – 4:00 | A simple but effective model predicting the effect of a karst conduit on an adjacent observation well by Fred Paillet |

| 4:00 – 4:20 | KIG Business-planning next meeting Spring 2014 |

| 4:20 -6:00 | **Busses take those that want to attend the HydroDays Banquet at the Savoy Experimental Watershed (SEW)** |
| 6:00 - ? | HydroDays Banquet of roast razorback and roast goat, in addition to many other tasty treats to satiate the full range of palates for which karstophiles are noted. We have arranged presentations by recent researchers at SEW. |

# THURSDAY, APRIL 28

| 8:00AM – 5:00PM | Field Trip Day 1—Geology and Karst Landscapes of the Buffalo National River, Northern Arkansas lead by Mark Hudson and Kenzie Turner, U.S. Geological Survey; and Chuck Bitting, National Park Service |

# FRIDAY, APRIL 29

| 8:00AM –5:00PM | Field Trip Day 2—Dominant Hydrogeologic Controls of Karst on the Buffalo National River—Float Trip |

## POSTERS

*Karst Mapping, Dye Tracing, and Geographic Information Systems*

Parallelism in karst development suggested by quantitative dye tracing results from Springfield, Missouri by Douglas R. Gouzie and Kati Tomlin

Geologic controls on karst landscapes in the Buffalo National River area of northern Arkansas: Insights gained from comparison of geologic mapping, topography, dye tracers and karst inventories by Mark R. Hudson, Kenzie J. Turner, Chuck Bitting, James E. Kaufmann, Timothy M. Kresse, and David N. Mott

Sinkholes identified from LiDAR in the Mill Creek area of the Buffalo National River, Arkansas by J. E. Kaufmann and A. T. Lingelbach

Investigation of the relation between surface-water losses in the Beaver Creek drainage and springs at the Kamas State Fish Hatchery, and implications for transmission of whirling by Larry Spangler

Survey of Springs Issuing from the Trinity Aquifer in the Vicinity of Northern Bexar County, Texas by Allan K. Clark, Robert R. Morris and Travis J. Garcia

Using GIS to identify cave levels and discern the speleogenesis of the Carter Caves karst area, Kentucky by Eric W. Peterson, Toby Dogwiler, and Lara Harlan

Surface Denudation of the Gypsum Plain, West Texas and Southeastern New Mexico by M. G. Shaw, K. W. Stafford, and B.P. Tate

*Geochemisty and Contaminant Transport of Karst Systems*

Evaluating the Stormwater Filters at Mammoth Cave, Kentucky by Ashley West, Rickard Toomey, Mike Bradley, and Tom Byl

Using labeled isotopes to trace groundwater flow paths in a northwestern Arkansas cave by K. J. Knierim, E. D. Pollock, and P. D.Hays

Interpreting a spring chemograph to characterize groundwater recharge in an urban karst terrain by Victor Roland, Carlton Cobb, Lonnie Sharpe, Patrice Armstrong, Dafeng Hui, and Tom Byl

Investigation of the fate of nitrate in the interflow zone of mantled karst by Jozef Laincz

Alternative approaches to dissolved organic matter characterization in karst aquifers by Terri Brown

*Karst Aquifer Systems*

Ten Relevant Karst Hydrogeologic Insights Gained from 15 Years of In Situ Field Studies at the Savoy Experimental Watershed by Van Brahana

Traps Designed to Document the Occurrence of Groundwater Stygofauna at the Savoy Experimental Watershed, Washington County, Arkansas by Justin Mitchell

Assessment of Sinkhole Formation in a Well Field in the Dougherty Plain, near Albany, Georgia by Debbie Warner Gordon

Detecting karst conduits through their effects on nearby monitoring wells by Fred Paillet and Terryl Daniels

Spring Hydrology of Colorado Bend State Park, Central Texas by K.W. Stafford, M.G. Shaw, and J.L. DeLeon

Analysis of long-term trends in flow from a large spring complex in northern Florida by J.W. Grubbs

*Karst Modeling*

Use of MODFLOW-CFP and MT3DMS to simulate karst flow in a laboratory karst analog model and the Woodville Karst Plain by Josue Gallegos, Bill X. Hu, Hal Davis, Eve Kuniansky, and Barclay Shoemaker

Numerical Evaluations of Alternative Spring Discharge Conditions for Barton Springs, Texas, USA by W.R. Hutchison and M.E. Hill

# SPELEOGENESIS
# Paleokarst of the USA: A Brief Review

By Arthur N. Palmer[1] and Margaret V. Palmer[2]
[1]Department of Earth Sciences, State University of New York, Oneonta, NY 13820-4015
[2]619 Winney Hill Road, Oneonta, NY 13820

## Abstract

Paleokarst consists of solutional features from a prior geomorphic phase that have been preserved by burial or by a substantial change in local environment. The two major paleokarst horizons in North America are of Early-Middle Ordovician (post-Sauk) age and Mississippian-Pennsylvanian (post-Kaskaskia) age. There are also several less extensive paleokarst zones. They all differ in detail but typically include remnants of surface karst features, caves, breccias, hypogenic porosity, and related mineral suites. In places they provide high-permeability zones significant to water supply, or serve as hosts to petroleum, ores, and later karst development. Studies of paleokarst give significant evidence for past geologic and hydrologic conditions, both surficial and deep seated.

## INTRODUCTION

Paleokarst is an important but often overlooked aspect of geology (Fig. 1). It is typically formed by lengthy exposure of soluble rock to karst processes, followed by burial beneath younger rocks or sediment, which preserve the karst features more or less intact. Karst features that are left as relics by changes in climate or hydrologic regime can also be considered paleokarst, although this use of the term is not so common. Paleokarst provides clues to former geologic and hydrologic conditions, as well as insight into global tectonics and changes in climate and sea level.

The major American paleokarst horizons are briefly described here, with attention to their physical characteristics and economic importance. Also described is the impact of paleokarst on later caves and karst, as well as the problems of recognizing and mapping paleokarst. For a broader perspective see books by James and Choquette (1988), which mainly concerns the USA, and the worldwide perspective by Bozák (1989), which includes chapters on the USA (M. Palmer and A. Palmer) and Canada (Ford).

## POST-SAUK PALEOKARST

The Sauk cratonic sequence (latest Precambrian – Early Ordovician) terminates upward mainly in carbonate formations that extend over most of the USA (Sloss, 1963). These strata were exposed to erosion because of relatively low sea level for up to tens of millions of years, during which an extensive karst surface developed. This surface was buried by Middle Ordovician strata, which also include many carbonates. The resulting paleokarst is extensive but poorly exposed at the surface. In the areas of greatest exposure, the surface is disturbed either by later tectonism (e.g., in the Appalachians) or obscured by weathering residuum (e.g., around the Ozark Dome). Most information on the unconformity is from mines and drillholes.

Figure 1. Example of paleokarst: Mississippian sinkhole still partly filled with in-place Pennsylvanian sediment, Bighorn Canyon in Wyoming. Cliff face is about 20 m high.

In the eastern USA, this stratigraphic interval was greatly deformed by all phases of the

Appalachian Orogeny, so it is often difficult to distinguish paleokarst from mechanical disruption of beds. Mussman et al. (1988) describe the paleo-karst in the upper Knox–Beekmantown Group of the southern Appalachians. They report erosional relief of more than 100 m, with filled sinkholes and caves extending to more than 65 m below the unconformity and intrastratal breccias down to 300 m. Paleosinkholes are filled with carbonate breccias and gravels with a fine detrital matrix. Caves are filled with breccia and laminated dolomite. Stratiform breccias containing dolomite clasts are present in intervals up to 35 m thick. Paleocaves appear to have formed by rapid conduit flow, but possibly also by dissolution in mixing zones (evidence is sparse). Further brecciation took place during deep burial, as did dissolution and dolomitization, and sulfide ore emplacement by migration of late Paleozoic warm saline basinal brines. This paleokarst is also well developed in parts of eastern Canada, with a relief up to 10 m that includes buried sinkholes and karren, as well as mineralized vuggy and cavernous porosity (Ford, 1989; Dix et al., 1998).

The most notable aspect of the post-Sauk paleokarst is its role in hosting lead-zinc ores (e.g., Tennessee, Missouri) and oil and gas fields (e.g., west Texas). It also provides high-permeability zones in carbonate aquifers, especially along the Cincinnati Arch in Tennessee and Kentucky. These are located in the top 60 m of the Knox Group, at and beneath the unconformity, along bedding-guided solution zones and fractured, vuggy dolomites and breccias. Where shallow, these zones provide much groundwater, and where deep they have been targets for injection wells. Artesian conditions are present in some areas.

The most extensive views of the post-Sauk paleokarst are in lead-zinc mines of Tennessee and Missouri (see Ohle, 1985; Sangster, 1988). But it is curious that a buried karst surface is not particularly clear. Most of the mines are located well below the unconformity, and their mineralized zones may be related to older paleokarst. Some breccias, mineralized or not, extend upward to the unconformity, and even above it. Are these breccias the indirect result of paleokarst, where subsidence propagated upward

from solution voids well after the overlying beds were deposited?

Breccias in the Tennessee mining districts form tabular bodies in a zig-zag pattern with irregular tops and bottoms, and their stratigraphic relationships are obscured by folding and faulting. The older breccia occupies the upper 150 m of the Knox Group beneath the unconformity. The younger breccia is of unknown age and in places extends upward across the unconformity. It formed around the earlier breccia and was apparently guided by it. The ores are present only in the younger breccia (Fig. 2). These features are not directly related to the paleokarst but were probably influenced by tectonic and hypogenetic processes. The lead-zinc ore fills spaces between angular carbonate clasts, and evidence for bedrock replacement is minimal, except in high-grade deposits. Regarding the origin of the breccias, Sangster (1988) notes their similarity to known paleokarst solution collapse breccias and the paucity of wall-rock alteration, both of which may relate to meteoric karst. (This viewpoint is controversial.)

Figure 2. Zinc ore (yellow sphalerite, medium gray in photo) from a mine in Ordovician carbonate breccia associated with the post-Sauk paleokarst, Jefferson City, Tennessee. White = saddle dolomite (slightly thermal). Pen for scale.

A great deal of information on the post-Sauk paleokarst is available from oil and gas wells. The Ellenburger Formation of western Texas is a typical example (Kerans, 1993). Most of the reservoir rock is brecciated and dolomitized (Fig. 3). There is considerable debate about the role of paleokarst in providing the reservoir porosity. Some attribute it almost entirely to paleokarst (e.g., Loucks and Handford, 1992). Others invoke deep fracture zones related to basement-rooted trans-tensional faults (e.g., Smith, 2009). According to the latter view, the breccias formed in spaces created by the faulting, and dissolution was performed mainly by hydrothermal fluids rising along faults. It is likely that both views are justified to varied degrees, depending on the local geomorphic setting and structural history.

Figure 3. Dolomite breccia below the post-Sauk unconformity, Ellenburger Formation. The white rinds are saddle dolomite (thermal), and the matrix is recrystallized internal dolomite sediment. This exposure is in the wall of a quarry in central Texas. Pen for scale.

## POST-KASKASKIA PALEOKARST

The post-Kaskaskia paleokarst is much more clearly defined than the post-Sauk and merits a more detailed description. The Kaskaskia sedimentary sequence extends from Middle Devonian to the end of the Mississippian (Sloss, 1963). Carbonates are abundant, especially near the top. These strata are host to the most extensive caves of the USA, as well as the best-exposed paleokarst (Figs. 1 and 4). Some Cenozoic caves, such as Wind and Jewel Caves in South Dakota, have inherited much of their pattern from these earlier features.

The widespread post-Kaskaskia erosion surface was well preserved by deltaic and transgressive Pennsylvanian deposits. Because of the onlappng pattern of strata, the exact time of burial varied from place to place. Valleys and karst depressions were filled, as were most of the late Mississippian caves. Paleokarst fill is most commonly sand and clay colored deep red by hematite.

Karst preservation was more complete than in the post-Sauk paleokarst, and exhumed relics of Mississippian sinkholes, fissures, and caves are well exposed in many areas. Erosional relief on the Mississippian strata ranged up to about 100 m, allowing extensive karst to develop in carbonate rocks. The paleokarst is best exposed in areas of Laramide uplift and diminishes in intensity eastward, where the Mississippian carbonates are overlain by thick detrital strata.

The easternmost extent of well-developed post-Kaskaskia paleokarst appears to be in northern Illinois, where the erosion surface extended down to Ordovician and Silurian carbonates (Plotnick et al., 2009). Clearly developed tubular caves up to 10 m high and 18 m wide were filled with clastic sediment, which fossil pollen show to be Middle Pennsylvanian.

Paleokarst is especially well developed in the Northern and Middle Rocky Mountains, where uplift near the end of the Mississippian was accelerated by the Antler Orogeny farther west. In this region, local relief on the paleokarst is typically about 20–40 m. In places its relief is sharp, with deep depressions and underlying caves. Elsewhere it is subdued and undulatory. In places there is evidence for buried surface stream valleys (Sando, 1988).

Figure 4. Post-Kaskaskia paleocaves exposed in the Bighorn Canyon, Wyoming. These and related caves are concentrated about 15 m below the top of the Madison Limestone (Mississippian) and are filled with poorly lithified Pennsylvanian sediments. Also see Figure 1.

Widespread breccias are also common in the early Meramecian carbonates of this area. The breccias are laterally correlative with evaporite deposits in structural basins (e.g., Williston Basin) and appear to represent collapse caused by intrastratal solution in exposed areas (Roberts, 1966; Sando, 1988). They consist mainly of angular, unstratified, and unsorted carbonate and chert fragments in a red silt-clay matrix. Breccias are present only in domal areas, so evaporite dissolution must have taken place during and/or after uplift. The post-Kaskaskia erosion surface truncates most of the breccias, so it is likely that these areas were relatively high long before Laramide uplift. Farther down the flanks of uplifts, strata were exposed to abundant circulation of meteoric groundwater only after large-scale Laramide tectonism.

Lead-zinc-silver mineralization has taken place along the post-Kaskaskia paleokarst in central Colorado, and paleovalleys and caves have also been mapped (Tschauder and Landis, 1985; DeVoto, 1988), as well as buried tower karst (Maslyn, 1976). Correlative breccias, breccia pipes, and air-filled pockets lined with calcite crystals are abundant in the Mississippian limestones of the Grand Canyon, Arizona (Troutman, 2004).

## OBSERVATIONS IN THE BLACK HILLS

The various aspects of the post-Kaskaskia paleokarst are most clearly displayed in caves of the Black Hills of South Dakota. On the basis of field mapping and petrographic study of Wind Cave and Jewel Cave (Palmer and Palmer, 1995, 2008), distinct stages of paleokarst and related cave development are described here in chronological sequence. Similar paleokarst exposures are also common in outcrops and caves in the Northern Rockies, although local details differ.

### 1. Early Diagenetic Karst Processes

The earliest phase of the post-Kaskaskia paleokarst in the future Black Hills involved the interaction between sulfates and carbonates in the Mississippian Madison (locally Pahasapa) Formation. Plastic flow, hydration of anhydrite, and dissolution of sulfates formed carbonate breccias in stratiform bodies and local discordant dikes, mainly in the middle and upper Madison (Fig. 5). These zones correlate with evaporites in the neighboring Williston Basin.

Many clasts were wedged apart and lifted against gravity by crystal growth. In places, fracturing was less displacive and produced mosaic breccias with calcite veins.

Meanwhile, sulfate reduction fostered the growth of pyrite and associated dolomite crystals in the walls of fractures and pores. The source of organic compounds as a reducing agent is uncertain, although migration from nearby petroleum deposits, deep-seated methane, and organics in connate water are all possible. Isotopic evidence has been disrupted by later mineral replacement. Saddle dolomite, an indicator of thermal conditions, is absent.

Figure 5. Early diagenetic breccia in Madison Limestone in Jewel Cave, South Dakota. Bedrock clasts are surrounded by a matrix of yellow-brown calcite coated with a thin layer of deep-burial scalenohedral calcite. Width of photo is approximately 25 cm.

The effect of the Antler Orogeny in the Black Hills area was limited to minor uplift and block faulting, which probably involved reactivation of Precambrian structures (Sando, 1988). During the middle and late Mississippian an increasing influx of meteoric water took place as the Madison Formation was gradually exposed at the surface.

Oxidation of pyrite (and possibly of hydrogen sulfide from sulfate reduction) produced sulfuric acid that attacked the carbonate bedrock and breccia clasts, producing porosity and secondary gypsum. Textures similar to those in active sulfuric acid caves are still present in places, and microscopic remnants of gypsum and anhydrite crystals and pseudomorphs are visible in thin sections. With the increasing influx of meteoric water and a rise in water table, the gypsum was gradually replaced by calcite through the common-ion effect. Iron oxides and fossil filaments of iron-oxidizing bacteria are abundant in the calcite and impart a distinct yellow-brown color to the calcite. Some of the oxides are pseudomorphic after pyrite. The calcite is rather uniform in texture, with poorly defined and irregular growth patterns, which is compatible with sulfate replacement, rather than deposition on the walls of water-filled voids. It also contains floating bedrock residue and sparse anhydrite inclusions. U/Pb dating of this calcite confirms a Mississippian-Pennsylvanian age (Palmer et al., 2009).

Dissolution of the carbonate bedrock by sulfuric acid produced voids up to ~2 m in diameter lined by calcite-cemented breccia, and with large brown calcite scalenohedra projecting into the openings. The openings were probably not extensive enough to be considered caves. Most have been almost entirely assimilated by later dissolution, but they can still be seen more or less intact where bedrock collapse has exposed them in the present caves.

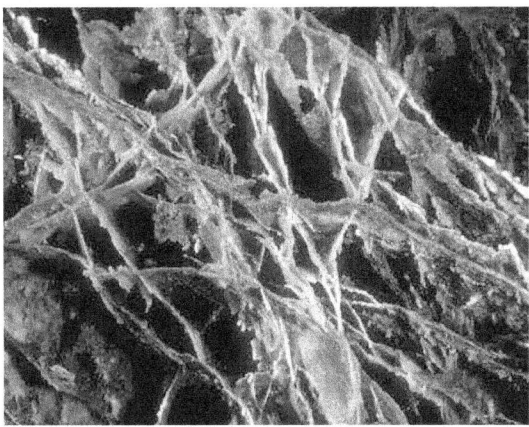

Figure 6. Boxwork in Wind Cave, South Dakota, is composed of veins of the yellow-brown calcite identical to the breccia matrix shown in Figure 5. The veins shown here project about 20 cm from the cave wall.

Closely spaced calcite veins developed in many of the dolomitic strata in the middle section of the Madison and project from the present cave walls as boxwork, especially in Wind Cave (Fig. 6). These veins show considerable scatter in orientation, even in the same location. The small scale, high spatial density, and diversity of fracture patterns required multi-directional stresses that were local, rather than regional, and hydration and plastic flow of sulfates was the likely cause. At first the fractures were filled with gypsum. The dolomitic bedrock was later altered by sulfuric acid, which had virtually no effect on the gypsum veins. Still later, the gypsum veins were replaced by calcite as the influx of meteoric water increased. The boxwork seen in the caves today has been accentuated by later removal of the bedrock and reinforcement of the calcite veins by accretion of calcite, aragonite, and internal carbonate sediment. This type of boxwork is rare elsewhere in the world. It also accompanies certain sulfide ores, which might suggest a hydrothermal origin; but the essential factor seems to be sulfur redox processes, rather than high temperature.

## 2. Late Mississippian Karst and Caves

Meteoric caves formed in local zones at an average of 15–30 m below the highs in the karst surface. Most were limited to the upper massive limestone of the Madison, above a prominent zone of bedded chert. A few solutional fissures extended to greater depths, but most were enlarged during the Cenozoic and do not retain their original form. Exposures in canyon walls and in the present caves show that the Mississippian caves were mainly low arched rooms or galleries up to roughly 8 m wide and high (Fig. 4). They are densest where the Madison contains overlying sinkholes and fissures, but a clear relation between the surface and subsurface features is difficult to distinguish in outcrops. The caves appear to have survived deep burial and Laramide exhumation with little modification.

Most of the depressions and fissures were formed by water infiltrating from small local catchments, with some of their growth aided by collapse into earlier voids. Sulfate-related breccias surround and underlie many of the depressions, and it is apparent that either dissolution of sulfates contributed to the collapse, or the depressions were guided by existing breccias. A combination of the two scenarios is likely.

The paleocaves resemble those that presently form in zones of mixing between fresh water and saline water. This process is common in coastal caves. Walls of paleokarst depressions show no evidence of flowing water (e.g., vadose flutes) and no clearly vadose passages, such as canyons, extend from them. Depressions and caves contain few if any coarse-grained fluvial deposits. Groundwater recharge must have been constrained to small local catchments.

## 3. Burial of the Mississippian Karst

All of the surface depressions and most of the caves in the future Black Hills area were completely filled with Pennsylvanian sediment of the Minnelusa Formation, which consists mainly of red sandstones with lesser amounts of shale, siltstone, carbonates, and evaporites. The Minnelusa directly overlies the carbonate rocks of the Madison Formation. Some paleocaves have only thin fine-grained deposits of this material on their floors, which implies that the caves were fairly isolated from the overlying depositional surface. Most of those exposed in canyon walls are blocked with Pennsylvanian sediment a short distance inside.

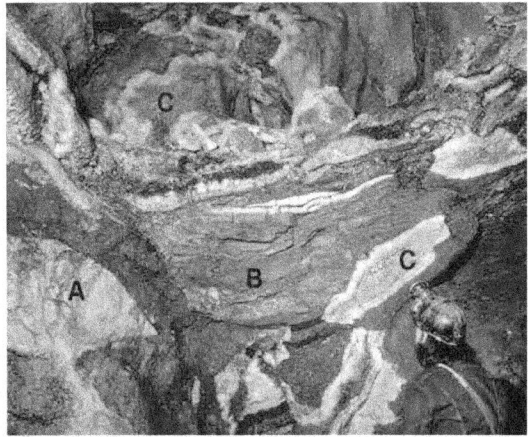

Figure 7. Poorly indurated Pennsylvanian fill in Jewel Cave. A = brecciated bedrock, B = red paleofill, C = Oligocene-Miocene calcite crust, partly fallen away. The sediment originally filled Mississippian paleocaves, and as those openings were enlarged after Laramide uplift, much of the sediment subsided into the growing voids. Some continues to do so today.

The earliest detrital cave sediments were autochthonous carbonate sands derived from weathering of overlying beds. These are overlain by laminated non-carbonate yellow clay and silt, and, in turn by poorly bedded red sand (Fig. 7) with clasts of carbonates and chert. In places, either or both of the first two sediment types is absent. The presence of a clearly defined sequence is important in determining whether the sediment is in place and not disrupted or redistributed by later tectonics or sapping into lower voids.

Sediment continued to be deposited from the Pennsylvanian through the Cretaceous, burying the paleokarst to depths up to 2 km. During deep burial, a thin coating of white scalenohedral calcite was deposited on the walls of solution pockets and caves that were not completely filled with sediment. It also lined pockets in the paleofill. U/Th dating of this calcite has so far been unsuccessful because of low U content.

## 4. Laramide Uplift and Cave Enlargement

Laramide uplift during the late Cretaceous and early Cenozoic allowed the sedimentary rocks to be removed from the central Black Hills. The truncated edges of these strata are exposed around the perimeter of the uplift. Recharge along this boundary fed substantial groundwater flow through the limestone, and eventually reached the situation seen today, where most of the Madison groundwater rises to surface springs through the overlying Minnelusa Formation along fracture zones, especially along anticlinal crests. The paleokarst was partly exhumed, and paleocaves were enlarged to the size seen in the present caves (Fig. 8). Not all the presently accessible cave passages had earlier precursors, but they follow the paleokarst breccias and paleocaves rather faithfully. Few cave passages are exposed in surface canyons, despite the dense array of passages seen in the accessible caves. This suggests that the caves are far from uniformly distributed and are concentrated in paleokarst zones and, one step further back, to bodies of Mississippian sulfates.

Caves show little evidence for rapid groundwater flow except for local incursions of coarse cobbles and boulders. Most of the cave enlargement took place in slow phreatic flow, where mixing of several sources could take place along the Mississippian solution pockets and caves, which afforded the greatest permeability. Presently accessible caves are highly constrained. They do not extend indefinitely down-dip into the heart of the Madison aquifer. Only a couple of passages in Wind Cave reach the water table. And only a few minor passages extend up-dip into areas where the overlying Minnelusa has been stripped off the limestone. No passages reach the base of the limestone (Jewel Cave occupies only the upper half of the Madison), and very few small domes and chimneys extend up to the base of the Minnelusa. On the other hand, the caves have their greatest extent along the strike of the beds, maintaining a somewhat narrow vertical range (Fig. 8). This pattern suggests that dissolution was concentrated in zones of converging groundwater – a combination of water sinking along the eroded edge of the limestone, infiltrating from above through the Minnelusa, and probably also rising from below. Each source would have had a different chemical character (e.g., $CO_2$ partial pressure), and the mixture would be more solutionally aggressive than any individual source.

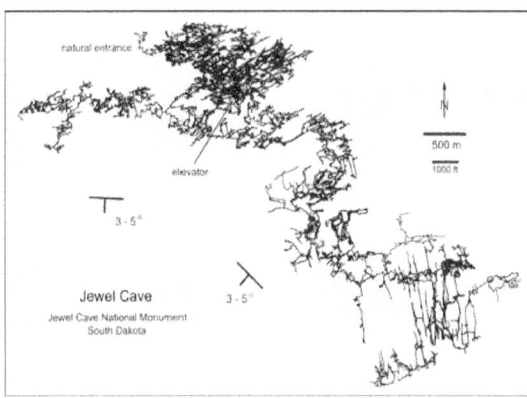

Figure 8. Map of Jewel Cave, South Dakota, as of 2003. Its NW-SE extension follows the mean strike of the Madison beds. The actual structure is more varied than indicated on the map. (Map courtesy of National Park Service.)

The early erosion of the Black Hills was rapid because of active tectonic uplift and a warm, humid climate with much runoff. The present caves formed late in the erosional history of the area, apparently around the end of the Eocene, when the topography had reached a stage nearly identical to that of today. Most cave walls show evidence that after their main phase

of enlargement they were exposed to prolonged weathering above the water table.

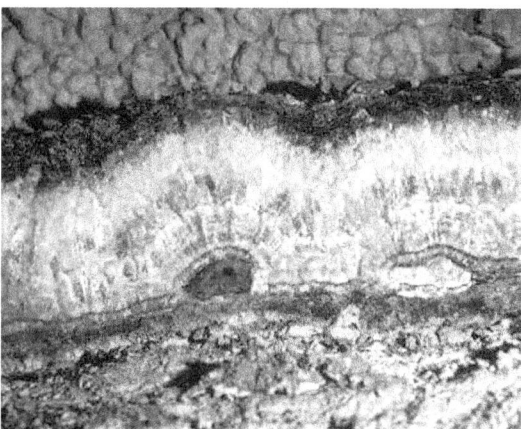

Figure 9. Cross section of Oligocene-Miocene calcite wall crust 15 cm thick overlying weathered dolomite in Jewel Cave. The crust was naturally broken when part of it fell away from the weathered surface.

## 5. Late Cenozoic Reburial and Exhumation

The weathered bedrock surfaces in the Black Hills caves were later coated with subaqueous calcite (Fig. 9). In Jewel Cave these crusts average about 15 cm thick. U/Pb dates on this crust give a range of 26 Ma at the base and 14.7 Ma at the outer edge. This validates the hypothesis that the crust was deposited during occlusion of the springs and burial of the caves by Oligocene-Miocene continental deposits. Since the mid-Miocene these deposits have been eroding from the Black Hills and the (presumed) original groundwater flow paths have been reinstated, while the caves have drained once more. Wind Cave is still draining and its lowest wall crusts are of Pleistocene age. The mid-Cenozoic topography around the caves is still being exhumed and is well preserved in today's semi-arid climate. This late-Cenozoic burial and re-exposure represents yet another phase of paleokarst development. The cumulative result of all these events, from mid-Mississippian onward, is a karst system that may be more complex in history and details than any other documented in the world.

## OTHER PALEOKARST HORIZONS

A few other American paleokarst zones should be mentioned. Skotnicki and Knauth (2007) describe an extensive paleokarst on the Mescal Limestone of middle Proterozoic age in central Arizona preserved by overlying quartzite and predating the intrusion of 1.1 Ga diabase sills. Weathering of overlying basalt led to silicification of the paleokarst. Dissolution of the carbonate led to accumulation of early diagenetic chert while the remaining carbonate underwent nearly complete replacement by secondary silica. In paleocaves, silicified collapse breccias, fills of sandstone and thinly laminated cave-floor siltstone, and flowstone indicate extensive cave development and filling during the karst event. These represent some of the best-preserved silicified karst and Precambrian cave flowstone ever described.

Paleokarst is locally significant in the Permian carbonates of New Mexico and west Texas. From drill-core and related data, Craig (1988) describes solutional voids up to 6 m in diameter that reveal distinct levels of paleokarst. His interpretation is that they formed during the Late Permian by freshwater-seawater mixing along the flanks of low-relief carbonate islands.

In the nearby Guadalupe Mountains, caves such as Carlsbad Cavern display early paleokarst features of both Permian and Laramide age that partly guided later caves that were formed mainly by sulfuric acid (Hill, 1987; Queen, 2009). These early features include fissures and small caves lined by large scalenohedral spar (Fig. 10). At the surface, Koša and Hunt (2005) have mapped many of the paleokarst fissures and breccias and their relation to faulting.

In Arkansas, deep-seated dissolution related to the Oachita Orogeny (late Paleozoic) has produced cavities lined by calcite crystals up to 1.5 m long (Brahana et al., 2009). Some have been intersected by presently active caves. The best-preserved cavities were apparently encountered by vadose cave streams, which dropped to the floors of the fissures without dissolving away the calcite crystals, although there has been some etching and mud coating by periodic flooding.

Tertiary-Quaternary fluctuations in sea level have periodically exposed the carbonates of Florida to karst development and burial. Buried paleokarst horizons have been mapped as far down as 90 m below sea level. In southern Florida a strata-bound horizon of cavernous porosity, called the "boulder zone," lies as much as 900 m below sea level. Rather than a relic of low sea level, this is probably the result of dissolution of sulfates.

The USA contains other paleokarst zones of local importance, but they are of limited extent.

Figure 10. Large calcite crystals lining a paleokarst fissure in Carlsbad Cavern, New Mexico. The calcite is apparently early Laramide in age (ca. 87–98 Ma; Lundberg et al., 2000). The fissure may be as old as Permian. Carbide lamp for scale = 12 cm tall.

## EFFECT OF PALEOKARST ON LATER CAVES

Several examples of how paleokarst can guide later cave development have been described here. Other less extreme examples are cited in this paper. Nearly all of these caves have a hypogenic origin – i.e., they represent rising water or deep-seated processes in which solutional aggressiveness is poorly related to meteoric groundwater flow. Ford (1995) gives field examples that show that hypogenic caves are more likely to be influenced by paleokarst than epigenic caves. Why? A paleokarst zone should be just as permeable to descending water as to rising water.

The tendency for hypogenic caves to follow paleokarst is partly due to the fact that both are most abundant (or at least most recognizable) in the semi-arid West. Although some epigenic caves in humid regions follow paleokarst horizons, examples are fewer and less clear. There is also an inherent reason why buried karst should favor hypogenic cave development. Epigenic caves are fed by water that collects at the surface and drains underground at specific points. These points are governed mainly by topography and distribution of rock types. Paleokarst zones have no preferred orientation relative to this recharge pattern. But rising groundwater concentrates along the most efficient routes – i.e., the widest available paths – with little regard for their location or orientation.

## MAPPING PALEOKARST

Note that this paper does not include a map showing the distribution of paleokarst. Such maps have been attempted in the past (e.g., Palmer and Palmer, 1989), but the results are so general and boundaries so vague that they can be misleading without extensive annotation.

But mapping the distribution and character of paleokarst is worthwhile. Besides its economic value, paleokarst provides a record of former geomorphic settings, climates, tectonics, and global sea-level stands. The difficulty of mapping paleokarst lies in its highly variable character and degree of development, and the fact that most of it is buried. Paleokarst horizons also overlap in many places, so their pattern is three-dimensional. Distinguishing paleokarst features of meteoric origin from those formed by deep-seated processes is a further complexity. Finally, it is often difficult to determine whether paleokarst is even present at all.

With these handicaps, a karst mapper finds it difficult to portray the presence and nature of paleokarst and to delineate boundaries. But even attempting to do so provides a great amount of insight and is a worthwhile goal.

## ACKNOWLEDGMENT

Many thanks to Drs. David Weary and Daniel Doctor, of the U.S. Geological Survey, Reston, Va., for their helpful reviews.

## REFERENCES

Bozák, P., ed., 1989, Paleokarst – A systematic and regional review: Elsevier and Academia, Amsterdam and Prague, 725 p.

Brahana, J.V., Terry, J., Pollock, E., Tennyson, R., and Hays, P.D., 2009, Reactivated basement faulting as a hydrogeologic control of hypogene speleogenesis in the southern Ozarks of Arkansas, USA, *in* Stafford, K., Land, L., and Veni, G., eds., Advances in hypogene karst studies: National Cave and Karst Research Institute, Carlsbad, N.M., Symposium 1, p. 99–110.

Craig, D.H., 1988, Caves and other features of Permian karst in San Andres Dolomite, Yates Field Reservoir, west Texas, *in* James, N.P., and Chouquette, P.W., Paleokarst: Springer-Verlag, New York, p. 342–365.

DeVoto, R., 1985, Late Mississippian paleokarst and related mineral deposits, Leadville Formation, central Colorado, *in* James, N.P., and Chouquette, P.W., Paleokarst: Springer-Verlag, New York, p. 278–305.

Dix, G., Robinson, G., and McGregor, D., 1998, Paleokarst in the Lower Ordovician Beekmantown Group, Ottawa Embayment: Structural control inboard of the Appalachian orogen: Geological Society of America Bulletin, v. 110, p 1046–1059.

Ford, D.C., 1989, Paleokarst of Canada, *in* Bosák, P., ed., Paleokarst – A systematic and regional review: Elsevier/Academia, Amsterdam/Prague, p. 337–363.

Ford, D.C., 1995, Paleokarst as a target for modern karstification: Carbonates and Evaporites, v. 10, p. 138–147.

Hill, C.A., 1987, Geology of Carlsbad Cavern and other caves in the Guadalupe Mountains, New Mexico and Texas: New Mexico Bureau of Mines and Mineral Resources, Socorro, NM, Bulletin 117, 150 p.

James, N.P., and Chouquette, P.W., 1988. Paleokarst: Springer-Verlag, New York, 416 p.

Kerans, C., 1993, Description and interpretation of karst-related breccia fabrics, Ellenburger Group, West Texas, *in* Fritz, R., Wilson, J., and Yurewicz, D., eds., Paleokarst related hydrocarbon reservoirs: SEPM (Society for Sedimentary Geology), New Orleans, Core Workshop 18, p. 181–200.

Koša, E., and Hunt, D.W., 2005, Growth of syn-depositional faults in carbonate strata: Journal of Structural Geology, v. 27, p. 1069–1074.

Lundberg, J., Ford, D., and Hill, C., 2000, A preliminary U-Pb date on cave spar, Big Canyon, Guadalupe Mountains, New Mexico, USA, *in* DuChene, H., and Hill, C., eds., The caves of the Guadalupe Mountains: Journal of Cave and Karst Studies, v. 62, no. 2, p. 144–146.

Loucks, R.G., and Handford, C.R., 1992, Origin and recognition of fractures, breccias, and sediment fills in paleocave-reservoir networks, *in* Candelaria, M., and Reed, C., eds., Paleokarst, karst-related diagenesis, and reservoir development: Examples from Ordovician-Devonian age strata of west Texas and the mid-continent: Permian Basin Section, Society of Economic Paleontologists and Mineralogists, Publication 92-33, p. 31–44.

Maslyn, R.M., 1976, Late Mississippian paleokarst in the Aspen, Colorado, area: MS thesis, Colorado School of Mines, Golden, CO, 96 p.

Mussman, W., Montañez, I., and Read, J., 1988, Ordovician Knox paleokarst unconformity, Appalachians, *in* James, N., and Chouquette, P., Paleokarst: Springer-Verlag, New York, p. 211–228.

Ohle, E.L., 1985, Breccias in Mississippi Valley-type deposits: Economic Geology, v. 80, p. 1736–1752.

Palmer, A.N., and Palmer, M.V., 1995, The Kaskaskia paleokarst of the Northern Rocky Mountains and Black Hills, northwestern USA: Carbonates and Evaporites, v. 10, no. 2, p. 148–160.

Palmer, A.N., and Palmer, M.V., 2008, Field guide to the paleokarst of the Black Hills, *in* Sasowsky, I., Feazel, C., Mylroie, J., Palmer, A., and Palmer, M., eds., Karst from recent to reservoirs: Karst Waters Institute, Special Publication 14, p. 189–220.

Palmer, A., Palmer, M., Polyak, V., and Asmerom, Y., 2009, Geologic history of the Black Hills caves, South Dakota, USA: International Congress of Speleology Proceedings, v. 2, p. 946–951.

Palmer, M.V., and Palmer, A.N., 1989, Paleokarst of the United States, *in* Bosák, P., ed., Paleokarst – A systematic and regional review: Elsevier and Academia, Amsterdam and Prague, p. 337–363.

Plotnick, R., Kenig, F., Scott, A., Glasspool, I., Eble, C., and Lang, W., 2009, Pennsylvanian paleokarst and cave fills from northern Illinois, USA: Palaios, v., 24, p. 627–637.

Queen, J.M., 2009, Geologic setting, structure, tectonic history and paleokarst as factors in speleogenesis in the Guadalupe Mountains, New Mexico and Texas, USA: Kerrville, Texas, Proceedings of 15th International Congress of Speleology, v. 2, p. 952–957.

Roberts, A.E., 1966, Stratigraphy of the Madison Group near Livingston, Montana, and discussion of karst and solution-breccia features: U.S. Geological Survey Professional Paper 526.B, 23 p.

Sando, W.J., 1988, Madison Limestone (Mississippian) paleokarst: A geologic synthesis, *in* James, N.P., and Chouquette, P.W., Paleokarst: Springer-Verlag, New York, p. 256–277.

Sangster, D.F., 1988, Breccia-hosted lead-zinc deposits in carbonate rocks, *in* James, N.P., and Chouquette, P.W., Paleokarst: Springer-Verlag, New York, p. 102–116.

Skotnicki, S., and Knauth, P., 2007, The Middle Proterozoic Mescal paleokarst, central Arizona, USA: Karst development, silicification, and cave deposits: Journal of Sedimentary Research, v. 77, p. 1046–1062.

Sloss, L.L., 1963, Sequences in the cratonic interior of North America: Geological Society of America Bulletin, v. 74, p. 93–114.

Smith, L.B., 2009, Upper Ordovician Trenton – Black River hydrothermal dolomite reservoirs of eastern North America, *in* Stafford, K., Land, L., and Veni, G., eds., Advances in hypogene karst studies: National Cave and Karst Research Institute, Carlsbad, N.M., Symposium 1, p. 99–110.

Troutman, T., 2004, Reservoir characterization, paleogeomorphology, and genesis of the Mississippian Redwall Limestone paleokarst, Hualapai Indian Reservation, Grand Canyon area, Arizona: MS thesis, University of Texas, Austin, 221 p.

Tschauder, R., and Landis, G., 1985, Late Paleozoic karst development and mineralization in central Colorado, *in* DeVoto, R., ed., Sedimentology, dolomitization, karstification, and mineralization of the Leadville Limestone (Mississippian), central Colorado: Field trip guidebook, Society for Economic Paleontologists and Mineralogists, Section 6, p. 89-91.

# KARST IN COASTAL ZONES

# Influence of Tides and Salinity on the Discharge at the Coastal Spring Creek Springs Group and the Connection to Discharge at the Inland Wakulla Springs, Wakulla County, Florida

By J. Hal Davis[1] and Richard J. Verdi[1]
[1]U.S. Geological Survey, 2639 North Monroe Street, Suite A-200, Tallahassee, FL 32303

## Abstract

The Floridan aquifer water budget in north-central Florida has been poorly described in the past because Spring Creek Springs Group, one of the major discharge points, is in a tidal estuary and discharge has been measured only a few times. To improve the understanding of the water budget, the U.S. Geological Survey and the Florida Department of Environmental Protection conducted a cooperative study to measure discharge in Spring Creek Springs Group.

The Spring Creek Springs Group is composed of 13 springs located in a tidal estuary in north-central Florida. In June 2007, the U.S. Geological Survey installed a gaging station to monitor stage, water velocity, precipitation, and conductance in Spring Creek, which connects these springs to the Gulf of Mexico. Salinity was calculated from conductance and used to determine the freshwater and saltwater components of flow. Spring flow varied from 2007 to 2009, but could be grouped into periods of similar characteristics, each lasting several months. There was no net outflow of freshwater from Spring Creek Springs Group for the three periods of June 2007 to March 2008, June to September 2008, and July to September 2009. In contrast, there was a net outflow of freshwater from the aquifer of about 800 ft$^3$/s for the periods from March to June 2008 and September 2008 to July 2009.

Salinity in Spring Creek Springs is strongly influenced by inflows into nearby sinkholes, especially Lost Creek Sink, which contribute flow to the springs. Flow into Lost Creek Sink is generally low, less than 50 cubic feet per second (ft$^3$/s), except during periods of heavy rainfall when inflows can briefly peak at 2,000 to 3,000 ft$^3$/s. This influx of freshwater causes the salinity in Spring Creek Springs to drop from near seawater values of 35,000 parts per million (ppm) to values as low as 5,000 ppm. Extended periods of low rainfall result in low flows into Lost Creek Sink and the increase of salinity in Spring Creek Springs to near seawater values.

When salinity increases in Spring Creek Springs, the increase in water density results in a higher equivalent freshwater head. This higher head exceeds the head in the more inland Wakulla Springs, causing increased flow at Wakulla Springs and decreased freshwater flow at Spring Creek Springs. When salinity is low, the equivalent freshwater head at Spring Creek Springs is lower than at Wakulla Springs, and relatively higher freshwater discharge shifts back to Spring Creek Springs. An estimated 9-inch rise in sea level since the 1930's may be causing long-term salinity increases in Spring Creek Springs and the cause of an observed increase in flow at Wakulla Springs over this same period.

# Revised Hydrogeologic Framework for the Floridan Aquifer System in the Northern Coastal Areas of Georgia and Parts of South Carolina

By Harold E. Gill (retired) and Lester J. Williams
U.S. Geological Survey, 3039 Amwiler Rd, Suite 130, Atlanta, GA 30360

## Abstract

The hydrogeologic framework for the Floridan aquifer system was revised for eight northern coastal counties in Georgia and five coastal counties in South Carolina (fig. 1) as part of a regional assessment of water resources by the U.S. Geological Survey (USGS) Groundwater Resources Program. In this study, selected well logs were compiled and analyzed to determine the vertical and horizontal continuity of permeable zones that make up the aquifer system, and define more precisely the thickness of confining beds that separate individual aquifer zones. The results of the analysis indicate that permeable zones in the Floridan aquifer system can be divided into (1) an upper group of extremely transmissive zones that correlate to the Ocala Limestone in Georgia and the Parkers Ferry Formation in South Carolina, and (2) a lower group of zones of relatively lower transmissivity that correlates to the middle part of the Avon Park formation in Georgia and updip clastic equivalent units of South Carolina (fig. 2). This new subdivision simplifies the hydrogeologic framework originally developed by the USGS in the 1980s and helps to improve the understanding of the physical geometry of the system for future modeling efforts. Revisions to the framework in the Savannah–Hilton Head area are particularly important where permeable beds control the movement of saltwater contamination. The revised framework will enable water-resource managers in Georgia and South Carolina to assess groundwater resources in a more uniform manner and help with the implementation of sound decisions when managing water resources in the aquifer system.

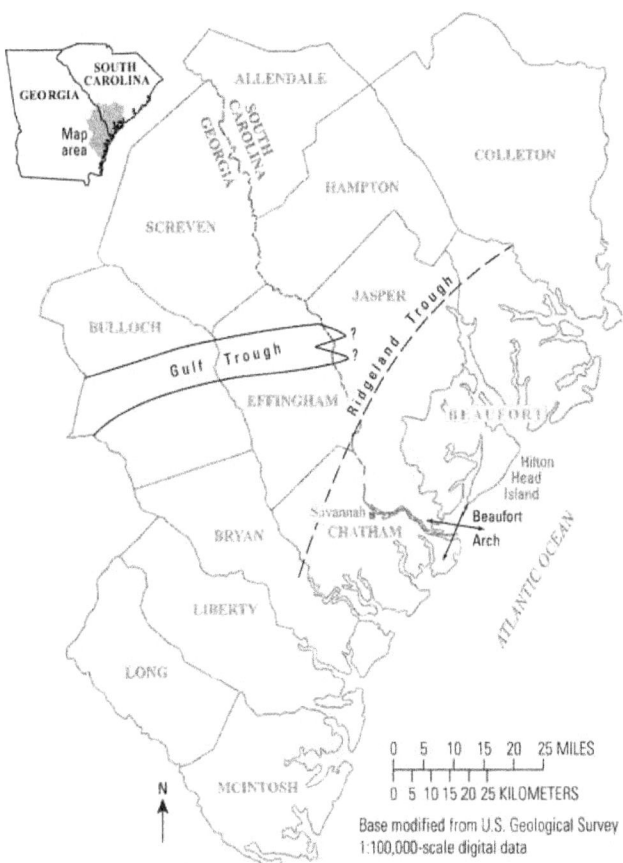

## REFERENCE

Williams, L.J., and Gill, H.E., 2010, Revised hydrogeologic framework of the Floridan aquifer system in the northern coastal area of Georgia and adjacent parts of South Carolina: U.S. Geological Survey Scientific Investigations Report 2010–5158, 103 p., 3 plates.

Figure 1. Study area in eight northern coastal counties of Georgia and five southern coastal counties of South Carolina. (Structural features: Gulf Trough from Applied Coastal Research Laboratory, 2002; Beaufort Arch from Clarke and others, 1990; modified from Williams and Gill, 2010).

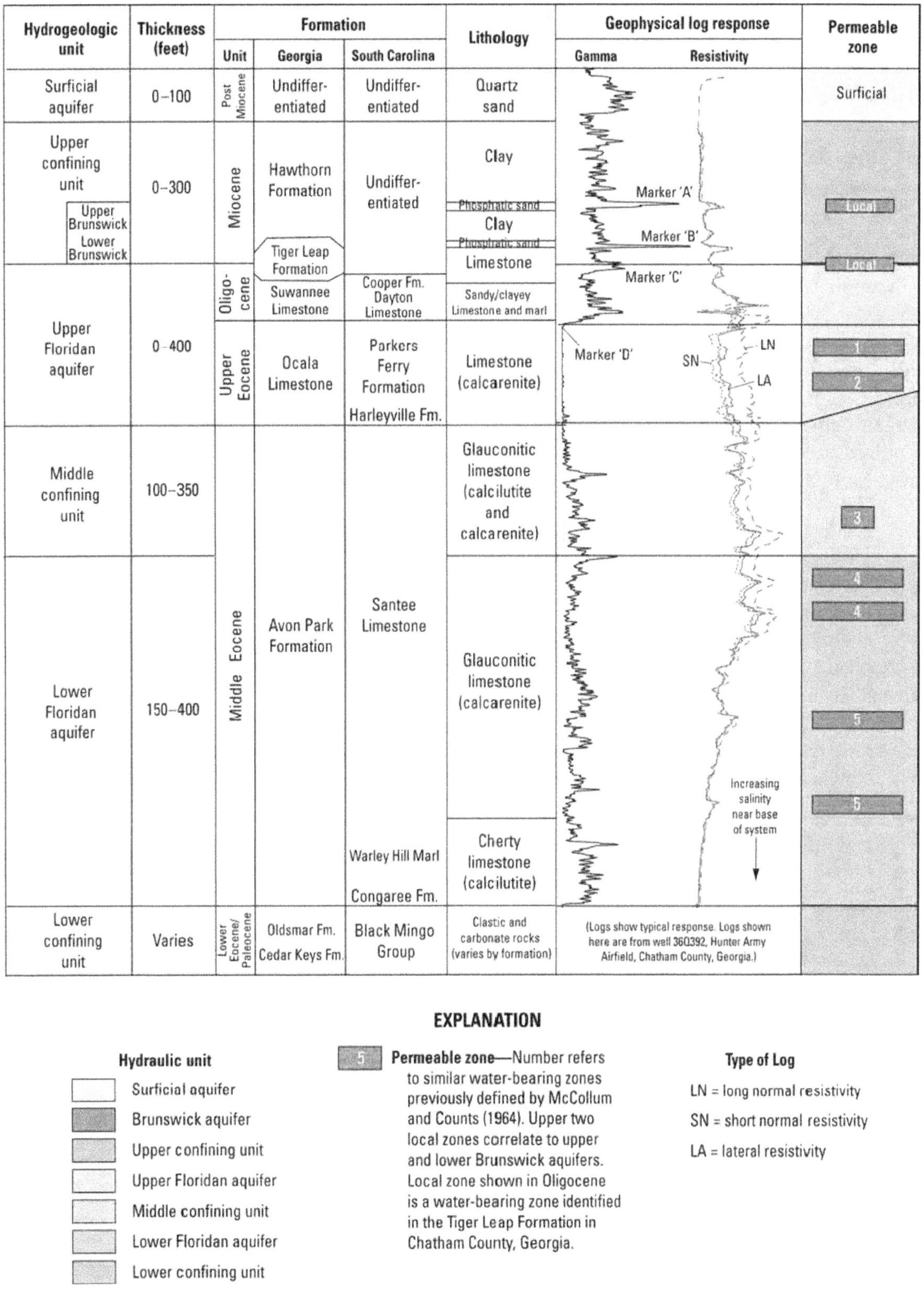

| Hydrogeologic unit | Thickness (feet) | Formation | | | Lithology | Geophysical log response | | Permeable zone |
|---|---|---|---|---|---|---|---|---|
| | | Unit | Georgia | South Carolina | | Gamma | Resistivity | |
| Surficial aquifer | 0–100 | Post Miocene | Undifferentiated | Undifferentiated | Quartz sand | | | Surficial |
| Upper confining unit — Upper Brunswick / Lower Brunswick | 0–300 | Miocene | Hawthorn Formation / Tiger Leap Formation | Undifferentiated | Clay / Phosphatic sand / Clay / Phosphatic sand / Limestone | Marker 'A' / Marker 'B' / Marker 'C' | | Local / Local |
| Upper Floridan aquifer | 0–400 | Oligocene | Suwannee Limestone | Cooper Fm. Dayton Limestone | Sandy/clayey Limestone and marl | Marker 'D' | | |
| | | Upper Eocene | Ocala Limestone | Parkers Ferry Formation / Harleyville Fm. | Limestone (calcarenite) | | LN / SN / LA | 1 / 2 |
| Middle confining unit | 100–350 | Middle Eocene | Avon Park Formation | Santee Limestone | Glauconitic limestone (calcilutite and calcarenite) | | | 3 |
| Lower Floridan aquifer | 150–400 | | | | Glauconitic limestone (calcarenite) | | Increasing salinity near base of system | 4 / 4 / 5 / 5 |
| | | | | Warley Hill Marl / Congaree Fm. | Cherty limestone (calcilutite) | | | |
| Lower confining unit | Varies | Lower Eocene/ Paleocene | Oldsmar Fm. Cedar Keys Fm. | Black Mingo Group | Clastic and carbonate rocks (varies by formation) | (Logs show typical response. Logs shown here are from well 36Q392, Hunter Army Airfield, Chatham County, Georgia.) | | |

## EXPLANATION

**Hydraulic unit**

- Surficial aquifer
- Brunswick aquifer
- Upper confining unit
- Upper Floridan aquifer
- Middle confining unit
- Lower Floridan aquifer
- Lower confining unit

**5** Permeable zone—Number refers to similar water-bearing zones previously defined by McCollum and Counts (1964). Upper two local zones correlate to upper and lower Brunswick aquifers. Local zone shown in Oligocene is a water-bearing zone identified in the Tiger Leap Formation in Chatham County, Georgia.

**Type of Log**

LN = long normal resistivity

SN = short normal resistivity

LA = lateral resistivity

Figure 2. Hydrogeologic units and confining beds of the Floridan aquifer system showing representative log response and location of permeable zones and mapping horizons (modified from Williams and Gill, 2010; Fm., formation)

# Horizontal Bedding-Plane Conduit Systems in the Floridan Aquifer System and Their Relation to Saltwater Intrusion in Northeastern Florida and Southeastern Georgia

By Lester J. Williams[1] and Rick M. Spechler[2]
[1]U.S. Geological Survey, 3039 Amwiler Rd, Suite 130, Atlanta, GA 30360
[2]U.S. Geological Survey, 12703 Research Parkway, Orlando, FL 32826

## Abstract

Acoustic televiewer (ATV) images, flowmeter, and borehole geophysical logs obtained from the open intervals of deep test wells were used to develop a revised conceptual model of groundwater flow for the Floridan aquifer system in northeastern Florida and southeastern Georgia. Borehole information was used to identify and map the types and distribution of highly-transmissive production zones in the Floridan aquifer system. The ATV images and flowmeter traverses indicate that water produced from most wells is largely derived from a system of highly-transmissive solution zones formed along bedding planes and major formational contacts. These "horizontal bedding-plane conduit systems" may locally influence the movement of brackish and saline water in the Floridan aquifer system.

A modified conceptual model of regional flow in the Floridan aquifer system is proposed that incorporates locally interconnected horizontal conduit systems within the largely porous matrix rock (fig. 1). Each of the conduit systems represents a highly-transmissive zone along which water can move preferentially through the aquifer system. These may or may not be laterally continuous across the area. Flow paths within the system are restricted vertically by local or regional confining units except where these are breached by collapse features or vertical fractures. Near major pumping centers, water probably moves preferentially along the horizontal conduits to reach the discharging well. The source of water moving into the transmissive open conduits is either derived from upward migration along vertical discontinuities in the rock or from diffuse leakage from adjacent porous rock units. Some trapped relict water in adjacent lower-permeability units may locally contribute to the higher chloride concentrations observed in some wells.

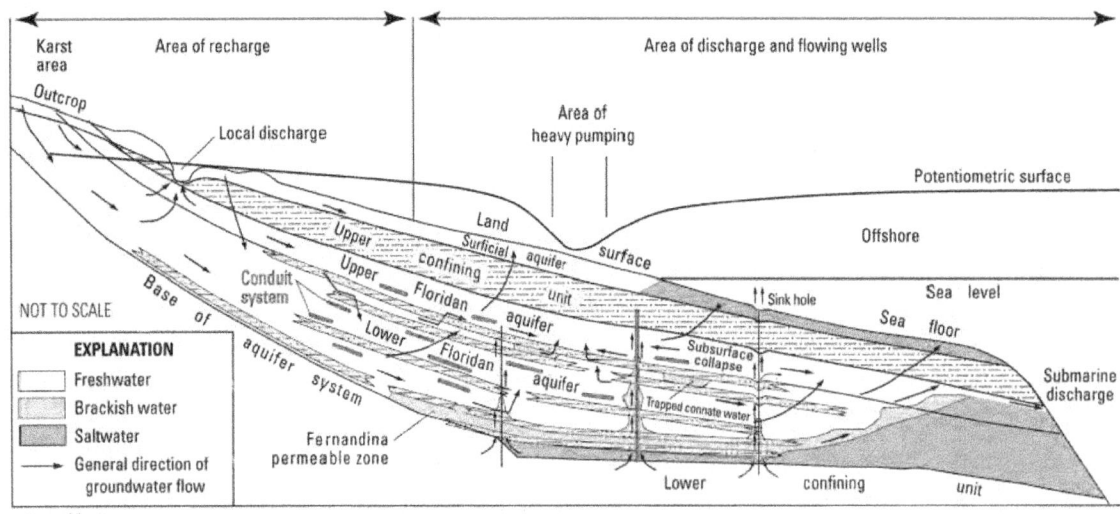

Figure 1. Conceptual model of for the Floridan aquifer system from the outcrop area to the offshore area (modified from Krause and Randolph, 1989; Spechler, 1994).

## REFERENCES

Krause, R.E., and Randolph, R.B., 1989, Hydrology of the Floridan aquifer system in southeast Georgia and adjacent parts of Florida and South Carolina: U.S. Geological Survey Professional Paper 1403-D, 65 p., 18 pl.

Spechler, R.M., 1994, Saltwater intrusion and the quality of water in the Floridan aquifer system, northeastern Florida: U.S. Geological Survey Water-Resources Investigations Report 92–4174, 76 p.

# GEOCHEMISTRY AND CONTAMINANT TRANSPORT OF KARST SYSTEMS

## Water-Quality Changes and Dual Response in a Karst Aquifer to the May 2010 Flood in Middle, Tennessee

By Michael W. Bradley and Thomas D. Byl

U.S. Geological Survey, Suite 100, 640 Grassmere, Nashville, Tennessee 37211

**Abstract**

During May 1 and 2, 2010, an area stretching from Memphis through Nashville received about 14 to 19 inches of rain. Extensive flooding occurred and the Cumberland River in Nashville was above flood stage from May 1 through May 7, 2010. Water-quality data were collected at a spring and an observation well at Tennessee State University in Nashville, Tennessee (figure 1) during the flood. Changes in groundwater quality were observed, including a dual response to the heavy rainfall at the spring and the well. The dual response observed in the water-quality data consists of more rapid changes in water quality at the spring and minimal water-quality changes in the well. The spring issues from the Ordovician Bigby-Cannon Limestone. The observation well is 202 feet deep with water-bearing zones in the lower Bigby-Cannon Limestone.

Water level elevations in the well and in the spring rose in response to the heavy rainfall (figure 2). The spring area was flooded by water from the Cumberland River and the water-level elevations of the spring match the elevation of the Cumberland River during the flood (figure 2). Groundwater elevations in the observation well are typically higher than the Cumberland River. However, during the flood, the gradient reversed and the level of the Cumberland River was higher than the groundwater elevations (figure 2).

Specific conductance records show the contrasting responses to rainfall and flooding at the spring and the well (figure 3). The specific conductance of water in the observation well shows little response to the flood event with values fluctuating from about 700 to 705 microSiemens per centimeter (figure 3). The specific conductance of water from the spring fluctuates rapidly over a range of about 300 to 500 microSiemens per centimeter during and after the flood (figure 3). The water quality in the spring responded quickly to the large amounts of precipitation. The water-quality parameters in the observation well, however, showed very little response to the flood event, even though water levels in the well rose about 23 feet (figures 2 and 3).

Imagery from USDA 2006 NAIP

| 0 | 625 | 1,250 | 2,500 Feet |
|---|-----|-------|-----------|

| 0 | 155 | 310 | 620 Meters |
|---|-----|-----|-----------|

EXPLANATION
Location of water-quality monitor

⚲    Observation well

⚲S   Spring

▲    Cumberland River

⚬    Location of supply wells

Figure 1. - Location of water-quality monitoring sites near Tennessee State University, Nashville, Tennessee

22

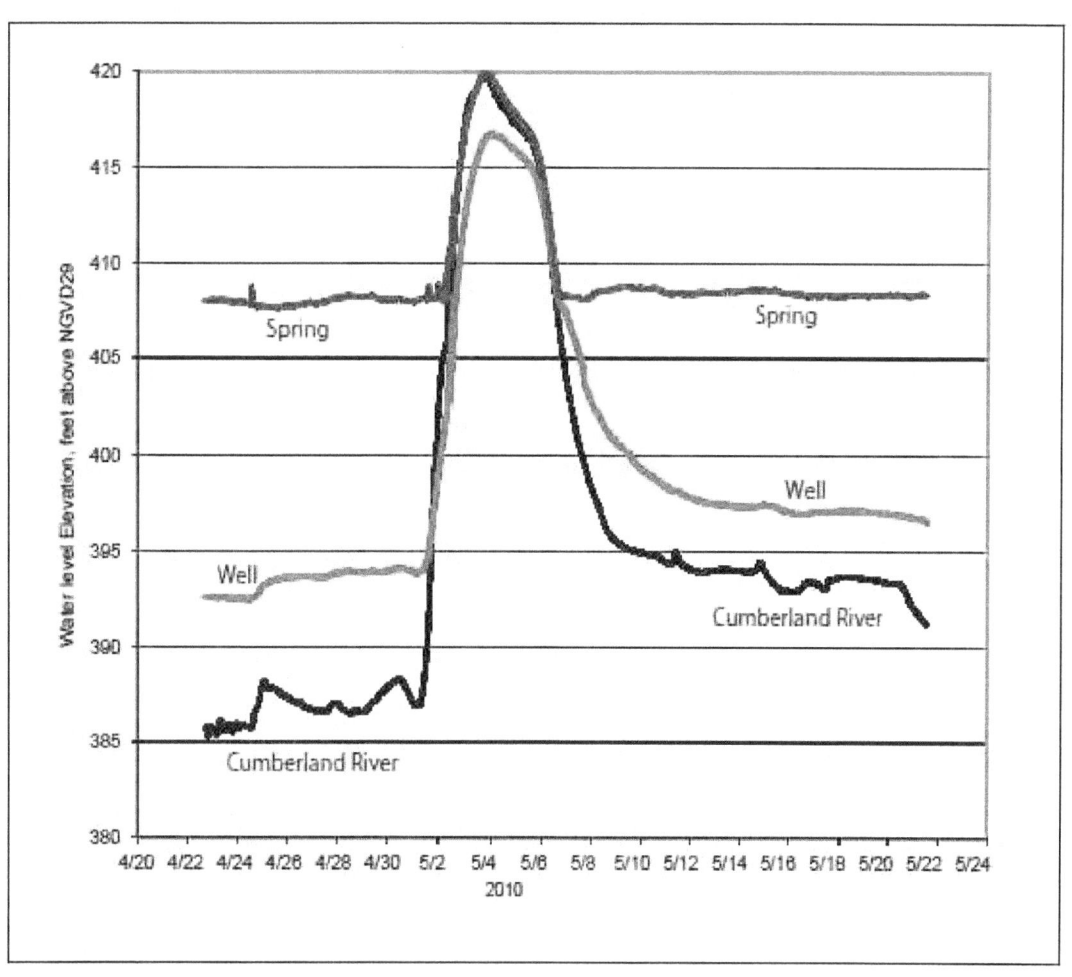

Figure 2. – Water-level elevations in the Cumberland River and a spring and observation well at Tennessee State University, Nashville, Tennessee during April 22 - May 24, 2010

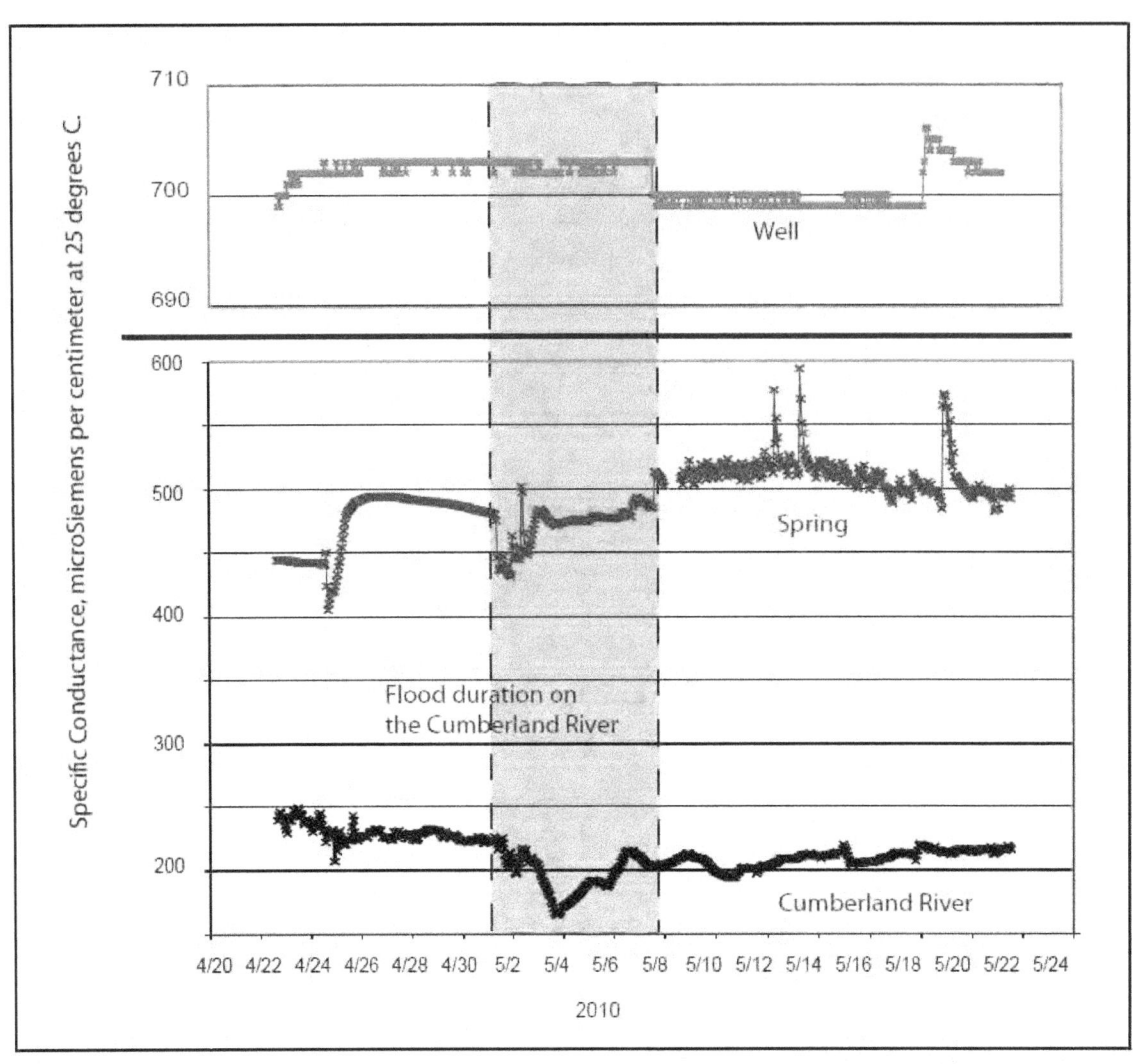

Figure 3. -- Specific conductance of water in the Cumberland River and a spring and observation well at Tennessee State University, Nashville, Tennessee during April 22 - May 24, 2010

# Interaction between Shallow and Deep Groundwater Components at Fay Spring in the Northern Shenandoah Valley Karst

By Daniel H. Doctor[1], Nathan C. Farrar[2], and Janet S. Herman[2]
[1]U.S. Geological Survey, 12201 Sunrise Valley Drive, MS 926A, Reston, VA, 20191
[2]University of Virginia, Dept. of Environmental Sciences, P.O. Box 400123, Charlottesville, VA 22904

## Abstract

The fractured carbonate rock aquifer of the Shenandoah Valley has been karstified to depths exceeding 300 meters (m). Springs in the region integrate flow through both shallow and deep flowpaths within solutional conduits which tend to be concentrated in areas of bedrock structural weakness, such as along lineaments and fault zones, and through fractured bedrock. In order to investigate the local dynamics of the aquifer, continuous monitoring of discharge, conductivity and temperature was combined with frequent geochemical sampling and dye-tracing at Fay Spring, located in Winchester, Va. This perennial spring shows rapid response in discharge following large rain events with a concomitant decrease in conductivity; however, snowmelt leads to an increase in conductivity, indicating recharge by surface water carrying seasonally applied road salt. Quantification of local, surface-water input to the spring was sought through a dye trace in July 2009. Rhodamine WT injected at the terminal sink point of an ephemeral stream approximately 1 kilometer west of Fay Spring provided evidence of initial rapid dye transit (greater than 500 m/day), yet overall low total mass recovery (12 percent after 5 weeks). Water samples were collected by an automatic sampler every six hours to daily during the trace period, and also from January-April, 2010 during snowmelt of the largest snowfall on record for the region. Chemographs revealed dilution in $NO_3$, Cl, $SO_4$, and Mg with most large rain events; however, Cl increased during peak snowmelt with little change in other ion concentrations. Principal components analysis (PCA) was used on major ions (Mg, $SO_4$, $NO_3$, Cl, Sr, and Si) in an effort to distinguish components of deeper, regional flow and shallow, local flow within the spring discharge. The PCA results show that two components cumulatively explain 72 percent of the variance of the data, permitting a mixing model comprised of three end-members. The end-members were representative of 1) a Ca-Mg-$HCO_3$ type water of relatively low $NO_3$, $SO_4$ and Cl concentrations, 2) an "agriculturally influenced" type water of high $NO_3$ with moderate $SO_4$ and Cl concentrations, and 3) an "urban influenced" type water with high Cl and moderate $SO_4$ and $NO_3$ concentrations. The results of the end-member mixing model indicate that Fay Spring is dominated by a regional flow component with elevated background levels of $NO_3$ (10-12 mg/L as $NO_3$), $SO_4$ (45-50 mg/L) and Cl (35-40 mg/L). The background concentrations of $SO_4$ and $NO_3$ are similar between summer and winter; however, Cl concentration is higher in the winter (up to 80 mg/L). The spring chemistry is noticeably diluted only after large rain events by water interpreted to be derived from shallow diffuse groundwater storage; direct sinking surface runoff is a lesser component. Overall, the impacts of rapid surface runoff are muted at Fay Spring compared with springs in other Appalachian karst regions. A combination of approaches including dye tracing and interpretation of natural geochemistry is helpful for elucidating the nature of discharge from karst springs in fractured rock carbonate aquifers.

## INTRODUCTION

The carbonate aquifer system of the northern Shenandoah Valley in Virginia and West Virginia is a fractured rock aquifer that has undergone karstification at depth far below the water table. Maximum depth of groundwater circulation is unknown; however, circulation through conduits exceeds depths of 300 meters (m) as evidenced by a small number of high-yield deep wells. For example, Cady (1936)

described a well in the City of Winchester (well no.W 161, drilled in 1931) that is 436 m deep and produced 560 liters/minute (L/min), or 148 gallons/minute (gal/min). Water bearing cavities intersected by this well are reported at depths of 30, 91, 213, and 335-365 m below the surface. Cady (1936) reported three other high-yield wells in the Shenandoah Valley at depths greater than 300 m below the surface that have intersected cavities, all drilled in Cambrian and Ordovician carbonate rocks. Recent drillers' accounts corroborate the older information. A well constructed approximately 12 km north of Winchester near Green Spring was mostly dry until a depth of 410 m, at which point the well hit a cavernous zone and produced 750 L/min (198 gal/min); a second well in the vicinity hit a cavernous zone at a depth of approximately 245 m, and emanated sulfurous water and black fine-grained sediment that caused the well to be abandoned (G. Payne, Payne Well Drilling, pers. comm., 2008).

McCoy and Kozar (2007) correlated high-yield wells and springs with cross-strike fractures and faults where upward-convergent groundwater flow may occur. Kozar and others (2007a,b) and Jones (1991, 1997) interpreted the groundwater system as being a solutionally modified fractured-rock carbonate aquifer, with interspersed solution conduits serving as primary drains for the karst groundwater system. Kozar and others (2007 a,b) point out that such conduits typically coincide with strike-parallel

thrust faults and cross-strike extensional or strike-slip faults. Conduit flow is evident from springs often found along high angle normal or reverse faults that cut across the dominant strike of bedding, or along thrust faults where carbonates are brought into contact with overlying shales. Springs are common where cross-strike faults intersect thrusts within carbonate units (Perry and others, 1979; Orndorff and Harlow, 2002).

The karst aquifer discharges to numerous artesian springs that range in flow rate over three orders of magnitude. Cady (1936) reported 95 springs with discharge ranging from 27 to 29,000 $m^3$/day (0.01 to 11.9 cfs), and an average flow of 363 $m^3$/day (0.15 cfs). Some larger springs exhibit a delayed response to rainfall events, with long hydrograph recession periods and muted chemical variability (Vesper and others, 2008). Spring discharge accounts for 60 percent to 97 percent of stream flow (Harlow and others, 2005; Nelms and Moberg, 2010).

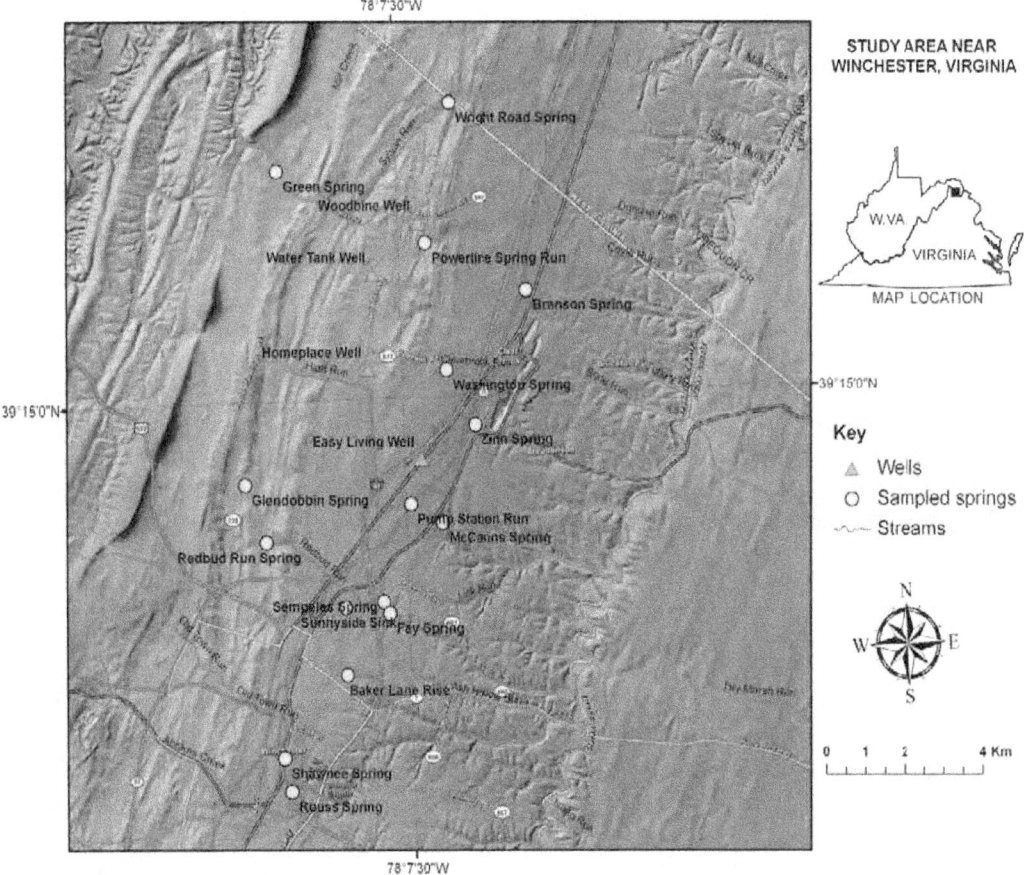

**Figure 1**. Map of study area near Winchester, Virginia showing sampling sites.

With the majority of the surface streamflow derived from karst aquifer discharge, questions remain as to the mechanisms and proportions of recent aquifer recharge expressed in the discharge at springs. Here, we investigated the interaction between shallow, recently recharged water at a stream sink and water derived from deeper, phreatic flow at one artesian spring site in the Shenandoah Valley.

## Study site: Fay Spring

Fay Spring is located within the City of Winchester, Virginia (fig. 1). The spring is situated on a northwest-trending strike-slip fault cutting Lower Ordovician dolomite and limestone (Orndorff and others, 2004). The spring is owned by the City of Winchester, and was once used as a part of the municipal water supply (Harlow and others, 2005).

The spring was instrumented by the Virginia Water Science Center of the U.S. Geological Survey for continuous monitoring of discharge, water temperature, and specific conductance from late April 2007 to October 2010. For the period of record, the mean discharge was 2,592 $m^3$/day (1.09 $ft^3$/s) and the mean temperature was 13.1 °C. The continuous record for this perennial spring shows a broad seasonal rise and fall in water temperature and discharge, punctuated by rapid increases in flow with concomitant decreases in conductivity (fig. 2). An increase in spring discharge corresponding to a decrease in conductivity generally indicates the rapid addition of less-mineralized water from the surface to the spring during recharge events, and is commonly observed in karst settings (White, 1988). At Fay Spring, the water temperature does not respond to discharge in as flashy a manner as conductivity, and relatively little temperature change is observed even during periods of large changes in flow (fig. 2). Conversely, conductivity does not show a pronounced seasonal trend like water temperature, and tends not to vary greatly when above the long-term mean value of 711 µS/cm (fig. 2). However, a winter precipitation event in January 2008 caused an unusual dramatic increase in conductivity, with no corresponding change in discharge or water temperature at the spring (fig. 2). This increase in conductivity was hypothesized to result from sinking surface

water carrying road salt to the spring, and may have been a delayed response to events ranging from several days up to two weeks earlier. In order to test this hypothesis, quantification of the influence of local, surface-water input to the spring was sought through a dye trace.

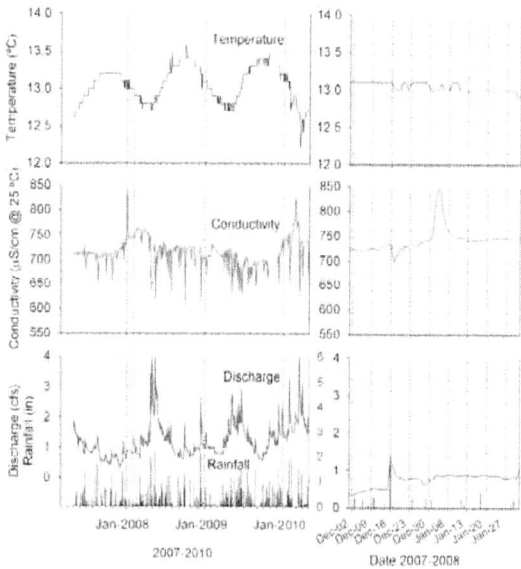

**Figure 2.** Continuous monitoring at Fay Spring, from 2007 to 2010. An unusual conductivity increase in January 2008 is highlighted in the green box and expanded in the right-hand panels. A similar, but more gradual conductivity excursion was observed during a major period of recharge as a result of above normal snowfall in 2010. Precipitation data obtained from National Weather Service Coop Station 449181 at Winchester, Virginia.

## Dye trace to Fay Spring

On June 30, 2009, approximately 2 kg of 20 percent Rhodamine WT dye solution was injected at the terminal sink point of Sunnyside Run, an ephemeral stream ~1 km west of Fay Spring (fig. 3). Water samples for dye and major ion analysis were collected by an automatic sampler with a frequency of every six hours to once per day during the trace period. Dye was recovered at two springs (Fay Spring and Sempeles Spring, fig. 1) located 300 m apart along a single fault, and within the channel of Redbud Run immediately upstream of its confluence with the spring run of Sempeles Spring; dye was not recovered at any of the other sampling sites.

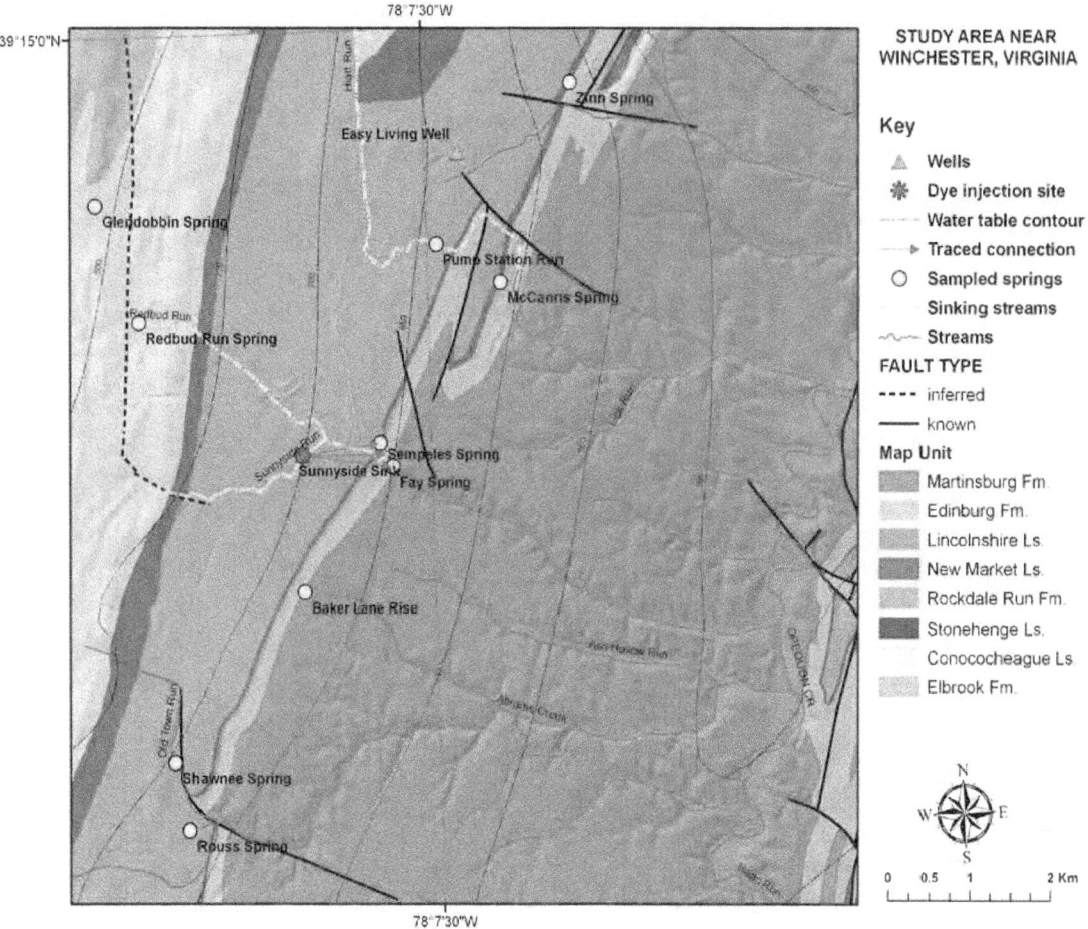

**Figure 3**. Geologic map of study area showing water table contours and traced connections.

The dye breakthrough curves at Fay Spring and Sempeles Spring are shown in figure 4. The results of this tracing test provided evidence of initial rapid dye transit (>500 m/day) and low overall mass recovery (~12 percent) that includes sampling at both springs over the course of five weeks. In spite of the low tracer recovery in the water samples, monitoring by passive activated charcoal samplers verified a positive recovery of the dye at each of the three sites. The amount of dye recovered at Sempeles Spring was approximately ten times greater than that at Fay Spring; however, Sempeles Spring was not instrumented for continuous monitoring, so our discussion focuses on Fay Spring. The amount of dye recovered at Redbud Run was less than at either spring based on the passive charcoal sampler results.

### Snowmelt monitoring at Fay Spring

Chemical sampling continued at Fay Spring throughout the winter and spring of 2009-2010.

The results of the continuous monitoring during this period revealed a gradual rise in conductivity following the recession of a rainfall event in late January. Then in early February 2010, the largest snowfall on record for the region occurred. Melting of this snow began on February 7, and during the melt period the rise in discharge was closely correlated to a rise in conductivity (fig. 5). Rapid increases in discharge occurred with rainfall both prior to and following the snowmelt period, and resulted in concomitant decreases in conductivity. However, during the large snowmelt event, only a gradual, damped discharge increase was observed that corresponded with a similar damped increase in conductivity.

Water samples were collected by an automatic sampler every six hours to daily from January-April 2010, and a complete record of daily samples was obtained during the snowmelt period. Chemographs revealed dilution in $NO_3$,

Cl, SO₄, and Mg with most large rain events; however, Cl increased during peak snowmelt with little change in other ion concentrations (fig. 6). In comparison to the results of chemical sampling during the summer, the background concentrations of $SO_4$ and $NO_3$ are similar in the winter; however, Cl concentration is higher in the winter (up to 80 mg/L). This added chloride component may be due to the impact of road salt in the winter months.

## Principal components analysis and estimation of end-members

We used principal components analysis (PCA) on major ions (Mg, $SO_4$, $NO_3$, Cl, Sr, and Si) in an effort to distinguish possible end-member components of deeper, regional flow and shallow, local flow that are mixed within the spring discharge. PCA has been shown to be an effective means for screening hydrochemical data in order to identify possible end-member compositions contributing to mixed samples (Christopherson and Hooper, 1992), and has previously been successfully applied in a karst aquifer setting (Doctor and others, 2006).

PCA reduces the entire dataset into a smaller set of factors that account for the greatest variance in the data. In this case, the PCA results from the winter samples show that two components cumulatively explain 72 percent of the variance of the data. The first component (54 percent) is most positively weighted on Mg and $NO_3$; the second component (18 percent) is most positively weighted on $SO_4$, and negatively weighted on Sr and Cl.

An end-member mixing analysis (EMMA) was performed using the results of the PCA. Since two principal components account for the majority of the variance of the data, only three end-members are necessary to account for the compositions of the mixtures, subject to the constraint that the proportions of all end-members in each observed mixture sum to 1 (Christopherson and Hooper, 1992). For the EMMA, it is assumed that end-members mix linearly and that the geochemical constituents behave conservatively. For this reason, the most non-conservative chemical parameters (Ca and $HCO_3$) were excluded.

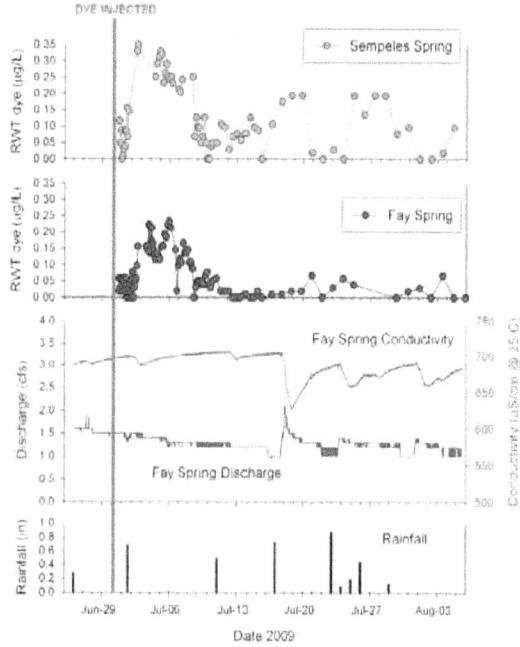

**Figure 4.** Dye recovery at Sempeles Spring and Fay Spring. Note pulsed recovery of dye released from shallow groundwater storage during later storm events.

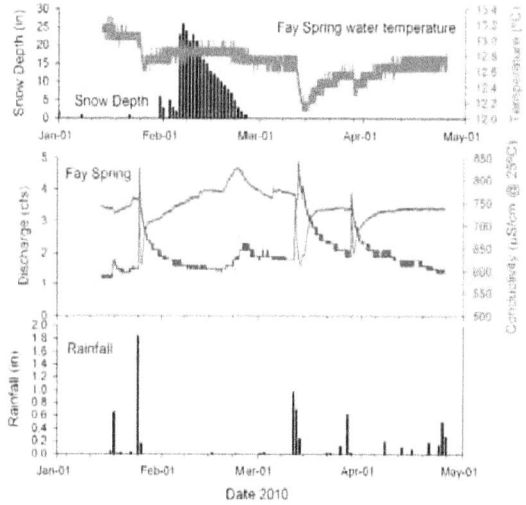

**Figure 5.** Record at Fay Spring before and after snowmelt in winter and spring 2010. Note the muted, gradual discharge increase in response to snowmelt versus the greater, more rapid discharge increase in response to rainfall events. Temperature decrease in response to rainfall is evident, though damped, and almost nonexistent in response to snowmelt.

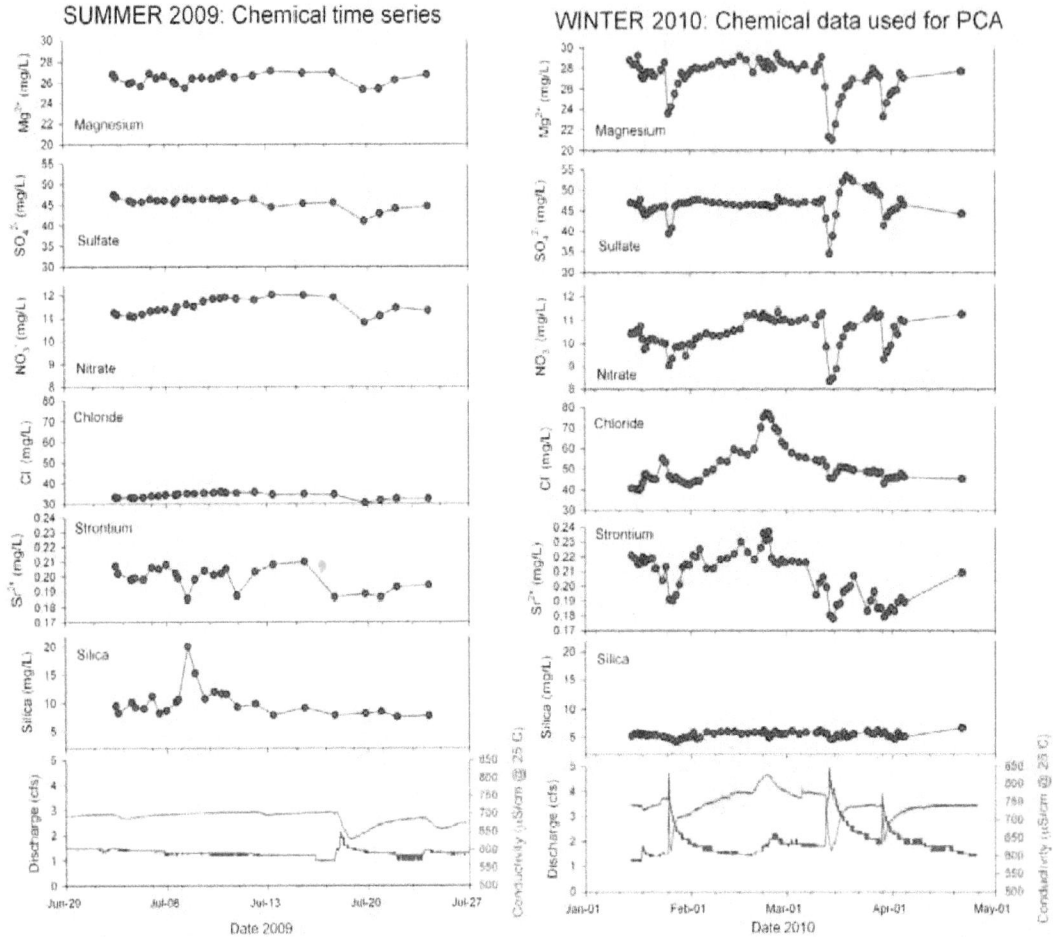

**Figure 6**. Major ion chemistry at Fay Spring during summer and winter events. Left panel shows chemographs during the dye-trace monitoring in July 2009; right panel shows chemographs during the snowmelt period of 2010. Note that the ranges of each chemical constituent are shown on equivalent vertical axes for both sampling periods. Except for chloride and silica, the chemistry varies around average background values that are remarkably consistent between the two seasons, despite nearly the full range of observed discharge at Fay Spring being represented.

Reducing the data matrix to a set of two factors allows projection of the transformed data into a 2-dimensional mixing space (fig. 7). This data projection is convenient for exploring the data cloud in relation to possible end-member compositions that are likewise projected into the principal components mixing space. The end-member compositions might be estimated by the extremes of the data cloud if candidate samples are not available, but this can result in ambiguous and non-unique solutions that may be unrealistic (Christopherson and Hooper, 1992). Therefore, it is best to obtain samples of candidate end-member sources according to a conceptual model of the system in order to test

the mixing model. In this study, samples of surrounding springs and wells were obtained periodically during the course of the monitoring at Fay Spring. These samples were projected into the same principal components space as the Fay Spring samples, and of those samples the three that fell at the outer extremes and enclosed the data cloud of the Fay Spring data were chosen as possible end-members (fig. 7).

The three estimated end-members were: 1) the composition of the well water at the Woodbine Assembly Church, 2) the composition of Washington Spring and 3) the ephemeral flow in Sunnyside Run (see fig. 1 for localities).

**Table 1**. Hypothesized geochemical end-member compositions used in the mixing model for Fay Spring.

| Hypothesized End-Members[1] | Cl | NO₃ | SO₄ | Mg | Si | Sr |
|---|---|---|---|---|---|---|
| **EM1: Woodbine Well (WW)**<br>low NO$_3$, low SO$_4$ , low Cl | 4.0 | 7.0 | 18.0 | 15.1 | 8.7 | 0.2 |
| **EM2: Washington Spring (WS)**<br>high NO$_3$, mod. SO$_4$ , mod. Cl | 18.0 | 21.1 | 22.4 | 20.5 | 8.0 | 0.4 |
| **EM3: Sunnyside Run (SR)**<br>mod. NO$_3$, high SO$_4$ , high Cl | 47.4 | 14.0 | 48.0 | 18.5 | 7.0 | 0.2 |

[1]Concentrations are in units of mg/L, or ppm; nitrate is as NO$_3$.

Table 1 shows the chemical composition of these hypothesized end-members. Each of these waters is a Ca-Mg-HCO$_3$ type composition, with an additional component of Cl, NO$_3$ and SO$_4$. The Woodbine Well end-member has the most dilute chemical composition in relation to the other end-members, with the lowest NO$_3$, SO$_4$ and Cl values among them. This well is located 10 km outside of the Winchester city limits in a pastoral setting of low development and moderate agricultural use. Washington Spring is also located outside of the city limits; however, it is surrounded by land used for cattle pasture and a dairy operation. Due to restricted access, the spring run was sampled approximately 500 m downstream of the rise pool; no tributary flows enter the spring run upstream of the sampling point. The composition of Washington Spring shows the highest measured NO$_3$, with moderately high SO$_4$ and Cl values. Sunnyside Run has its source within the city limits, and flows through an industrial park; it therefore represents an urban influence with the highest Cl and SO$_4$ of the end-members. It is important to emphasize that listing of these hypothetical end-members does not necessarily mean that the actual water from these locales is being expressed at Fay Spring; rather, water samples from these locales exhibit a chemistry of a type that satisfies the end-member mixing model, and are feasible within the conceptual model of the aquifer system.

Assuming simple linear mixing among the end-members and conservative behavior among the chemical constituents, the proportions of the end-members in each of the mixed samples were subsequently calculated. The results of the mixing calculation are shown in time series in figure 8. The shifting end-member (EM) proportions are evident in the different sizes of the bars that comprise each sample collected at Fay Spring. One can observe large increases in the proportion of EM1 during rainfall events, with a limited increase during snowmelt. This shift is reflective of more dilute water affecting Fay Spring during these events. It is noteworthy that the overall trend is a gradual decrease in the proportion of this dilute end-member as baseflow discharge gradually increases in between discrete events. Also, decreases in water temperature accompany increases in the proportion of EM1 during rainfall events (fig. 5).

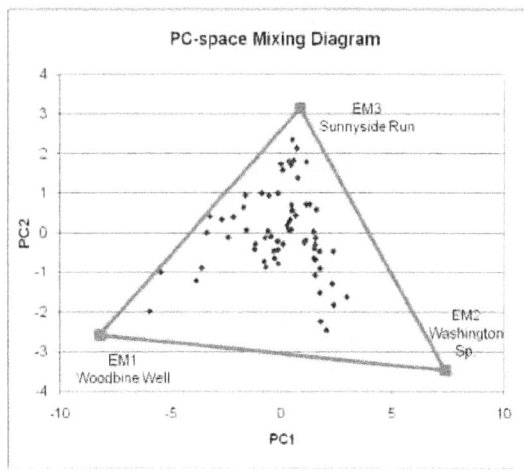

**Figure 7**. Principal components mixing-space diagram indicating Fay Spring samples fully enclosed by the three hypothesized end-members.

During the snowmelt period, the proportion of the high-Cl end-member (EM3) is observed to rise and fall, primarily at the expense of the higher NO$_3$ end-member (EM2). If Cl is indeed a tracer of recent recharge from the surface during snowmelt, then this behavior may be reflective of the input of recently recharged snowmelt water. The high NO$_3$ component represented by EM2 appears to provide the background chemical composition within the spring discharge, with shifting proportions between the low-Cl, low-$NO_3$ composition of EM1 and high-Cl, high-$SO_4$ composition of EM3 reflective of the dynamic changes in aquifer components expressed at the spring.

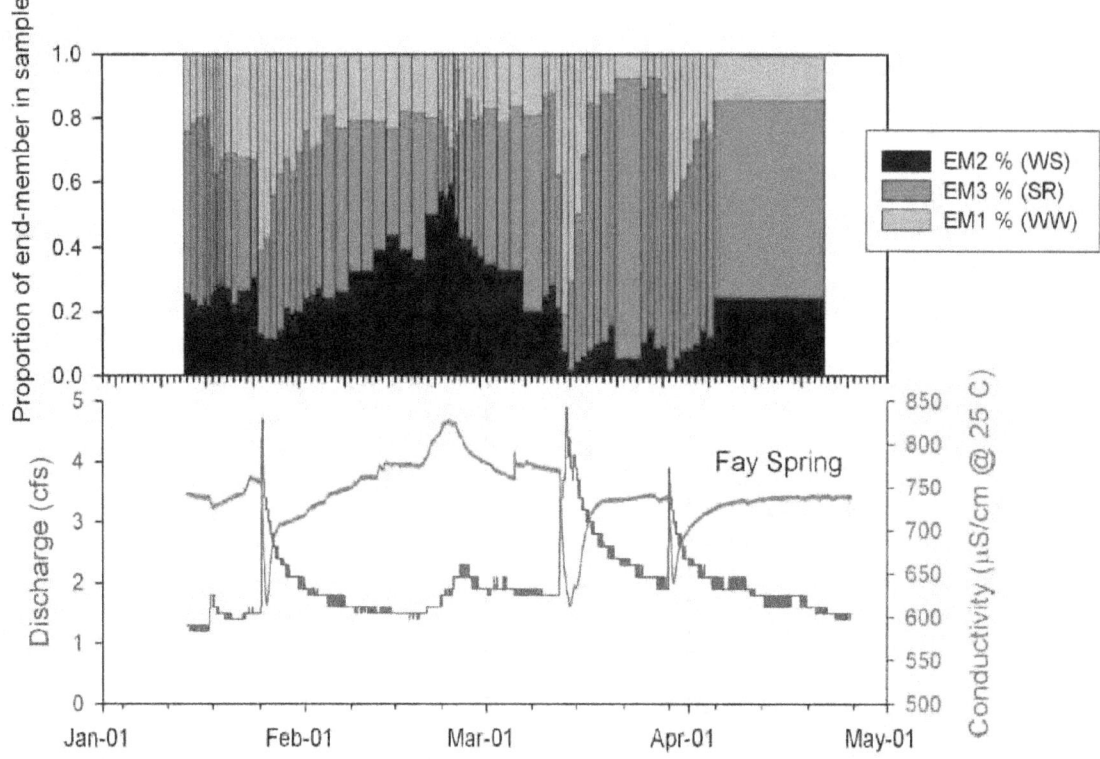

**Figure 8**. Estimated end-member proportions in the Fay Spring samples collected during the winter and spring of 2010.

## DISCUSSION

In many karst areas, focused recharge from sinking surface runoff into the karst aquifer comprises a large proportion of the discharge at springs. In the Shenandoah Valley, however, the karst aquifer seems to have a large component of deeper, phreatic flow in its discharge. In spite of the evidence of sinking surface runoff rapidly impacting Fay Spring through conduit-dominated flowpaths documented by the dye trace results, the amount of dye recovered was low, and repeated pulses of dye recovery during subsequent storm events indicate significant dye retention within the shallow portions of the aquifer.

### Chemical variability at Fay Spring

Frequent sampling of ambient geochemistry combined with continuous monitoring of discharge, temperature and conductivity also reveals functioning of the aquifer. First, only rainfall events above a threshold magnitude cause a discharge response at Fay Spring. This threshold is not well defined, and is dependent upon antecedent conditions, but appears to be in the range of a 0.4 in (10 mm) rainfall event. When a rainfall event of sufficient magnitude does cause a response at Fay Spring, the pattern is generally one of conductivity decrease that is a nearly direct reflection of the discharge hydrograph, and a muted or negligible change in water temperature. Second, only during brief periods in winter does the conductivity rise with an increase in discharge, yet during these periods the water temperature has also shown little change.

Based upon geochemical sampling in both summer and winter seasons, Fay Spring seems to be dominated by a diffuse flow component with elevated background levels of $NO_3$

32

(10-12 mg/L as $NO_3$), $SO_4$ (45-50 mg/L) and Cl (35-40 mg/L) that is noticeably diluted only after large rain events. The majority of the background at Fay Spring is a similar water type as that sampled at Washington Spring (EM2). Greater chemical variability was observed during larger discharge events in the winter and spring than in the summer, but chemical variability was relatively muted overall. For example, conduct-ivity values at Fay Spring do not drop below 600 µS/cm, even during the largest events. Sinking surface runoff might be a reasonable source of the water that causes dilution of the spring chemistry during peak flows, but this component does not dominate the spring discharge. The dye trace resulted in a demonstrable conduit connec-tion between the sinking surface stream of Sunnyside Run and the spring, yet the overall mass recovery was low, and a large proportion of that recovery occurred with subsequent storm events long after the initial dye breakthrough.

During the snowmelt period of 2010, Fay Spring showed a higher conductance and temperature than the water sinking at Sunnyside Run throughout the melt period. It is possible that another, as yet unidentified, source of sinking surface runoff affects Fay Spring.

**Mixing of end-member waters at Fay Spring**

An end-member mixing analysis (EMMA) using samples collected from nearby sources was performed. The chemical composition of Fay Spring was modeled as a mixture among three end-members using the compositions shown in table 1, and all the Fay Spring samples fell inside the range of compositions generated by mixing among these three end-member types. EM1, represented by the Woodbine Assembly well, reflects karst groundwater that has been impacted least by $NO_3$, Cl, or $SO_4$. Our preliminary interpretation is that this water type represents a component within relatively shallow fractures and solutional voids. During larger rain events, this water is mobilized into conduits as connections are made among normally isolated solutional pockets. We observe temperature decreases at Fay Spring after the snowmelt in 2010, but these temperature responses are quite damped (at most 0.6 °C), indicating the cooler component is probably shallow groundwater and

not sinking surface water. EM2, represented by Washington Spring, reflects karst groundwater that has likely been impacted by agricultural activities with the highest $NO_3$ concentrations of the end-members and moderate Cl and $SO_4$ concentra-tions. Being a rather large spring with flow on the order of that at Fay Spring, this spring can be interpreted as part of the broader regional conduit flow system, comprised itself of mixed sources. EM3 is the most impacted by the urban environment, and is represented by the ephemeral sinking surface runoff of Sunnyside Run. When sampled during the snowmelt period, it had high Cl and $SO_4$ concentrations with moderately high $NO_3$ concentration; however, it is expected that the composition of this end-member would not be constant through time.

The changing proportions of the hypothesized geochemical end-members shown in time series illustrate dilution of the spring water across hydrographs in response to large rainfall events; however, during winter the spring water is impacted by a high Cl component instead. It is difficult to define the actual sources of water responsible for these contrasting effects, because the composition of any single end-member is unlikely to be constant in time. For example, although Sunnyside Run (EM3) showed high Cl values during the snowmelt period in which samples were collected, this ephemeral flow may show a different composition in another season. Further sampling is needed to account for this seasonal variability.

In conclusion, a combination of approaches including dye tracing and interpretation of natural geochemistry is helpful for elucidating the nature of discharge for karst springs in fractured carbonate regions. In this case study, the chemistry of Fay Spring appears to show dominance of a deeper aquifer component at base flow, and dominance of shallow ground-water as opposed to sinking surface water runoff during events. The geochemistry of the deeper component indicates impacts by both agricultural and urban activities, resulting in elevated background levels of $NO_3$ and Cl in the spring discharge greater than that observed in groundwater from a well farther from the urban center.

## ACKNOWLEDGMENTS

We thank George Harlow and Roger Moberg of the USGS Virginia Water Science Center (VWSC) for installing and maintaining the continuous monitoring station at Fay Spring. Jim Lawrence of Opequon Watershed, Inc. provided valuable field assistance and use of equipment. Chemical analyses were performed by Mike Doughten at the USGS National Research Program (NRP) lab in Reston, Va. Helpful insight and discussions were provided by Niel Plummer (USGS, NRP) and David Nelms (USGS, VWSC). Finally, we thank Clearbrook Fire & Rescue and Whit Wagner for assisting with the dye trace. Partial funding for this project was provided to N.C. Farrar through the Harrison Undergraduate Research Award from the University of Virginia.

## REFERENCES

Cady, R.C., 1936, Ground-water resources of the Shenandoah Valley, Virginia. Virginia Geological Survey Bulletin 45. 137 pp., 2 plates.

Christopherson, N., and Hooper, R.P., 1992, Multivariate analysis of stream water chemical data: the use of principal components analysis for the end-member mixing problem. Water Resources Research, vol. 28, no. 1, p. 99-107

Doctor, D.H., Alexander, E.C., Jr., Petric, M., Kogovsek, J., Urbanc, J., Lojen, S., and Stichler, W., 2006, Quantification of karst aquifer discharge components during storm events through end-member mixing analysis using natural chemistry and stable isotopes as tracers, Hydrogeology Journal, vol. 14, p. 1171-1191

Harlow, G.E. Jr., Orndorff, R.C., Nelms, D.L., Weary, D.J., and Moberg, R.M., 2005, Hydrogeology and Ground-Water Availability in the Carbonate Aquifer System of Frederick County, Virginia, U.S. Geological Survey Scientific Investigations Report 2005-5161, 30 p.

Jones, W.K., 1991, The carbonate aquifer of the Northern Shenandoah Valley of Virginia and West Virginia, in Kastning, E.H., and Kastning, K. M., editors, Proceedings of the Appalachian Karst Symposium, Radford Virginia, March 23-26, 1991, p. 217-222.

Jones, W.K., 1997, Karst hydrology atlas of West Virginia, Special Publication no. 4, Karst Waters Institute, Charles Town, West Virginia, 111 pp.

Kozar, M.D., McCoy, K.J., Weary, D.J., Field, M.S., Pierce, H.A., Schill, W.B., and Young, J.A.,

2007a, Hydrogeology and Water Quality of the Leetown Area, West Virginia. U.S. Geological Survey Open File Report 2007-1358, 99 pp., 6 appendices.

Kozar, M.D., Weary, D.J., Paybins, K.S. and Pierce, H.A., 2007b, Hydrogeologic setting and ground-water flow in the Leetown area, West Virginia, U.S. Geological Survey Scientific Investigations Report 2007-5066, 80 pp.

McCoy, K.J., and Kozar, M.D., 2007, Use of sinkholes and specific capacity distributions to assess vertical gradients in a karst aquifer, Environmental Geology, DOI: 10.1007/s00254-007-0889-1.

Nelms, D.L. and Moberg, R.M., 2010, Hydrogeology and Groundwater Availability in Clarke County, Virginia, U.S. Geological Survey Scientific Investigations Report 2010-5112, 119 p.

Orndorff, R.C. and Harlow, Jr., G.E., 2002, Hydrogeologic Framework of the Northern Shenandoah Valley Carbonate Aquifer System. Proceedings of the U.S. Geological Survey Karst Interest Group, Shepherdstown, West Virginia, August 20-22, 2002 (E. Kuniansky, ed.), Field Trip Guide, USGS Water Resource Investigations Report WRI-02-4174, p 81-89.

Orndorff, R.C, Weary, D.J. and Parker, R.A., 2004, Geologic Map of the Winchester Quadrangle, Frederick County, Virginia. U.S. Geological Survey Open-File Report 03-461, 1 plate.

Perry, L.D., Costain, J.K., and Geiser, P.A., 1979, Heat flow in western Virginia and a model for the origin of thermal springs in the folded Appalachians. Journal of Geophysical Research, v. 84 no. B12, p. 6875-6883.

Vesper, D.J., Grand , R.V., Ward, K. and Donovan, J.J., 2009, Geochemistry of a spring-dense karst watershed located in a complex structural setting, Appalachian Great Valley, West Virginia, USA. Environmental Geology, vol. 58, no. 3, p. 667-678.

White, W.B, 1988, Geomorphology and Hydrology of Karst Terrains, New York: Oxford University Press, 464 p.

# Aqueous Geochemical Evidence of Volcanogenic Karstification: Sistema Zacatón, Mexico

By Marcus O. Gary[1], Daniel H. Doctor[2], and John M. Sharp, Jr.[3]

[1] Zara Environmental LLC, 1707 West FM 1626, Manchaca, TX 78652

[2] U.S. Geological Survey, 12201 Sunrise Valley Drive, MS 926A, Reston, VA 20192-0002

[3] Department of Geological Sciences, Jackson School of Geosciences The University of Texas, Austin, TX 78712-0254

## Abstract

The Sistema Zacatón karst area in northeastern Mexico (Tamaulipas state) is limited to a relatively focused area (20 km$^2$) in a carbonate setting not prone to extensive karstification. The unique features found here are characteristic of hydrothermal karstification processes, represent some of the largest phreatic voids in the world, and are hypothesized to have formed from interaction of a local Pleistocene magmatic event with the regional groundwater system. Aqueous geochemical data collected from five cenotes of Sistema Zacatón between 2000 and 2009 include temperature (spatial, temporal, and depth profiles), geochemical depth profiles, major and trace ion geochemistry, stable and radiogenic isotopes, and dissolved gases. Interpretation of these data indicates four major discoveries: 1) rock-water interaction occurs between groundwater, the limestone matrix, and local volcanic rocks; 2) varying degrees of hydrogeological connection exist among cenotes in the system as observed from geochemical signatures; 3) microbially-mediated geochemical reactions control sulfur and carbon cycling and influence redox geochemistry; and 4) dissolved gases are indicative of a deep volcanic source. Dissolved $^{87}Sr/^{86}Sr$ isotope ratios (mean 0.70719) are lower than those of the surrounding Cretaceous limestone (0.70730-0.70745), providing evidence of groundwater interaction with volcanic rock, which has a $^{87}Sr/^{86}Sr$ isotope ratio of 0.7050. Discrete hydraulic barriers between cenotes formed in response to sinkhole formation, hydrothermal travertine precipitation, and shifts in the local water table, creating relatively isolated water bodies. The isolation of the cenotes is reflected in distinct water chemistries among them. This is observed most clearly in the cenote Verde where a water level 4-5 meters lower than the adjacent cenotes is maintained, seasonal water temperature variations occur, thermoclines and chemoclines exist, and the water is oxic at all depths. The surrounding cenotes of El Zacatón, Caracol, and La Pilita show constant water temperatures both in depth profile and in time, have similar water levels, and are almost entirely anoxic. A sulfur ($H_2S$) isotope value of $\delta^{34}S = -1.8$ ‰ (CDT) in deep water of cenote Caracol, contrasted with two lower sulfur isotopic values of sulfide in the water near the surface of the cenote ($\delta^{34}S = -7$ ‰ and -8 ‰ CDT). These $\delta^{34}S$ values are characteristic of complex biological sulfur cycling where sulfur oxidation in the photic zone results in oxidation of $H_2S$ to colloidal sulfur near the surface in diurnal cycles. This is hypothesized to result from changes in microbial community structure with depth as phototropic, sulfur-oxidizing bacteria become less abundant below 20 m. Unique microbial communities exist in the anoxic, hydrothermal cenotes that strongly mediate sulfur cycling and likely influence mineralization along the walls of these cenotes. Dissolved $CO_2$ gas concentrations ranged from 61-173 mg/L and total dissolved inorganic carbon (DIC) $\delta^{13}C$ values measured at cenote surfaces ranged from -10.9 ‰ to -11.8 ‰ (PDB), reflecting mixed sources of carbon from carbonate rock dissolution, biogenic $CO_2$ and possibly dissolved $CO_2$ from volcanic sources. Surface measurements of dissolved helium gas concentrations range from 50 nmol/kg to 213 nmol/kg. These elevated helium concentrations likely indicate existence of a subsurface volcanic source; however, helium isotope data are needed to test this hypothesis. The results of these data reflect a speleogenetic history that is inherently linked to volcanic activity, and support the hypothesis that the extreme karst development of Sistema Zacatón would likely not have progressed without groundwater interaction with the local igneous rocks.

# The Influence of Land Use and Occurrence of Sinkholes on Nitrogen Transport in the Ozark Plateaus in Arkansas and Missouri

By Timothy M. Kresse[1], Phillip D. Hays[1], Mark R. Hudson[2], James E. Kaufman[3]
[1]U.S. Geological Survey, 401 Hardin Road, Little Rock, AR, 72211
[2]U.S. Geological Survey, Box 25046, Denver Federal Center, Denver, CO, 80225-0046
[3] U.S. Geological Survey, 1400 Independence Road, Rolla, MO 65401-2602

## Abstract

The presentation will focus on results from a 4-year study that occurred in 3 phases. Phase 1 included a data mining exercise that resulted in over 850 sites (wells and springs) with nitrate-N concentrations for 8 counties in northwest Arkansas. Statistical analysis showed a strong correlation to increasing nitrate-N and increasing cleared (agricultural) land use. Using only spring data, rather than combined springs and wells, resulted in better correlation and higher $r^2$ values. The Ozarks are divided into three hydrologic units: the Western Interior Plains confining system of the Boston Mountains, the Springfield Plateau aquifer, and the Ozark aquifer (Salem Plateau). The Springfield Plateau aquifer was observed to have statistically significant greater mean $NO_3$-N concentrations than the other two units. Using this information, only springs were selected for sampling in the Buffalo River watershed in areas with outcropping Boone Formation (Springfield Plateau aquifer) and in areas with dominantly agricultural land use for the phase 2 study. By controlling the influence of these variables, shown to be dominant variables affecting nitrate-N concentrations in the phase 1 study, approximately half of the fifty-six springs sampled for the study were in areas with a high density of sinkholes and the other half in areas devoid of mapped sinkholes to test the influence of karst features on nutrient transport. Nitrate-N concentrations were significantly higher in areas with high sinkhole density using non-parametric statistics. Because an inspection of the data revealed a strong correlation between nitrate-N concentration and agricultural land use using regression analysis, an analysis of covariance was used to test the level of importance of sinkhole density as a secondary variable. Results demonstrated that at a 10 percent significance level, nitrate-N concentrations were higher for increasing agricultural land use for springs in areas of high sinkhole density. Results from the first two phases were compared to available data from the Ozark Scenic Riverways in southeastern Missouri (phase 3), where data in regard to mapped sinkholes and defined spring recharge areas are much more detailed than that from Arkansas. Using water quality available for 9 springs in this area showed a strong relation ($r^2 = 0.86$) between nitrate-N concentrations and agricultural land use, with a lower, but positive relation between nitrate-N concentrations and sinkhole density ($r^2 = 0.19$). The data set was too small for use of multivariate analysis, but indicates that the occurrence of sinkholes influences nitrate transport, although possibly to a lesser degree than percentage of agricultural land use. Results from this study provide the Buffalo River Park Service and other land use mangers with a potentially strong management tool for identifying areas within a given watershed that are more vulnerable to nutrient sources based on the occurrence of sinkholes. An interesting finding from this study is that nitrate-N concentrations exceed regional mean concentrations where agricultural land use exceeds approximately 30-40 percent of total land use. This suggests that shifting of land use from dominantly forested to agricultural land use exceeding 30-40 percent may result in elevated $NO_3$-N concentrations and consequently may increase the potential for degradation of groundwater resources. With further corroboration, this finding additionally could be an important criterion for guiding land-management planning.

# Seasonal Carbon Dynamics in a Northwestern Arkansas Cave: Linking Climate and Cave Conditions

By Erik D. Pollock[1], Katherine J. Knierim[2], and Phillip D. Hays[3]

[1]University of Arkansas Stable Isotope Lab, Department of Biological Sciences, University of Arkansas, Fayetteville Arkansas 72701

[2]Environmental Dynamics Program, University of Arkansas, Fayetteville, Arkansas 72701

[3] U.S. Geological Survey, Arkansas Water Science Center, Fayetteville, Arkansas 72701

## Abstract

The transport of carbon (inorganic and organic and solid, aqueous, and gaseous phases) was examined in a simple, well-characterized cave system to understand the seasonal variations and meteorological effects on carbon cycling. Soil- and cave-gas and water samples were collected monthly from August 2008 until August 2009 and cave atmosphere carbon dioxide was continuously monitored from April 2009 until September 2009. Water samples were analyzed for the concentration and isotopic composition ($\delta^{13}C$) of dissolved inorganic and organic carbon. Gas samples were analyzed for carbon dioxide concentration and isotopic composition ($\delta^{13}C$). Seasonal variations were apparent in cave carbon dioxide concentration, ranging from near atmospheric concentrations in the winter (580 parts per million) to over 4,200 parts per million in the summer. Carbon dioxide $\delta^{13}C$ ranged from -10.79 per mil (‰) in the winter to -23.18‰ in the summer. Water sample dissolved inorganic carbon concentrations ranged from 2.9 to 6.9 milligrams per liter with $\delta^{13}C$ values from -24.19 to -9.47‰. Dissolved organic carbon ranged from 1.1 to 4.4 milligrams per liter with $^{13}\delta C$ values from -29.78 to -24.71‰. Cave-atmosphere carbon dioxide concentration followed a seasonal pattern with high concentrations in the summer, indicating soil gas (~10,000 parts per million) infiltration into the cave air, and low concentrations in the winter as atmospheric carbon dioxide (~380 parts per million) ventilated the cave. Surface air temperature was positively correlated with carbon dioxide concentrations in the cave. The carbon isotopic composition of both the dissolved carbon species and carbon dioxide did not follow a simple seasonal pattern. The system varied from more negative carbon isotope values in the summer, representative of soil gas input, to more positive values during the spring and fall. However, during the winter a more negative excursion occurred in dissolved inorganic carbon and carbon dioxide isotopic compositions. This excursion is interpreted as the dynamic mixing of soil air and atmospheric air.

## INTRODUCTION

The transport of carbon (inorganic and organic and solid, aqueous, and gaseous phases) was examined in a simple, well-characterized cave system to understand the seasonal variations and meteorological effects on carbon (C) cycling. In karst systems, the well-developed interaction between surface and groundwater and soil, cave, and atmosphere gas masses and the interplay of biotic and abiotic processes provide multiple, bidirectional pathways for C movement between organic and inorganic pools resulting in a system of considerable complexity (Fairchild and others, 2006). Thorough understanding of C processing in these systems is critical to the development of effective land-use and agricultural management practices, as well as protection of sensitive karst ecosystems (Davis and others, 2000). Important inorganic processes (such as bedrock dissolution and mineral precipitation) and biogeochemical processes (such as nutrient attenuation) are dependent on the quantity and quality of C in these distinct pools (Spötl and others, 2005; Winston, 2006). For example, calcite precipitation in caves is controlled by the concentration gradient between dissolved inorganic carbon (DIC) in infiltrating water and carbon dioxide ($[CO_2]$) in the cave air (Baldini and others, 2006). The $[CO_2]$ in the cave atmosphere is a function of the relative contribution of soil gas and atmospheric $CO_2$ (Spötl and others, 2005). In one study on the interflow zone (the soil zone immediately above the epikarst surface), dilution decreased solute concentration during storm events, but biological processing also contributed to lower nitrate ($NO_3^-$) and dissolved organic carbon (DOC) concentrations (Winston, 2006). Although water residence times were shorter during storm events, an influx of labile organic matter to the epikarst zone may have increased rates of denitrification and decomposition of organic matter (Winston, 2006).

For this study, air and water samples from a small, well-characterized cave and the overlying soil in northwestern Arkansas were analyzed for aqueous C species and gaseous $CO_2$ to better understand the movement of C on a seasonal time scale.

## STUDY SITE

The study site is located in the southern part of the Ozark Plateaus Province, which receives an average of 112 centimeters (cm) of precipitation annually (Adamski and others, 1995). According to the National Oceanic and Atmospheric Administration's (NOAA) National Climatic Data Center for the northwestern Arkansas division for the years 1895-2009 precipitation peaks in May at an average of 13.8 centimeters per month and a second, lower maximum typically occurs in October with 10.5 centimeters per month. Annual average air temperature is 14.1°C (NOAA, 2009).

Jack's Cave (36°17'16"N, 93°40'10"W) is representative of the physical and chemical hydrogeology of caves and karst in northwestern Arkansas. The morphology of most caves in the Ozarks comprises single passages controlled by joints that terminate in narrowing sediment-filled rooms (Taylor and others, 2009). The area is capped by Mississippian limestone of the St. Joe Limestone Member of the Boone Formation and other members of the Boone Formation. The entrance to the cave is a fracture in the Middle Ordovician Kings River Sandstone Member of the Everton Formation, and Jack's Cave is developed in an underlying dolomite unit of the Everton Formation. The soil is a very gravelly silt loam and thickness varies depending on the slope (National Cooperative Soil Survey, 2008).

Meteoric water recharges the shallow groundwater system at the site. In Jack's cave,

infiltrating water collects and flows along the cave passage towards the southwest. Direct hydraulic connections were observed at the study site between the soil and cave, but two dye traces failed to identify the discharge point(s) of the cave (Knierim, 2009).

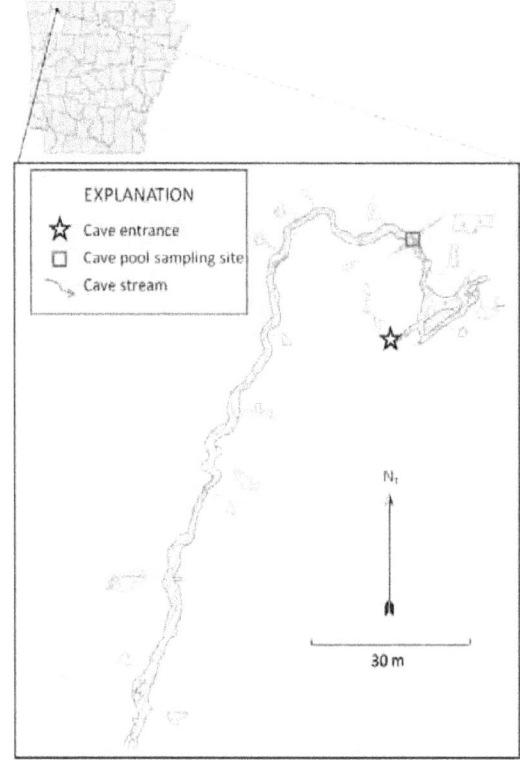

Figure 1. The location and cave map of Jack's Cave showing the cave pool sampling site. Soil lysimeters were installed at the surface approximately 75 m southwest of the cave entrance.

## METHODS

Two soil lysimeters were placed upslope from the cave at depths of 80 and 90 cm. A continuous $CO_2$ monitor was placed in the cave approximately 20 meters (m) from the entrance. Water was sampled from lysimeters and a small cave pool located near the continuous monitor (fig. 1). Water and gas samples were collected approximately monthly from August 2008 through August 2009.

Water samples were collected in 40-milliliter (mL) precombusted total organic carbon (TOC) vials and filtered through 0.45-micrometer (µm) filters. DOC samples were treated with 40 microliters of 3.6 M sodium azide, and all samples were stored at 4°C until analysis. Gas samples were collected in precombusted 100-mL serum vials purged with helium. Water samples were analyzed for the concentration and isotopic composition ($\delta^{13}C$) of DIC and DOC at the Colorado Plateau Stable Isotope Laboratory using continuous flow isotope ratio mass spectrometry (IRMS), TOC-IRMS (±0.75‰; St-Jean, 2003). [DIC] and [DOC] were analyzed using an Aurora IO TOC analyzer (accuracy ±0.1 milligram per liter (mg/L); St-Jean, 2003). Discrete samples of [$CO_2$] were measured monthly using a Vaisala CARBOCAP ® handheld $CO_2$ meter (accuracy ±1.5% of the range + 2% of the reading; Vaisala, 2008), and a LI-COR 840 $CO_2/H_2O$ gas analyzer (accuracy ± 1.5% of the reading; LI-COR, Inc., 2003) was installed in the cave for continuous measurements. Gas samples were analyzed for [$CO_2$] and $\delta^{13}C$-$CO_2$ at the University of Arkansas Stable Isotope Laboratory using gas chromatography-combustion IRMS (±0.56‰). Additionally, samples of TOC in soil (n=12) and carbonate in bedrock (n=5) were analyzed for carbon isotope ratios. Soil samples were collected at approximately 15-cm intervals from the holes in which the lysimeters were installed. Bedrock samples were collected from the St Joe Limestone Member of the Boone Formation above the cave and from dolomite of the Ordovician Everton Formation within the cave.

All isotope ratios are reported using delta notation ($\delta$) in per mil (‰) relative to VPDB (Vienna Peedee Belemnite). The $\delta$ notation indicates the ratio of the heavier to the lighter isotope ($^{13}C/^{12}C$) relative to VPDB (Clark and Fritz, 1997).

Daily temperature and precipitation data from NOAA's weather station at Huntsville, AR (located approximately 23 km southwest of the study location; NOAA, 2010) were averaged for the 14-day period prior to and including the date of sample collection. The 14-day averages eliminated short-term fluctuations in temperature and precipitation.

## RESULTS

Seasonal changes occurred in C species in soil and cave gas and water C isotope ratios and

concentrations. Soil gas [$CO_2$] ranged from 1,400 parts per million (ppm) (April 2009) to 19,450 ppm (July 2009) (data not shown). The [$CO_2$] in the atmosphere of Jack's Cave ranged from 4,200 ppm in the summer to near outside atmospheric concentrations (580 ppm) in the winter (fig. 2). Continuous data were averaged to provide daily readings, and the concentrations recorded by the continuous monitor were in good agreement with discrete monthly concentrations. Continuous data only were recorded for a part of the study period (April 2009 to September 2009) because of equipment problems. The daily averages recorded high-frequency variation of [$CO_2$] (fig. 3); daily changes of several hundred ppm in [$CO_2$] were common (fig. 3). The cave air temperature was nearly constant over the study period (fig. 2), and high frequency changes in $CO_2$ concentration were not related to precipitation events (fig. 3).

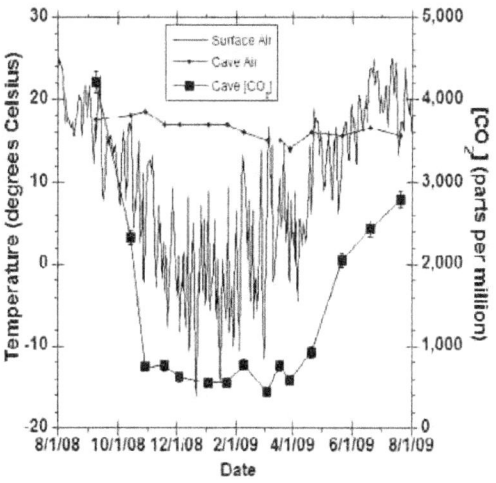

Figure 2. Cave air carbon dioxide concentration ([$CO_2$]) and cave air temperature over the course of the study compared to surface air temperature from August 2008 to August 2009.

The aqueous carbon concentrations in Jack's Cave did vary though not in a simple seasonal pattern. DOC concentration ranged from 1.1 to 4.4 milligrams per liter (mg/L) and DIC concentration ranged from 2.9 to 6.9 mg/L (data not shown). The highest DOC concentration was recorded in January 2009 and the lowest concentration was in December 2008 (fig. 4).

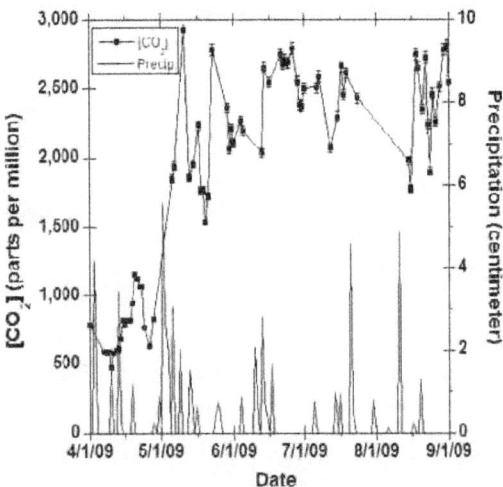

Figure 3. Daily average cave air carbon dioxide concentration ([CO$_2$]) recorded by the continuous monitor and precipitation from April, 2009 to September, 2009.

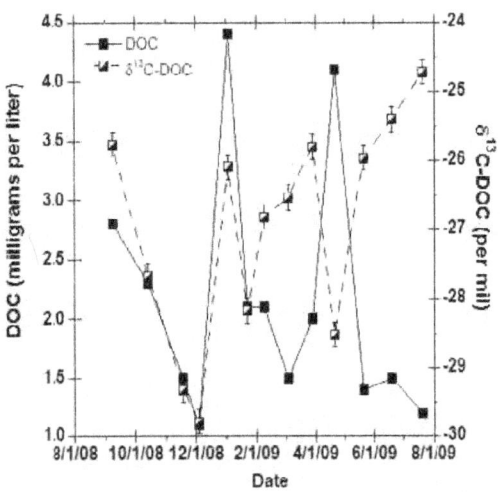

Figure 4. Dissolved organic carbon (DOC) concentration and isotopic composition of DOC ($\delta^{13}$C-DOC) from September 2008 to August 2009 in Jack's Cave.

The isotopic ratios of both the cave air CO$_2$ and the DIC showed similar but not identical patterns. The cave-air isotopic composition of CO$_2$ ($\delta^{13}$C-CO$_2$) ranged from -23.18‰ in the summer to -10.79‰ in the winter (fig. 5). The data exhibit bimodal negative periods. During the warm season, the CO$_2$ isotopic compositions show a long period of more negative isotope ratios with values all less than -17‰. During the cool season, a second period of more negative ratios occurred, from January through

March 2009, with consistent $\delta^{13}$C values of approximately -15.6 ±0.3‰ VPDB. The C isotope ratios of DIC in Jack's Cave varied from -24.19 to -9.47‰ VPDB (fig. 5). More negative isotopic ratios were observed during the warm season. A period of negative isotope ratios also occurred during the winter, separated by a preceding and a following more positive periods (fig. 5). The negative excursion is similar to that observed in the $\delta^{13}$C-CO$_2$ of the cave-air. The difference between the $\delta^{13}$C-CO$_2$ and the $\delta^{13}$C-DIC was variable (-5‰ to 12‰) (fig. 5).

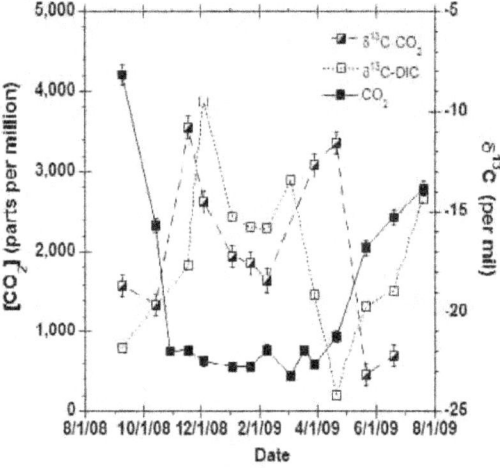

Figure 5. The isotopic composition ($\delta^{13}$C) of dissolved inorganic carbon (DIC) and carbon dioxide (CO$_2$) in Jack's Cave compared to [CO$_2$] from September, 2008 to August 2009. The solid lines represent equilibrium enrichment between DIC and CO$_2$ at 15°C.

The isotopic composition of DOC ($\delta^{13}$C-DOC) ranged from -29.78 to -24.71‰ (fig. 4) and showed a similar but inverse pattern to that of the cave CO$_2$. During the cool season, a period of more positive isotope ratios was observed from February to April. However, the timing of the cool-season light ratio period for $\delta^{13}$C DOC response was delayed relative to the $\delta^{13}$C-CO$_2$ response (February for DOC compared to January for CO$_2$).

Bedrock samples from the overlying St. Joe Limestone Member of the Boone Formation had a C isotope ratio of -1.36‰ ±0.01 (n=2). Dolomite samples from the Everton Formation collected within the cave had C isotope ratio of -3.7‰ ±0.43 (n=2). Soil samples collected from the lysimeter holes had an average $\delta^{13}C$ of -25.15‰ ±0.44 (n= 9).

## DISCUSSION

The [$CO_2$] in the cave displayed a seasonal signature as recorded by monthly readings. A highly variable [$CO_2$] was measured by the continuous monitor. The [$CO_2$] was lowest during the colder months and was correlated to surface air temperature. As surface air temperature increased, [$CO_2$] increased exponentially ($r^2 = 0.82$, data not shown) for the monthly data. The decrease in [$CO_2$] during periods of cooler surface temperatures has been observed in other cave systems (Baldini and others, 2008; Spötl and others, 2005). Seasonal temperature variations change the density contrast between the relatively static cave air and the outside air, which causes ventilation patterns to alternate seasonally. During the cooler months, dense, isotopically heavy, low [$CO_2$] surface air is pulled into the cave. As the surface temperature warms, this ventilation shuts down. Without this ventilation the cave air is dominated by $CO_2$-rich, isotopically more negative soil gas as a result of direct gas diffusion. The relation of the daily reading for [$CO_2$] to surface air temperature was less clear ($r^2= 0.45$, data not shown). The changes in daily [$CO_2$] also were not explained by changes in precipitation ($r^2= 0.04$). For some settings, the rewetting of soils has been shown to cause the release of a pulse of $CO_2$ (Xiang and others, 2008). Other research suggests that high or low soil water content can retard the production of $CO_2$ (Davidson and others, 1998). As such, the high frequency variation in the daily readings is likely a result of changes in the soil temperature and moisture, which are not currently resolved by our data.

The concentrations of DOC (fig. 4) and DIC did vary though not in clear seasonal pattern. The isotopic composition of DIC is a mixture of organically derived soil-gas $CO_2$ and bedrock carbonate, and varies seasonally. High soil gas contributions of relatively negative (approximately -22‰) $CO_2$ during the warm season decreased $\delta^{13}C$-DIC values (fig. 5). Cool season $\delta^{13}C$-DIC ratios were higher, showing an increase in the relative bedrock (-1.36‰ to -3.7‰) contribution to the DIC pool.

The difference between the $\delta^{13}C$-DIC in the cave pool and $\delta^{13}C$-$CO_2$ in the cave air showed that the system is in isotopic disequilibrium; the difference in the two isotopic compositions is variable (fig. 5) and generally less than the -7.77‰ that would be predicted by the relevant fractionation factor (Mickler and others, 2004). This indicates that the cave-air $CO_2$ is largely derived from soil gas during the warm season. The time required for isotopic equilibration for the DIC-$CO_2$ system is on the scale of hours (Zhang and others, 1995), sufficiently rapid for equilibrium to be established in the relatively low-flow conditions found in Jack's Cave. The lack of equilibrium suggests that the cave air is not from evolved DIC, but is from direct input of soil $CO_2$.

Winter [$CO_2$] were lower and $\delta^{13}C$-$CO_2$ values were heavier in Jack's Cave than during warmer periods (fig. 2); this is indicative of a change in the cave air ventilation as relative surface-cave air density values changed with cooling temperature (Spötl and others, 2005). The reversal allowed for surface air, relatively $CO_2$-poor (approximately 380 ppm) and isotopically heavy (approximately -7.7‰) to enter the cave. However, the heaviest $\delta^{13}C$-DIC values were not consistently observed during the period with the lowest [$CO_2$] (figs. 2 and 5). Both the $\delta^{13}C$-$CO_2$ and $\delta^{13}C$-DIC values in the cave were slightly lighter between January and February (fig. 5). The respiration of an organic C pool would provide a source for the lighter isotopic ratios observed in the cave. The addition of soil gas to the cave air would explain the higher values; however no increase in the [$CO_2$] was observed (soil [$CO_2$] averaged 6,880 ppm over the period). The concentration and isotopic ratios of the DOC support the idea of organic respiration. Biological processing of DOC would preferentially use the labile, isotopically higher C (Santruckova and others, 2000). This would have two effects on the DOC pool: (1) the pool itself would shrink and (2) the remaining pool would become progressively enriched in the heavier isotope ($^{13}C$) (Garten and others, 2000). Over the same time period that the cave-air $\delta^{13}C$ $CO_2$ exhibits a negative departure, DOC concentration is decreasing and the isotopic composition is becoming higher. The change in DOC occurs during the cooler and drier portion of the year when the water

41

movement between the surface and the cave is lower and less able to transfer mass. The soil system is also less able to provide those materials in dry, cold, low-productivity conditions, which decrease surface biological processing and subsequent respiration. However, the reduction in the DOC content is not sufficient to account for the more negative isotope ratios alone; insufficient C exists in the organic C pool. The $\delta^{13}$C-$CO_2$ in the winter is a mixture of atmospheric, soil gas and respired DOC, and the respired DOC accounts for the smallest contribution to cave $CO_2$.

## CONCLUSION

The [$CO_2$] in the cave displayed not only a seasonal signature, but also a highly variable daily signal. The [$CO_2$] in the atmosphere of Jack's Cave ranged from 4,200 ppm in the summer to near atmospheric levels (580 ppm) in the winter. During the cooler months, dense, isotopically heavy, low-$CO_2$ concentration surface air was pulled into the cave. The high frequency variation in the daily readings is likely a result of changes in the temperature and soil moisture, which are not currently resolved by our data. The difference between the $\delta^{13}$C-DIC in the cave pool and $\delta^{13}$C-$CO_2$ in the cave air showed that the system is in isotopic disequilibrium. The lack of equilibrium suggests that the cave air is not from evolved DIC, but is from direct input of soil $CO_2$. The $\delta^{13}$C-$CO_2$ and $\delta^{13}$C-DIC values in the cave were slightly lighter between January and February. The addition of soil gas to the cave air would explain the higher values; however no increase in [$CO_2$] was observed. The respiration of an organic C pool would provide a source for the lighter isotopic ratios observed in the cave. However, the reduction in the DOC content is not sufficient to account for the more negative isotope ratios alone; insufficient C exists in the organic C pool. The $\delta^{13}$C-$CO_2$ in the winter is a mixture of atmospheric, soil gas, and respired DOC, and the respired DOC accounts for the smallest contribution to cave $CO_2$.

## REFERENCES

Adamski, J.C., Petersen, J.C., Freiwald, D.A., and Davis, J.V., 1995, Environmental and hydrologic setting of the Ozark Plateaus study unit, Arkansas, Kansas, Missouri, and Oklahoma: U.S Geological Survey Water-Resources Investigations Report WRI 94-4022, 69 p.

Baldini, J.U.L., McDermott, F., Hoffmann, D.L., Richards, D.A., and Clipson, N., 2008, Very high-frequency and seasonal cave atmosphere pCO₂ variability; implications for stalagmite growth and oxygen isotope-based paleoclimate records: Earth and Planetary Science Letters, v. 272, p. 118-129.

Baldini, J. U. L., Baldini, L.. M., McDermott, F. and Clipson, N., 2006, Carbon dioxide sources, sinks, and spatial variability in shallow temperate zonecaves: Evidence from Ballynamintra Cave, Ireland: Journal of Cave and Karst Studies, v. 68, no. 1, p. 4–11.

Clark, I., and Fritz, P., 1997, Environmental Isotopes in Hydrogeology: Boca Raton, N.Y., Lewis Publishers.

Davidson, E. A., Belk, E., Boone, R. D., 1998, Soil water content and temperature as independent or confounded factors controlling soil respiration in a temperate mixed hardwood forest: Global Change Biology, v. 4, p. 217-227.

Davis, R. K., Brahana, J. V., and Johnston, J. S., 2000, Ground water in northwest Arkansas: Minimizing nutrient contamination from non-point sources in karst terrane: Arkansas Water Resources Center, Publication Number MSC-288, http://www.uark.edu/depts/awrc/pdf_files/MSC/MSC-288.pdf.

Fairchild, I.J., Smith, C.L., Baker, A., Fuller, L., Spötl, C., Mattey, D., and McDermott, F., 2006, Modification and preservation of environmental signals in speleothems: Earth-Science Reviews, v. 75, p. 105-153.

Garten, C.T., Cooper, L.W., Post, W.M., and Hanson, P.J., 2000, Climate controls on forest soil C isotope ratios in the southern Appalachian Mountains: Ecology, v. 81, p. 1108-1119.

Knierim, K.J., 2009, Seasonal variation of carbon and nutrient transfer in a northwestern Arkansas cave [M.S. thesis]: Fayetteville, University of Arkansas, 141 p.

LI-COR Biogeosciences, Inc., 2003, LI-840 CO₂/H₂O Analyzer Instruction Manual: Lincoln, NE, LI-COR, Inc., p. 111.

Mickler, P.J., Banner, J.L., Stern, L., Asmerom, Y., Edwards, R.L., and Ito, E., 2004, Stable isotope variations in modern tropical speleothems; evaluating equilibrium vs. kinetic isotope effects: Geochimica et Cosmochimica Acta, v. 68, p. 4381-4393.

National Cooperative Soil Survey, 2008, Custom Soil Report for Madison County, Arkansas: United States Department of Agriculture, Natural Resources Conservation Service, p. 1-16.

National Oceanic and Atmospheric Administration, 2009, Arkansas Northwest Division 01, 1895-2009, http://www7.ncdc.noaa.gov/CDO/CDODivisionalSelect.jsp, [accessed March 3, 2010].

National Oceanic and Atmospheric Administration, 2010, Huntsville, AR (033544), http://www4.ncdc.noaa.gov/cgi-win/wwcgi.dll?wwDI~StnSrch~StnID~20000776, [accessed September 15, 2010].

Santrucková, H., Bird, M.I., and Lloyd, J., 2000, Microbial processes and carbon-isotope fractionation in tropical and temperate grassland soils: Functional Ecology, v. 14, p. 108-114.

Spötl, C., Fairchild, I.J., and Tooth, A.F., 2005, Cave air control on dripwater geochemistry, Obir Caves (Austria): Implications for speleothem deposition in dynamically ventilated caves: Geochimica et Cosmochimica Acta, v. 69, p. 2451-2468.

St. Jean, G., 2003, Automated quantitative and isotopic ($^{13}$C) analysis of dissolved inorganic carbon and dissolved organic carbon in continuous-flow using a total organic carbon analyser: Rapid Communications in Mass Spectrometry, v. 17, p. 419-428.

Taylor, D. S., Goodwin, D. P., Bitting C. J., Handford, R., and Slay, M., 2009, The Ozark Plateaus: Arkansas, *in* eds. Palmer, A. N. and Palmer, M. V., Caves and Karst of the USA: Huntsville, Ala., National Speleological Society, Inc., p. 172-178.

Vaisala, 2003, Vaisala CARBOCAP ® Hand-Held Carbon Dioxide Meter GM70 User's Guide: Finland, Vaisala, 68 p.

Winston, B., 2006, The biogeochemical cycling of nitrogen in a mantled karst watershed: [M.S. thesis]: Fayetteville, University of Arkansas, 98 p.

Xiang, S., Doyle, A., Holden, P.A., Schimel, J.P., 2008, Drying and effects on C and N mineralization and microbial activity in surface and subsurface California grassland soils. Soil Biology and Biochemistry, v. 40, p. 2281-2289

Zhang, J., Quay, P.D., Wilbur, D.O., 1995, Carbon fractionation during gas-water exchange and dissolution of $CO_2$: Geochimica et Cosmochimica Acta, v. 59, p107-114.

# Characterization of Urban Impact on Water Quality of Karst Springs in Eureka Springs, Arkansas

By Renee Vardy

Environmental Dynamics, 113 Ozark Hall, University of Arkansas, Fayetteville, AR 72701

**Abstract**

Shallow karst aquifers in northwest Arkansas are highly vulnerable to urban land use; many springs have shown contamination from dissolved constituents and by indicator bacteria. Eureka Springs, Arkansas is located on the karst terrain of the Ozark Plateau and more than 100 springs have been identified within the city limits. Through collaborative efforts to characterize spring water quality and potentially identify contaminant sources, the U. S. Geological Survey, University of Arkansas, City of Eureka Springs, and Arkansas Department of Environmental Quality have partnered in three recent studies. For these studies, spring water samples were collected and analyses were conducted for Organic Wastewater Constituents (OWC's), indicator bacteria (*E. coli* and fecal coliform), selected geochemical water quality parameters, and flow. OWC's have been detected in both stream sites and one of seven springs. Indicator-bacteria colony counts in the springs ranged from 2 (estimated(e)) to 500(e) cfu (colony forming units)/100mL with an average of 115 cfu/100mL for *E. coli*, and from 8(e) to 130 cfu/100mL with an average of 59 cfu/100mL for fecal coliform. Dissolved nitrate, and chloride concentrations in the springs are greater than natural background levels, with averages of 2.75 mg/L and 11.42 mg/L respectively. These data suggest the influence of urban contaminants in the karst aquifer, highlighting the vulnerable character of the karst hydrologic system.

## INTRODUCTION

Accounting for approximately 25% of worldwide water needs, karst aquifers are a vital water source for many areas of the world (Ford and Williams, 2007). When urbanized areas encroach on karst recharge areas, these aquifers are extremely vulnerable to degradation (Jiang et al., 2009). As urban populations continue to grow, karst areas will be subjected to greater development pressures, and potentially urban contaminants. Protection of karst aquifers from impacts of urbanization is imperative if the water needs of millions of people are to be met, which will be increasingly difficult as urbanization expands (Showers et al., 2008).

Karst aquifers are composed of a framework of interconnected conduits, fractures, voids and pores formed by dissolution of the host rock resulting in a complex flow system with intimate connection between surface water and groundwater. Urban contamination is a common source of degradation of many karst aquifers, not only in the southern Ozarks (Adamski, 1997), but also throughout the world (Jiang et al., 2009; Murray et al., 2007; Zhao et al., 2010). These aquifers are especially vulnerable to contamination for several reasons. For one, rather than being composed of a fine-grained medium capable of effective filtration, as is the case for granular-media aquifers, the open conduits in karst aquifers typically allow unimpeded, unfiltered channel-type flow. In addition, karst areas generally have relatively thin soils, providing minimal filtration between surface runoff and the aquifer (Ford and Williams, 2007). Sinkholes and other solutional features also provide direct connections between surface runoff and aquifer. Underground, leaky storage tanks or damaged piping can lead to direct contamination of the aquifer. These are just a few of the many contamination pathways and sources; others include surface runoff, septic-system effluent, road salt, automotive fluids, household products and industrial chemicals. (Davis et al., 2000; Zhao et al., 2010; Liu et al., 2006; Panno et al., 2001).

As the world population nears 7 billion, the finite water resources available will be stressed to accommodate this growing population. By 2025, close to five billion people are expected to live in urban areas, nearly double the 1995 statistic (United Nations, 1997). The amount of water required for these urban populations would significantly increase, escalating in conjunction with projected agricultural demands. This leads to a dilemma, to serve the basic domestic water needs, or to serve the agriculture food source requirements (Postel, 2000). Areas of the world, stressed by water usage today, will only experience increasing turmoil as the struggle between food production, urban water needs, and monetary gain of water rights sales continues (Postel, 2000). Owing to the overwhelming trend of increasing population (United Nations, 2009) with access to finite resources, even the small, currently unusable, water sources cannot be ignored. The restoration and development of these abused and unused water sources will be increasingly important as the present finite water resources are allocated. The project described in the following pages has been designed to identify sources of urban contamination, suggest options to alleviate the problems, and to restore the shallow karst aquifer to a usable condition in hopes of reestablishing a viable water resource for the city of Eureka Springs. Additionally, it is hoped this case study will provide valuable information on the structure and vulnerability of karst aquifers in similar settings elsewhere.

## Study Area

Eureka Springs is a city of ~2,500 residents, in northwestern Arkansas (Figure 1). Of the known springs, none have acceptable water quality to serve as a public drinking water supply. Recent studies have shown that many of the springs continue to be contaminated by

nitrate, organic wastewater constituents (OWC's), and indicator bacteria (Hays, personal communication; Eureka Springs Springs Committee, personal communication).

Eureka Springs was founded in 1879 as a resort town known for the healing powers of its springs. Although earliest water quality analyses show little uniqueness to the water, tourists were drawn to the city and its economy flourished as bathhouses, hotels, and other attractions opened to accommodate the guests. By late 1879, the population was over 10,000 and it was the fourth largest city in Arkansas. Today the population has dropped and tourism is the towns largest industry (Eureka Springs Historical Society, 2010). Although there are other attractions within the city that draw tourists, the springs are an integral part of the allure. At their current levels of contamination, signs warn of the danger of drinking the water. Several of the springs are even considered public hazards by the Arkansas Department of Environmental Quality owing to their high indicator bacteria counts. Iron gates now restrict access to many springs and none of them are suitable for potable consumption without treatment.

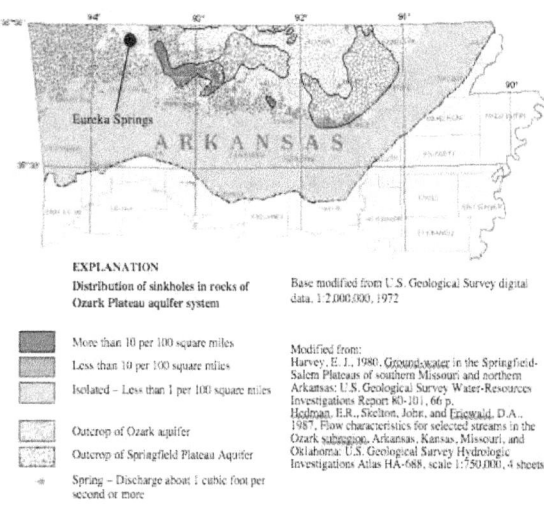

EXPLANATION

Distribution of sinkholes in rocks of Ozark Plateau aquifer system

Base modified from U.S. Geological Survey digital data, 1:2,000,000, 1972

More than 10 per 100 square miles
Less than 10 per 100 square miles
Isolated – Less than 1 per 100 square miles

Outcrop of Ozark aquifer
Outcrop of Springfield Plateau Aquifer

Spring – Discharge about 1 cubic foot per second or more

Modified from:
Harvey, E. J., 1980, Ground-water in the Springfield-Salem Plateaus of southern Missouri and northern Arkansas: U.S. Geological Survey Water-Resources Investigations Report 80-101, 66 p.
Hedman, E.R., Skelton, John, and Freiwald, D.A., 1987, Flow characteristics for selected streams in the Ozark subregion, Arkansas, Kansas, Missouri, and Oklahoma: U.S. Geological Survey Hydrologic Investigations Atlas HA-688, scale 1:750,000, 4 sheets

Figure 1: Distribution of sinkholes and aquifer outcrops in northwestern Arkansas (Renken, 1998)

| Table 1. Water Quality Attributes Indicating Anthropogenic Impacts on Springs in Eureka Springs, Arkansas | | |
| --- | --- | --- |
| Pathogens and Microbes: | Range | Average |
| E. coli | <1 to 1200 cfu/100 mL | 115 cfu/100 mL |
| Fecal Coliform | 8 to 130 cfu/100 mL | 59 cfu/100 mL |
| Total Coliform | 6 to 38,360 cfu/ 100 mL | 2,782 cfu/100 mL |
| Dissolved Constituents: | Range | Average |
| Nitrate | 0.63 to 4.63 mg/L | 2.75 mg/L |
| Chloride | 2.34 to 34.60 mg/L | 11.42 mg/L |
| Organic Wastewater Constituents: | # of Detections | |
| Pesticides | 1 | |
| Synthetic Musks | 2 | |
| UV Light Protectant | 1 | |
| Fragrance/Pharmaceutical Additive | 1 | |
| Food Additive | 1 | (Hays, personal communication; Eureka Springs Springs Committee, personal communication) |
| Flame Retardant | 4 | |
| Dye/Resins/Asphalt | 1 | |
| Topical Insect Repellent | 2 | |

For this area, available background values show springs in the study area are impacted by an external contaminant source(s). Nearby studies of pristine springs have shown nitrate concentrations are < 1 mg/L, typically about 0.3 mg/L (Adamski, 1997). In the study site, nearly all nitrate samples analyzed are above 1 mg/L (Table 1). Also, OWC's have been detected in one of seven springs and both stream sites at low flow. Acceptable levels for many OWC's have not been established. Detections of OWC's in northwest Arkansas are primarily associated with wastewater treatment plant (WWTP) effluent (Galloway et al., 2004). OWC's have been detected in areas of low population density in northwest Arkansas, but they are less prevalent than in sites receiving WWTP effluent. In Eureka Springs, OWC's are detected both above and below the WWTP, as well as in one spring at low flow. Bacteria also are a common contaminant in karst systems in northwest Arkansas (Davis et al., 2000). Their presence can also be highly variable in streams and springs in the area. Low impact areas, even within the study area, can display values as low as 2 cfu/100 mL E. coli, and 8 cfu/100 mL fecal coliform

(Table 1). Other sites within the study area show counts of as many as 1200 cfu/100 mL E. coli, and 38,360 cfu/100 mL total coliform (Table 1). These data suggest that the springs and karst aquifer in Eureka Springs is influenced by urban contamination.

**Hydrologic Framework**

Eureka Springs is located in an area of thinly mantled karst terrain on the Ozark Plateau, a 48,000 $mi^2$ area covering parts of Arkansas, Missouri, Kansas and Oklahoma (Imes and Emmett, 1994). The city lies on a steep escarpment between the Springfield and the Salem Plateau. The geologic formations are the Mississippian aged St. Joe Limestone member of the Boone Formation and older, Ordovician dolomites (Haley et al., 1993; Renken, 1998). These two units host the two regional aquifers in the area. Springs in this area have discharge variable with season and location generally ranging from <1 to 5 cfs.

The St. Joe member of the Boone Formation is included in the Springfield Plateau aquifer. The St. Joe is a relatively pure phase limestone, with little chert. It is known throughout northwest Arkansas as a prominent cave former. Caves and karst in the St. Joe are largely aligned with solutionally enlarged fractures, which are arranged orthogonally and potentially were caused by regional uplift. The resulting karst terrain allows surface runoff to quickly mix with groundwater with little filtration (Ford and Williams, 2007, Winter et al., 1998; Davis et al., 2000).

The Chattanooga Shale comprises the Ozark confining unit, which separates the Springfield Plateau aquifer from the deeper Ozark aquifer. The Ordovician aged Cotter and Jefferson City Formations (Ozark aquifer) lie beneath the Chattanooga. Many prominent springs in the area flow from the Ozark aquifer. Locally, water from the Springfield Plateau aquifer passes though the Ozark confining unit by way of fractures. These fractures, along with direct infiltration into exposed Cotter and Jefferson City Formations, cause the Ozark aquifer to also be susceptible to contamination.

## Historical Studies

Several studies provide an excellent water quality database and hydrogeologic framework for Eureka Springs. The most significant of these are an extensive geologic mapping project by Purdue and Miser (1916), and an exfiltration study by Husain and Aley (1981).

Although several other geologic maps of the area are available, none are as detailed as the original map produced by Purdue and Miser. In this map localized faults and fractures are delineated along with precise locations of outcrops. This aids in analyzing potential flow within the aquifer, and suspected mixing between aquifers along fracture zones.

The exfiltration study was designed to identify leakage in aging sewer lines throughout the city. It was funded by a group of concerned citizens, who would later become the Eureka Springs Springs Committee, through a grant from the EPA. In this study performed by the

Comprehensive Planning Institute, optical brighteners, chemicals found in most laundry detergents, were utilized to identify sewage effluent and classify spring contamination levels. Dye tracing, as well as knowledge of the geology of the aquifer-containing units and the structures present, were used to delineate the recharge areas for each of the springs (Husain and Aley, 1981). These data have proven to be a great resource for comparisons over the last 30 years.

## Recent Studies

Since the exfiltration study in 1981, several other studies have been conducted on the springs of Eureka Springs. These include bacterial investigations, student dye tracing, hydrologic investigations, general geochemistry, and organic wastewater constituent sampling (Eureka Springs Springs Committee, personal communication; Husain and Aley, 1981; Brahana et al., 1993; Hays, personal communication).

### Indicator Bacteria

Many different types of bacteria are found in spring water systems, several of which are not dangerous for human consumption. Those associated with wastewater and more specifically, fecal matter, can be harmful to human health. Several types of fecal bacteria are present in the digestive tracts of mammals, and passed to the outside environment through excrement. Indicator bacteria, easily identifiable bacterial species that are associated fecal matter, are often used to signal fecal input (Davis et al., 2005; Carson et al., 2001). At the study site, indicator bacteria analysis is used to signify the presence of fecal matter and thus potentially more dangerous bacterial species.

In Eureka Springs, indicator bacteria have been recorded for various springs since 2008 (Figure 2). The USGS in conjunction with the Springs Committee initiated sampling for the common indicator bacteria *E. coli*, fecal coliform, and total coliform. Counts for these bacteria range from <1 to 1,200 cfu/100 mL (n=38, average=100 cfu/100 mL) for *E. coli*, 8 to 130 cfu/100 mL (n=9, average=59 cfu/100mL) for fecal coliform and from 6 to 38,360 cfu/100 mL (n=31, average=2,782 cfu/100 mL) for total coliform (Table 1, Figure 2; Springs Committee, personal communication; Hays, personal communication).

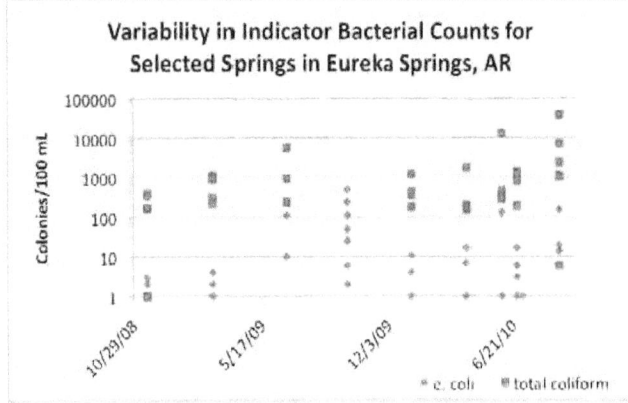

Figure 2: Indicator bacteria concentrations for selected springs, 2008-2010 (Eureka Springs Springs Committee, personal communication)

Bacteria counts in these springs vary significantly with precipitation patterns, spring location, and spring aquifer source. Precipitation increases runoff, which in turn causes a flush of sediment and bacteria from the surface into the aquifer or potentially from storage within in the aquifer (Davis et al., 2005). In a study done on *E. coli* growth and survival bacteria were observed to survive in a dormant form in the aquifer for up to four months. Therefore, if the precipitation events flushing fresh bacteria into the karst system occur more often than every four months, a viable bacterial community likely is present in the aquifer at all times, capable of being flushed out into the spring during rain events (Davis et al., 2005).

Many of the spring orifices' are altered by decorative additions meant to showcase the pride of Eureka Springs, their springs. The construction of houses, businesses, roadways, parking lots, and drainage structures in springs recharge areas, has caused a general decrease in flow (Eureka

Springs Springs Committee, personal communication) as well as increased the potential for contamination from urban sources, such as wastewater. As a prominent source of bacteria, wastewater is a likely culprit of the high indicator bacteria counts in several of the springs.

Wastewater in the city has two major potential sources: leaking sewer lines, or improperly functioning septic systems. Much of the city's sewer infrastructure was designed and installed prior to 1900. These original structures were constructed from vitrified clay piping. Many of these pipes have been replaced, but several original sections within the city limits are still in use. These aged pipes are suspected of leakage in some areas of the city and could be contributing wastewater input to the springs. On one occasion, preliminary dye tracing discovered connection between a toilet and spring within minutes of dye introduction (Hays, personal communication).

The other major source of potential wastewater contamination in the springs are improperly functioning septic systems. Properly designed and functioning septic systems require the presence of soils to disperse the nutrient and bacteria-rich wastewater evenly to aid in degradation (McQuillan, 2004). In the karst terrain at Eureka Springs, soils are generally thin leading to little filtration of the wastewater between the septic leach lines and the open conduits of the karst aquifer.

Distinct chemical differences are observable between the Springfield Plateau and Ozark aquifers and the springs that recharge from each (Figure 4). Many small springs discharge from the Springfield Plateau aquifer, which on average show higher bacteria counts than their Ozark aquifer counterparts. This is likely due to the large outcrop of the Springfield Plateau aquifer unit in the central part of town allowing increased exposure to potential contamination sources. The Ozark aquifer also shows impacts of bacterial contamination in many springs, but the degree of contamination is not as great as the Springfield Plateau aquifer, due to the presence of the Ozark confining unit that separates the two aquifers and the relatively small outcrop of the Ozark aquifer in the center of the city (Purdue and Miser, 1916).

**Organic Wastewater Constituents**

Organic Wastewater Constituents (OWC's) are a relatively new set of chemical parameters to be examined in water systems because the ability to detect them at the concentrations in which they typically are found in the environment has only recently been developed. This group of chemicals includes any substance that is passed into wastewater after its intended use including antibiotics, pharmaceuticals, hormones, personal care products, caffeine, over the counter medications, and many others. Although the long-term effect of these products beyond their intended use requires further research, the potential effects of exposure to OWC's include reproductive impairment, carcinogenic effects, and the eventual development of antibiotic resistant bacteria (Galloway et al., 2004; Kolpin et al., 2002).

In northwest Arkansas, OWC's have been detected in many streams (Galloway et al., 2004). Common OWC's found in the area are caffeine, flame-retardants, byproducts of pharmaceutical degradation, personal care products, pesticides, and insect repellents. The presence of these constituents in northwest Arkansas leads to concern of their presence at Eureka Springs. In September 2009, the USGS, Eureka Springs Springs Committee, and the Arkansas Department of Environmental Quality analyzed water samples for OWC's in Eureka Springs. Of the 62 OWC's analyzed, 12 were found in at least one of the sampling sites (Hays, personal communication). Although the potential effects of exposure to OWC's are uncertain, in Eureka Springs, their presence could impact a much larger area than Eureka Springs (Galloway et al., 2004). Leatherwood Creek, the main stream draining Eureka Springs, is a tributary to

the White River. The White River, a nationally protected water body and municipal water supply to thousands, is located downstream of Eureka Springs. Not only could contamination of these springs lead to a decrease in the livelihood of Eureka Springs by discouraging tourism, but it could also affect a major regional water supply.

**General Geochemistry**

This study utilizes three separate sets of available geochemical data for selected springs in the study area. The first data set includes total nitrogen, ammonia, nitrate, total phosphorous, and reactive phosphorous for selected springs from 2008 to present (Figure 3).

In Eureka Springs a majority of nitrogen detections in groundwater are in the form of dissolved nitrate. Since 2008, the values found have all been under the Environmental Protection Agency maximum contaminant level for drinking water of 10 mg/L (EPA, 2009), but they have also almost all been above the pristine levels of >1 mg/L (Adamski, 1997). Basin, Harding, and Sweet springs show the highest levels of nitrate of any sampled in this dataset. These springs are all located, and recharged in the central part of town, potentially affected by wastewater effluent. They are also all sourced from the Springfield Plateau aquifer. Magnetic Spring is consistently one of the more pristine springs. It is sourced from the Ozark aquifer with a largely undeveloped recharge area on the outskirts of town. Leatherwood Creek is the main water body draining the central part of town and has also historically shown relatively low nitrate values (Eureka Springs Springs Committee, personal communication).

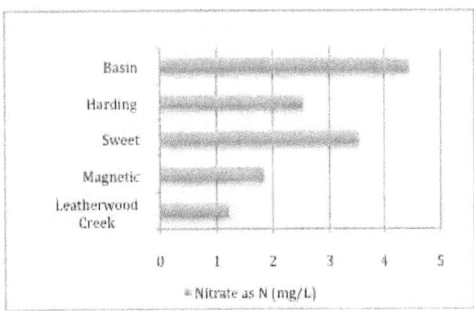

Figure 3A: Average nitrate concentration for selected springs and Leatherwood Creek 2008-2010

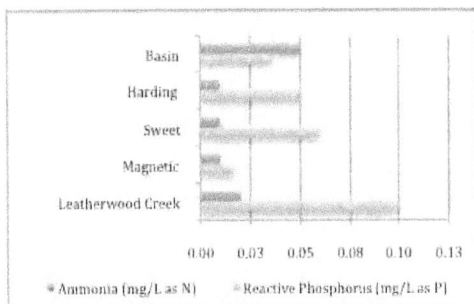

Figure 3B: Average ammonia and reactive phosphorous concentrations for selected springs and Leatherwood Creek 2008-2010

Other contaminants of concern, ammonia and reactive phosphorous, are also detected in area springs and streams. Basin spring shows the highest ammonia concentration at 0.05 mg/L. As with nitrate, a potential source of ammonia is wastewater effluent leaking from damaged underground piping. Phosphorous, also potentially from wastewater effluent, is a contaminant of major concern in northwest Arkansas (Soerens et al., 2003; Benson et al., 2008).

A second geochemical data set was collected simultaneously with the OWC samples. This includes data from 7 springs and 2 stream sites within the study area. Parameters included in this analysis included major cations and anions, metals and nutrients (Figure 4).

Through analysis of these geochemical data, much important information is gained. The characteristic Ozark aquifer spring, Magnetic, displays a relatively high concentration of $Mg^{2+}$, as a reflection of its dolomitic host rock. The Springfield Plateau aquifer springs, with their limestone host rock, show high $Ca^{2+}$ concentration as in the Ozark aquifer, but lack $Mg^{2+}$. The Ozark aquifer also has generally lower $Cl^-$ and nitrate concentrations than the Springfield Plateau aquifer, potentially indicating less affect from urban sources. This reflects the greater isolation from the anthropogenic activities on the surface.

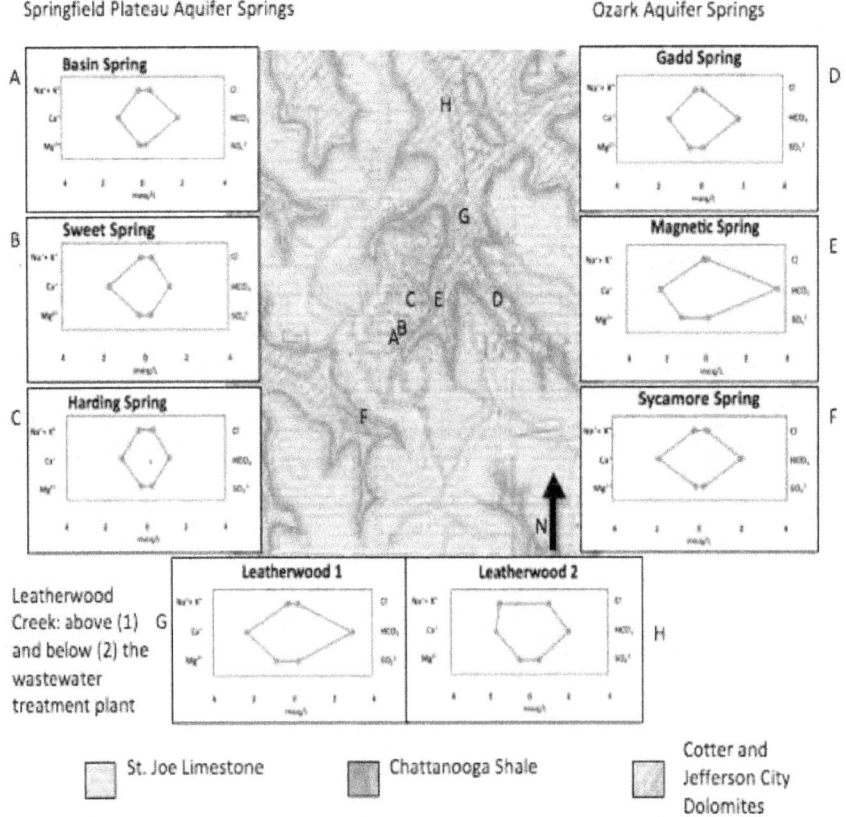

Figure 4: Geochemistry of selected spring and stream sites (Hays, personal communication; Purdue and Miser, 1916)

Ozark aquifer springs vary in chemistry, depending on location. Sycamore Spring is an Ozark aquifer spring, but it appears to be influenced by Springfield Plateau aquifer as indicated by the concentrations of $Na^+$ and $SO_4^{2-}$. The major dissolved constituents look similar to that of the Springfield Plateau aquifer springs. This apparent mixing suggests a fault or other open connection across the Chattanooga Shale near this location.

An integrated mixing of surface runoff and the springs can be observed in Leatherwood Creek prior to the addition of WWTP effluent. Below the WWTP outlet there is a distinct change in the chemistry noted by elevated levels of $Cl^-$, $Na^+$, and $SO_4^{2-}$. These are all urban contaminants common in wastewater effluent (Figure 4; McQuillan, 2004).

The final set of geochemical data was gathered Fall 2010. This project was designed to analyze nitrate, chloride, conductivity, and temperature though a rainfall event in order to characterize the "flush" of urban contaminants in the karst aquifer system. As suspected, nitrate varied with water level (Figure 5B). With the flush of water from the storm pulse, nitrate levels were elevated. Conductivity also showed an influence by the water level pulse, but in a different way. Prior to the nitrate surge, a conductivity pulse occurred as the dissolved solid laden aquifer water was forced out by the incoming runoff. Chloride showed no recognizable response to the storm event and the concentrations did not suggest anthropogenic chloride input.

**Flow Data**

Numerous spring discharge measurements have been recorded in Eureka Springs, and most reflect low-flow discharges of several liters per minute. Several are known to go dry during the summer months, whereas others have perennial flow. After intense precipitation, many of these springs show obvious increases in discharge. From known

records, no long term continuous monitoring of the springs in Eureka Springs has been undertaken.

During the fall 2010, the water level of Magnetic Spring was monitored throughout a storm event. In this analysis two transducers were set up within the Magnetic Spring pool along with a control transducer, which was used to correct the spring stage data for changes in barometric pressure. Several precipitation events were recorded over the 10-day span the transducers were in place. The average spring response lag time was approximately

A.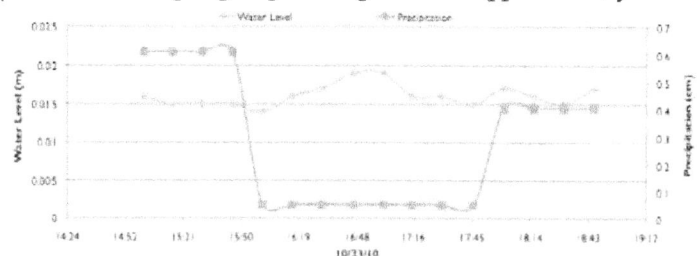

B.

C.

D.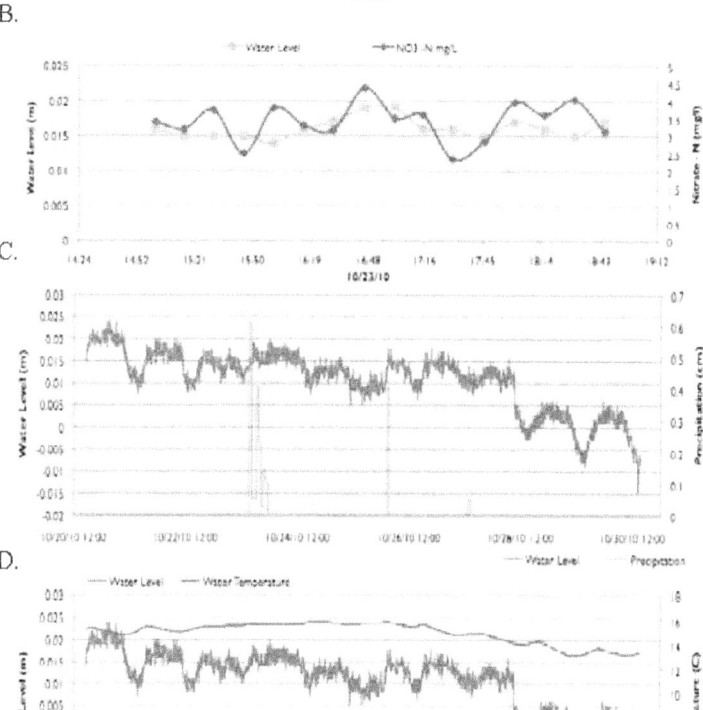

Figure 5:
A. Spring stage variation in response to precipitation 10/23/10.
B. Nitrate variation in response to spring stage 10/23/10.
C. Spring stage variation in response to precipitation for the period 10/20/10-10/30/10
D. Spring stage variation and water temperature for the period 10/20/10-10/30/10

51

1 hour (Figures 5A and 5C). A diurnal fluctuation in temperature and flow was recognized as well (Figures 5C and 5D). The diurnal fluctuation in flow showed an increase in flow at night relative to daylight. The recharge area of Magnetic Spring is heavily forested and thus, evapotranspiration is suspected as the cause for this periodic signal.

## Future Work

Within the study area there are many pieces of the puzzle that are missing. Community groups are providing an excellent base of information, but gaps are present in the data. A continuous suite of flow, indicator bacteria, and water chemistry data from representative springs would lend immense amounts of data on the variations of the springs throughout the year, relative to contaminant sources.

Isotopes from water and nitrate can also be helpful in determining sources. The two elements in water, hydrogen and oxygen, have varying isotopic ratios that can signify the evaporation exposure, and season of precipitation of the recharge water. Also, as nitrogen and oxygen are incorporated into nitrate, the original isotopic signatures of these elements are retained. These signatures distinguish whether the nitrate was originated from sewage, soil organic matter, agriculture, or the atmosphere (Clark and Fritz, 1997; Cascotti et al., 2002, Steffy and Kilham, 2004). Thus, invaluable information on contaminant sources is available through isotopic studies of Eureka Springs.

## Conclusion

Although there have been several excellent studies centered on Eureka Springs, there are still many basic questions left unanswered. Answering these questions is imperative to the preservation and restoration of this valuable groundwater resource. When the city was founded, the springs were the crowning jewel, and now their impurity is an embarrassment. The citizens of Eureka Springs are dedicated to the preservation and restoration of the springs. With the appropriate resources, these springs can once again be the pride of Eureka Springs.

## REFERENCES

Adamski, J., 1997, Nutrients and pesticides in ground water of the Ozark Plateaus in Arkansas, Kansas, Missouri, and Oklahoma, U. S. Department of the Interior, U. S. Geological Survey, Water-Resources Investigations Report 96-4313.

Benson, V. W., Baffaut, C., Robinson, D., Intarapapong, W., Farrand, T., Rogers, W., & Nam, L., 2008, Upper White River Watershed Integrated Economic and Environmental Management Project, Food and Agricultural Policy Research Institute, FAPRI–MU Report #09-08

Brahana, J. V., Leidy, V. A., Lindt, J., & Hodge, S. A., 1993, Hydrogeologic Data for Carroll County, Arkansas, U. S. Geological Survey, Open-file Report 93-150.

Carson, A., Shear, B., Ellersieck, M., Asfaw, A., 2001, Identification of Fecal Escherichia coli from Humans and animals by ribotyping, Applied and Environmental Microbiology, Apr. 2001, p. 1503–1507.

Casciotti, K., Sigman, D., Galanter Hastings, M., Bohlke, J., Hilkert, A., 2002, Measurement of the Oxygen Isotopic Composition of Nitrate in Seawater and Freshwater Using the Denitrifier Method, Analytical Chemistry, 74, 4905-4912.

Clark, I., and Fritz, P., 1997, Environmental Isotopes in Hydrogeology, CRC Press, New York.

Davis, R., and Brahana, J., Johnston, J., 2000, Groundwater in northwest Arkansas: Minimizing nutrient contamination from non-point sources in karst terrane, Arkansas Water Resources Center, Publication No. MSC-288.

Davis, R. K., Hamilton, S., and Brahana, J. V., 2005, Escherichia Coli survival in mantled karst springs and streams, northwest Arkansas Ozarks, USA, Journal of American Water Resources Association. vol. 41:6

Environmental Protection Agency, 2009, National Primary Drinking Water Standards, EPA 816-F-09-004

Eureka Springs Historical Society, accessed 10/27/2010, Historical, http://www.eurekasprings.com/history/

Ford, D., and P. Williams, 2007, Karst Hydrogeology and Geomorphology, Wiley.

Galloway, J. M., Haggard, B. E., Meyers, M. T., & Green, W. R., 2004, Occurrence of Pharmaceuticals and Other Organic Wastewater Constituents in Selected Streams in Northern

Arkansas, 2004, U. S. Geological Survey, Scientific Investigations Report 2005-5140.

Haley, B., Glick, E., Bush, W., Clardy, B., Stone, C., Woodward, M., and Zachry, D., 1993, Geologic Map of Arkansas, Arkansas Geological Commission and U. S. Geological Survey, scale 1:500,000

Husain, K., and Aley, T., 1981, An Exfiltration Study for Eureka Springs, AR, Comprehensive Planning Institute.

Imes, J.L., and Emmett, L.F., 1994, Geohydrology of the Ozark Plateaus Aquifer System in parts of Missouri, Arkansas, Oklahoma, and Kansas: U.S. Geological Survey Professional Paper 1414–D, 127 p.

Jiang, Y., Wu, Y., and Yuan, D., 2009, Human Impacts on Karst Groundwater Contamination Deduced by Coupled Nitrogen with Strontium Isotopes in the Nandong Underground River System in Yunan, China, Environ. Sci. Technol. 2009, Vol. 43, No. 20.

Kolpin, D. W., Furlong, E. T., Meyer, M. T., Thurman, E. M., Saugg, S. D., Barber, L. B., and Buxton, H. T., 2002, Pharmaceuticals, Hormones, and Other Organic Wastewater Contaminants in U.S. Streams, 1999-2000: A National Reconnaissance, Environ. Sci. Technol. 2002, 36, 1202-1211

Liu, C.,Li, S., Lang, Y., Xiao, H., 2006, Using $\delta^{15}N$- and $\delta^{18}O$-Values To Identify Nitrate Sources in Karst Ground Water, Guiyang, Southwest China, Environmental Science and Technology, vol. 40:6928-6933.

McQuillan, D., 2004, Ground-water quality impacts from on-site septic systems, Proceedings, National Onsite Wastewater Recycling Association, 13th Annual Conference Albuquerque, NM, November 7-10, 2004.

Murray, K. E., Straud, D. R., & Hammond, W. W., 2007, Characterizing groundwater flow in a faulted karst system using optical brighteners from septic systems as tracers, Environmental Geology, vol. 53:769–776

Panno, S. V., Hackley, K. C., Hwang, H. H., and Kelly, W. R., 2001, Determination of the sources of nitrate contamination in karstsprings using isotopic and chemical indicators, Chemical Geology 179:113–128

Postel, S., 2000, Enter an Era of Water Scarcity: The Challenges Ahead, Ecological Applications, Vol. 10:4, pp. 941-948Purdue, A. H. and Miser, H. D., 1916, U. S. Geological Survey Atlas, Folio 202.

Renken, R. A., 1998, Groundwater Atlas of the United States: Arkansas, Louisiana and Mississippi, U.S. Geolgical Survey, Hydrologic Atlas 730-F,

Showers, W., Genna, B., Mcdade, T., Bolich, R., & Fountain, J. C., 2008, Nitrate Contamination in

groundwater on an urbanized dairy farm, Environmental Science and Technology, vol. 42, no. 13.

Soerens, T. S., Fite, E. H., and Hipp, J., 2003, Water Quality in the Illinois River: Conflict and Cooperation between Oklahoma and Arkansas, Diffuse Pollution Conference Dublin 2003

Steffy, L., and Kilham, S., 2004, Elevated $\delta^{15}N$ in stream biota in areas with septic tank systems in and urban watershed, Ecological Applications, 14(3), pp. 637-641.

United Nations, 1997, World Urbanization prospects: the 1996 revision, New York, NewYork, 1997.

United Nations, 2009, World Population to Exceed 9 Billion by 2050: Developing Countries to Add 2.3 Billion Inhabitants with 1.1 Billion Aged Over 60 and 1.2 Billion of Working Age, United Nations Press Release.

Winter, T. C., Harvey, J. W., Franke, O. L., & Alley, W. M., Ground Water And Surface Water A Single Resource, USGS Circular 1139, http://pubs.usgs.gov/circ/circ1139/

Zhao, M., Zeng, C., Liu, Z., Wang, S., 2010, Effect of different land use/land cover on karst hydrogeochemistry: A paired catchment study of Chenqi and Dengzhanhe, Puding, Guizhou, SW China, Journal of Hydrology 388 (2010) 121–130

# The Relation between Dissolved Oxygen and other Physicochemical Properties in Barton Springs, Central Texas

By Barbara J. Mahler[1] and Renan Bourgeais[2]
[1]U.S. Geological Survey, 1505 Ferguson Lane, Austin, TX 78754
[2] Ecole Nationale du Génie de l'Eau et de l'Environnement de Strasbourg, 1 Quai Koch, B.P. 61039, 67070 Strasbourg, France

## Abstract

Dissolved oxygen (DO) is a critical component of karst water quality and is necessary for the health and reproduction of aquatic biota at karst springs. To better understand the relation between DO and other physicochemical properties, 6 years of daily DO data collected during January 2004–April 2010 from Barton Springs in Austin, Tex., were analyzed. Barton Springs is the only known habitat of the endangered Barton Springs salamander, *Eurycea sosorum*, whose probability of survival decreases by 50% at a DO concentration of less than 3.4 mg/L. For the 6 years examined, DO ranged from 4.0 to 8.5 mg/L. A two-segment multiple linear regression model was developed using DO, spring discharge (Q), and temperature (T), with a breakpoint of Q=70 ft$^3$/s separating the segments of the model. Q explained most of the variability in DO when Q was less than 70 ft$^3$/s (positive correlation with Q) ($r^2$=0.88), and T explained most of the variability in DO when Q was greater than 70 ft$^3$/s (inverse correlation) ($r^2$=0.67). When only the data collected after a change to an optical DO sensor (November 2006) were considered, an improved relation between DO and a combination of Q and T was obtained when Q was more than 70 ft$^3$/s ($r^2$=0.86). The relations between DO, Q, and T did not hold, however, in the hours to days following storm recharge. For example, DO was observed to increase by as much as 55 percent in 18 hours following one storm, and was closely correlated with turbidity and specific conductance, indicating that simple mixing with recharging surface water likely controls the DO concentration. These results indicate that species that depend on sufficient amounts of DO in Barton Springs for survival, such as the Barton Springs salamander, might be under increased stress if discharge decreases, temperature increases, or other processes that reduce DO concentrations intensify.

# Geochemistry, Water Sources, and Pathways in the Zone of Contribution of a Public-Supply Well in San Antonio, Texas

By Lynne Fahlquist, MaryLynn Musgrove, Gregory P. Stanton, and Natalie A. Houston
U.S. Geological Survey, 1505 Ferguson Lane, Austin, TX 78754-4733

## Abstract

In 2001, the National Water-Quality Assessment (NAWQA) Program of the U.S. Geological Survey initiated a series of studies on the transport of anthropogenic and natural contaminants (TANC) to public-supply wells in representative aquifers of the United States with the goal of better understanding source, transport, and receptor factors (such as well field management practices) that affect contaminant movement to public-supply wells (PSWs). One of the selected TANC regional study areas is within the San Antonio segment of the Edwards aquifer (hereinafter the regional aquifer), an important water resource in a rapidly urbanizing region in south-central Texas. Like many karst aquifers, the Edwards aquifer responds rapidly to changes in hydrologic conditions and is highly susceptible to contamination. A local-scale study area was delineated on the basis of a flow-modeled zone of contribution to a selected PSW and its associated well field in the larger study area. The local-scale study area is representative of the regional aquifer, including unconfined (recharge) and confined zones of the aquifer. Results for a variety of geochemical constituents and tracers were compared for groundwater samples from regional wells, the well field (wellhead samples), nearby monitoring well clusters, an overburden well, and selected depths within the selected PSW collected under varying pumping conditions. Temporal samples that were collected from the selected PSW and selected monitoring wells in response to a recharge event also were compared.

Although hydrogeologic zones of preferential flow were determined for the selected PSW, groundwater samples from different hydrogeologic zones were not geochemically distinct. The geochemistry of groundwater from the selected PSW was similar to that of confined groundwater samples collected from other PSWs throughout the regional aquifer. These results indicate that water from the selected PSW and its associated well field is generally representative of the regional confined aquifer. All samples collected from the well field and specifically those collected from different depths in the selected PSW under a variety of pumping conditions were relatively homogeneous and well-mixed for numerous geochemical constituents, with the notable exception of tritium/helium-3 age tracers.

Geochemical and isotopic data are useful tracers of recharge, groundwater flow, fluid mixing, and water-rock interaction processes affecting water quality at the well field. Results for stable isotopes of hydrogen and oxygen for groundwater samples indicated that groundwater is meteoric in origin. Sources of dissolved constituents to Edwards aquifer groundwater and the selected PSW include dissolution of and geochemical interaction with overlying soils and calcite and dolomite minerals that compose the aquifer. Molar ratios of magnesium to calcium and strontium to calcium in carbonate groundwater typically increase along flow paths; and strontium isotope values in Edwards aquifer groundwater typically decrease along flow paths, approaching values similar to those of the Cretaceous limestone aquifer rocks. Molar ratios of magnesium to calcium and strontium to calcium, along with strontium isotope values, were used to assess relative extents of water-rock interaction associated with shorter and longer flow paths.

One of the other well field PSWs and four selected monitoring wells were sampled several times in response to a storm event that recharged the regional aquifer. The measured temporal variations in water chemistry were consistent with the observation that response to individual recharge events in the confined aquifer is likely attenuated by mixing processes along regional flow paths, unless flow paths intersect conduits.

Anthropogenic sources that might influence groundwater quality at the selected PSW include septic systems, leakage from municipal water and wastewater systems, and general industrial, commercial, or residential use of pesticides and volatile organic compounds. Constituents of concern for the long-term sustainability of the groundwater resource include anthropogenic organic compounds and the nutrient nitrate. While the urban San Antonio environment is a likely source of anthropogenic chemicals of concern to the selected PSW, the data collected indicate a few organic chemicals of concern are widely distributed in the regional Edwards aquifer at low concentrations, which is indicative of widely distributed sources. These are the pesticide atrazine, its degradate deethylatrazine, the drinking water disinfection byproduct trichloromethane (chloroform), and the solvent tetrachloroethene (PCE). Detected concentrations of all organic constituents regionally and at the well field of the selected PSW were low (<1 µg/L), consistent with previous NAWQA findings. The frequent detection of these anthropogenic chemicals of concern indicates that the entire aquifer is susceptible to contamination, and water quality is likely affected by anthropogenic activities.

Nitrate concentrations for samples collected during this study had a similar median value (1.95 mg/L) compared to regional confined groundwater samples collected in previous studies (1.67 mg/L) and were below drinking water standards. Nitrogen isotope results from PSWs in the well field and monitoring wells indicate that soil organic nitrogen is the dominant source of nitrate in the regional aquifer. A comparison with historic nitrogen isotope values, however, suggests that inputs of nitrate from biogenic sources might have increased during the past 30 years (1980–2010).

Apparent groundwater ages (determined using a piston flow model) for samples from the well field and monitoring wells ranged from 0.8 to 41.3 years, with a median of 16.6 years. Although a piston-flow model might not be valid for karst aquifers, model ages provide a reference point for comparison. Results for depth-dependent samples from the selected PSW in San Antonio yielded a range of apparent ages (0.8 to 21.7 years) but with no consistency with respect to depth or stratigraphic unit contributing to the PSW. Selected geochemical constituents that might provide independent information about groundwater residence time or the extent of effect from anthropogenic activities were compared with piston flow model ages. Results of these comparisons were not consistent with the range of apparent groundwater ages. Mixing processes in this aquifer are complex; whereas age tracers might provide insight into the generally young nature of the groundwater, they might not readily allow for distinguishing the mixing history of water at a PSW or be well-suited to distinguish relatively small differences in groundwater age. Nonetheless, groundwater apparent age interpretations indicating that water supplied to the selected PSW in San Antonio is young, and these interpretations were consistent with particle tracking results from a companion hydrogeologic modeling study.

# Temporal Stability of Cave Sediments

By Eric W. Peterson, and Kevin Hughes
Illinois State University, Department of Geography-Geology, Campus Box 4400, Normal, IL 61790

## Abstract

Sediments within cave systems have been examined concerning source, mineralogy, and transport potential. While sediments in the thalweg are very mobile, the entrainment potential of the sediment piles has not been examined. Over the course of nine months, sediment piles in a Missouri cave were sampled to determine the stability of the piles and of the sediment properties. Sediment cores were analyzed for dry bulk density ($\rho_d$), porosity ($n$), volumetric wetness ($\theta$), organic content ($O.C.$), hydraulic conductivity ($K$), and sediment particle size distribution. Observational evidence, deposited sediment and deformation of previous sample holes, suggests that despite elevated flows, the sediment piles were not mobilized over the course of the nine months. Physical properties remained constant at each location, but varied among the various locations. Dry bulk density values ranged between 1.2 g/cm$^3$ to 1.5 g/cm$^3$. Porosity values were 0.42 to 0.57. Volumetric wetness showed similar variation ranging from 0.38 to 0.53. Organic content had the highest variation among the parameters ranging from 1.24 percent at one site to 4.84 percent at another. Hydraulic conductivity ranged from $2.13 \times 10^{-7}$ m/s to $3.10 \times 10^{-7}$ m/s.

## INTRODUCTION

Cave sediment work has focused on many different themes. White (1977) focused on finding the source of sediment. Herman and others (2007) examined the sediment mineralogy of the suspended sediment. Murray and others (1993) determined the sediment age and rate of accumulation. Granger and others (2001) and Anthony and Granger (2004) employed the dating of cave sediments to develop histories of cave systems. Krekeler and others (1997) examined the role of sediment in landscape evolution and what the sediments can reveal about surface-weathering conditions. Engel and others (1997) analyzed cave sediments to gain information about reversals of the geomagnetic field. Springer (2002) used cave sediments for paleoflood reconstruction.

Examining the mobility of sediment within the thalweg facies, Dogwiler and Wicks (2004) reported that fluviokarst systems can transport up to 85 percent of the substrate during bankfull conditions. In the systems they investigated, flows capable of transporting the $d_{50}$ and $d_{85}$ sized particles occurred at intervals of 2.4 and 11.7 months, respectively. Whereas Dogwiler and Wicks (2004) examined the sediment within the channel, they did not examine the sediment piles adjacent to the cave streams.

Sediment piles serve as zones of low discharge and high storage, for water, solutes, and bacteria. Peterson and Wicks (2003) analyzed the physical and hydraulic properties concluding that the sediment possessed hydraulic properties similar to the matrix and that the cave sediments were an extension of the bedrock for modeling purposes. Lines of evidence exist suggesting that the sediment piles are periodically entrained. White (1988) states that if cave streams were incapable of flushing sediment from the conduits, then the conduits would quickly clog. The large number of caves with traversable passages stands as a testament to the fact that flushing occurs. Herman and others (2008) reported that flow thresholds had to be exceeded to mobilize the sediment in the sediment piles. Thus, the sediment is mobile and can be reworked.

Hence, the question of frequency of entrainment for sediment piles needs to be examined. This work examines the stability, both temporally and spatially, of cave sediments. While, the work done by Peterson and Wicks (2003) examined the sediment piles, their work was a one-time sampling and did not examine the potential temporal changes. This work examines the role of high-flow events and the possibility that the hydraulic properties could change after the sediments are disturbed.

## FIELD AREA

The investigation centered on Berome Moore Cave of the Moore Cave System of the

Perryville Karst Area located in Perry County, Missouri (fig. 1) The strata of the Perryville Karst Area are comprised of three stratigraphic units of Middle Ordovician age. The basal unit is the St. Peter Sandstone, a well-sorted quartz arenite, marking the lower limit of karstification (Panno and others, 1999). Unconformably overlying the St. Peter Sandstone is the Joachim Dolomite; a yellow-brown, silty dolomite containing limestone interbeds and minor shale (Panno and others, 1999) with intense cavern development. The Joachim Dolomite is about 76 meters thick in the study area (Panno and others, 1999). Overlying the Joachim Dolomite, the Plattin Formation is even bedded, fine-grained to sublithographic limestone that has intervals of chert nodules and thin beds of shale (Dean, 1977; Martin and Wells, 1966). This formation is about 106 meters thick and forms most of the bedrock of the eastern portion of the Perryville Karst Area (Panno and others, 1999). The Plattin Formation forms a geomorphic surface where there is intense sinkhole formation. Overlying the Plattin Formation, are thick loess deposits (Panno and others, 1999). Geochemical analysis of the cave sediment reveals that the sediment composition is primarily silicates and are loess derived (unpublished data); entering the conduit system through the many sinkholes.

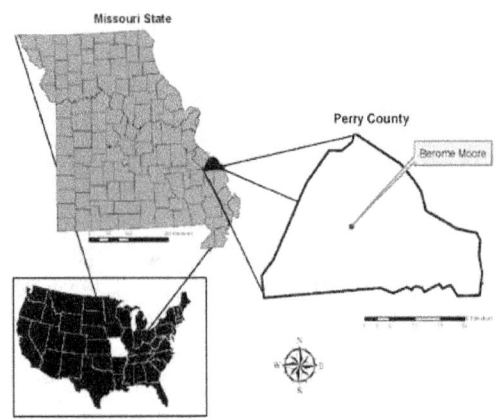

Figure 1: Location of the Moore Cave System, Perry County Missouri

Sinkholes of the Plattin Formation serve as discrete groundwater recharge points, allowing surface runoff and transported sediments to enter the cave system. After penetrating the Plattin Formation, the water and sediments enter the karst aquifer. Flow through the Moore Cave system is dominated by conduit flow that has been steady as indicated by small scallops along the main cave stream.

## METHODS AND PROCEDURES

### Sediment collection

Sediment piles from four sites were chosen within Berome Moore Cave for sampling. The Waterfall Passage site was located at the edge of a shallow plunge pool below a perennial waterfall. After entering the plunge pool, the water flows out to two small streams that eventually merge into a single stream. Collection was from a sediment pile between the two streams emerging from the plunge pool. Hydraulically, the site represented a waterfall. The Middle Main Stream Passage occurred within a conduit with a constant flow of water. In many places, the channel facies was almost non-existent; existing only as a laminated clay layer on top of bedrock. Collection occurred from a point bar along the main channel. Hydraulically, the site represented steady flows of moderate to high discharge. The Drum Passage has water pools and a small stream that was only centimeters across and millimeters deep. During high flow events, the main cave stream has the ability to overflow its banks and backflood the Drum Passage area. Sediment was collected from a point bar. Hydraulically, the site represented a point of stagnation with occasional high flows and backflooding. The No Name Passage was located in a small side conduit with a diameter of less than two meters. The roof of the conduit had fine sediment and plant debris deposited on the edges and along the apexes of dome structures. The deposits indicated that the conduit experienced pipe-full conditions and that the water source carried surface materials into the system. The collection site was a sediment pile on the edge of the conduit. Hydraulically, the site represented pipe-full flow conditions.

During each sampling trip, three cores were collected from each location to determine dry bulk density ($\rho_d$), porosity ($n$), volumetric wetness ($\theta$), and organic content ($O.C.$). A bulk density sampler was used to collect individual sediment cores of known volume. Each core was wrapped in aluminum foil, placed in a collection bag, and stored in a cooler with ice to preserve water content. Three additional sediment cores were collected in 1.4 cm internal diameter collection tubes to determine hydraulic conductivity (K). Finally, sediment for grain-size analyses was collected with a hand-trowel and placed into sample bags. The sample size was small, six cores per site, to accommodate the small area and immobility of the sediment piles. If more samples were collected, then the integrity of sediment piles would have been destroyed and the temporal objectives of this study would have been lost.

### Sediment Properties

Dry bulk density values were determined using the procedure presented by Blake (1986). Porosity (*n*) was determined from the formula:

$$n = 1 - \frac{\rho_d}{\rho_p} \tag{1}$$

where $\rho_d$ is the dry bulk density and $\rho_p$ is the sediment particle density, which for this work a value of 2.65 g/cm$^3$ was used. A sediment particle density of 2.65 g/cm$^3$ represents clays and quartz (Blake, 1986), which is the dominant mineralogical composition of the sediment as determined by SEM analysis. The volumetric wetness ($\theta$) was determined for each core using the method presented by Gardner (1986). The organic content (*O.C.*) of the sediment was determined following the method presented by Schulte and Hopkins (1996).

### Determination of Grain size distribution

For each sample, the grain-size distribution was determined using a combination of sieve and hydrometer analysis (ASTM D 422). The sieve and hydrometer data were used to generate a cumulative frequency curve.

### Determination of Hydraulic Conductivity

Hydraulic conductivity (K) was determined by conducting a falling head test on the intact sediment cores that were collected using a plastic tube with a known cross-sectional area. Hydraulic conductivity was found using the method presented by Lee and Fetter (1994).

## RESULTS

Six sampling events occurred over a nine month period from February to October 2006 (fig. 2). With the exception of the first sampling trip, all cores were successfully collected and analyzed. During the first sampling trip, all of the cores collected for the sediment property analyses from the Drum Passage were destroyed during transport out of the cave.

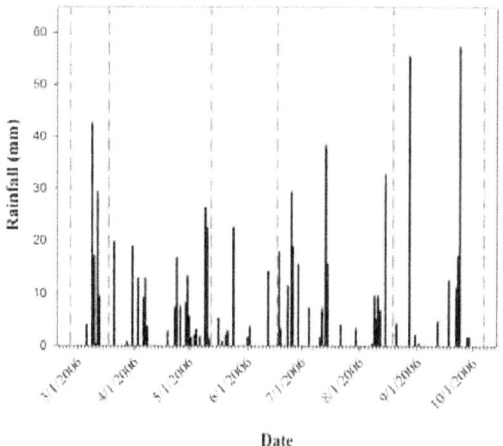

Figure 2: Distribution of daily rainfall (black bars) in Perry County (National Climatic Data Center, 2011). Gray dashed lines indicate sample collection dates.

Over the course of this study, the sediment piles appeared to be permanent fixtures of the cave. During each consecutive sampling trip, a thin layer of sediment, primarily clay, had been deposited in past sampling holes, indicating that deposition was active at each of these sites. High flow events occurred as a result of rain events (fig. 2). The presence of turbid pools of water located on shelves between three to five meters above the cave stream base level attests to the elevated flows. Even with the higher flows, the only noticeable alteration to the sediment piles was plastic deformation, as indicated by the elongation of past sampling points along the axis parallel to stream flow.

Each sediment pile was underlain by gravel sized particles. Overlying the gravel were layers of varying lithologies, but primarily fine grained, silt and clay sized, particles. One layer of note was an organic rich layer composed largely of forest litter from a conifer forest, predominant land-use for the area is agriculture. This layer was another indicator that the sediment piles had been immobile for quite some time. The top layer of all piles was a clay layer. This top clay layer would, with its strong electrostatic attraction, armor the sediment piles and make them hard to move. Sampling was conducted on the layers overlying the gravel. Although multiple layers were present, the collection methods for analysis of the physical properties and hydraulic conductivity did not allow for individual analysis of the layers. Thus, the results are representative of the material overlying the gravel as whole.

## Sediment Properties

### Bulk Density Porosity, Volumetric Water Content, and Organic Carbon Content

Waterfall Passage (fig. 3a) had a $\rho_d$ that ranged from 1.16 g/cm$^3$ to 1.25 g/cm$^3$ over the course of this study. Dry bulk density standard deviations ranged from 0.01 g/cm$^3$ to 0.08 g/cm$^3$. The Middle Main Stream Passage (fig. 3a) has $\rho_d$ values that ranged from 1.46 g/cm$^3$ to 1.53 g/cm$^3$ with a standard deviation that ranged from 0.01 g/cm$^3$ to 0.06 g/cm$^3$. The $\rho_d$ for sediment at the Drum Passage ranged from 1.32 g/cm$^3$ to 1.38 g/cm$^3$, with a standard deviation that ranged from 0.02 g/cm$^3$ to 0.22 g/cm$^3$ (fig. 3a). Dry bulk density for the No Name Passage ranged from 1.15 g/cm$^3$ to 1.29 g/cm$^3$ with a standard deviation ranging from 0.01 g/cm$^3$ to 0.23 g/cm$^3$ (fig. 3a).

Porosity of the sediment at the Waterfall Passage (fig. 3b) ranged from 0.53 to 0.56 with a standard deviation range from 0.00 to 0.02. Middle Main Stream Passage (fig. 3b) had porosity ranging from 0.42 to 0.45 with a standard deviation ranging from 0.00 to 0.02. Drum Passage sediments had a porosity range of 0.48 to 0.49 with a standard deviation ranging from 0.00 to 0.08 (fig. 3b). At No Name Passage, porosity ranged from 0.51 to 0.57 with a standard deviation ranging from 0.01 to 0.09 (fig. 3b).

Volumetric wetness at the Waterfall Passage (fig. 3c) ranged from 0.50 to 0.53 with the standard deviations ranging from 0.01 to 0.03. Middle Main Stream Passage had a $\theta$, ranging from 0.38 to 0.43, with standard deviations ranging from 0.00 to 0.02 (fig. 3c). For Drum Passage, $\theta$ ranged from 0.43 to 0.47 with a

standard deviation ranging from 0.01 to 0.06 (fig. 3c). The $\theta$ at No Name Passage ranged from 0.46 to 0.49 with a standard deviation ranging from 0.00 to 0.03 (fig. 3c).

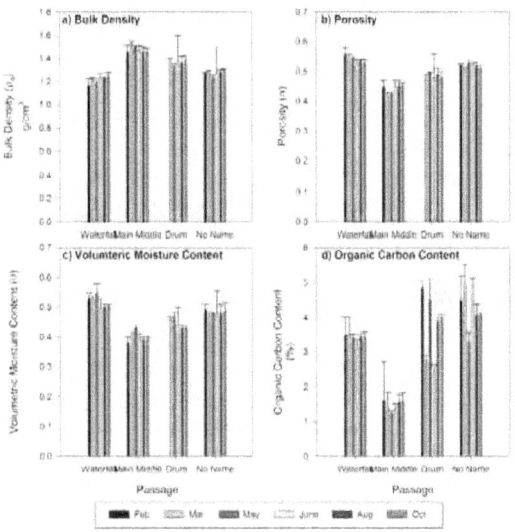

Figure 3: Sediment properties at the four sampling locations. a) Dry Bulk Density b) Porosity c) Volumetric wetness d) Organic Carbon Content

Compared to the other properties, the organic carbon content exhibited a greater range of values at each site (fig. 3d). At the Waterfall Passage, O.C. ranged from 3.12 to 3.53 percent with a standard deviation range of 0.07 to 0.52 percent. O.C. for the Middle Main Passage ranged from 1.24 to 1.61 percent with a standard deviation range of 0.07 to 1.11 percent. The Drum Passage experienced the greatest range in O.C. with values from 2.45 to 4.84 percent and a standard deviation range of 0.11 to 0.62 percent. No Name Passage exhibited some of the highest O.C. values ranging from 3.28 to 4.84 percent with standard deviation values ranging from 0.05 to 0.73 percent.

For each site and each sediment property an ANOVA was performed to test whether the values of the sediment properties changed from sampling event to sampling event (Table 1). A quick examination of figure 3 would suggest that the individual sediment property values are rather stable at each location. This is confirmed by the ANOVA analyses (Table 1). With the exception of the $\theta$ at the Water Fall Passage and Middle Main Passage, and the O.C. at the Drum Passage and No-Name Passage, the values were statistically similar for all sampling events. Note that the $\rho_d$ and *n* values did not statistically vary over the duration of the study for each location.

The values for the individual sediment properties vary among the sample locations (figs. 3 and 4). Waterfall Passage had the lowest $\rho_d$ value with the Main Middle Passage having the highest $\rho_d$ value. Given the relationship between $\rho_d$ and porosity, Waterfall Passage had the highest porosity value with the Main Middle Passage having the lowest porosity value. The $\theta$ values had a similar relationship as porosity. The *O.C.* values exhibited some of the larger differences among the sites. Drum Passage and No Name Passage had higher *O.C.* than Waterfall Passage and Middle Main Passage. Middle Main Passage had considerably less *O.C.* than the other three locations. To test whether the variation in the individual properties were statistically different among the locations, an ANVOA was performed for each property. Among the locations, there is significant variation for each of the sediment properties (Table 2).

**Grain Analysis**

Sediment particle size distributions reveal similar distributions among the four sites and show relatively stable particle size distributions at the sites over time (fig. 5). A noticeable exception is that the Drum passage has a smaller percentage of sand size particles, 12 percent as compared to the other sites that have between 17 percent at Middle Main to 21percent at No Name. At all locations clay sized particles comprised 2.5 percent or less of the overall material. No Name had the highest amount at 2.5 percent, followed by Middle Main (2 percent), Waterfall (1.8 percent), and Drum (1.6 percent).

Sieving of sediment samples revealed that more than geosediments composed the sediment piles. Plant debris, pine needles, and the shells of small mollusks were observed with the sediment. These materials account for the higher *O.C.* values measured for the sediments.

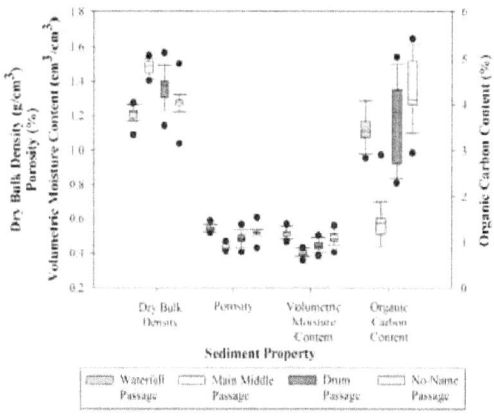

Figure 4: Comparison of the sediment properties among the four passages. The ends of the boxes represent the 25th and 75th percentiles with the solid line at the median; the error bars depict the 10th and 90th percentiles and the points represent the 5th and 95th percentiles.

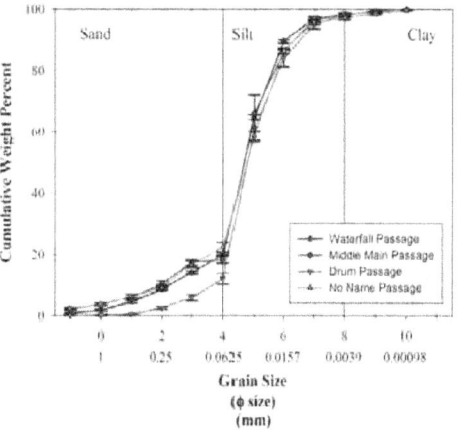

Figure 5: Grain size distribution curves for the sediment particles at the four sampling locations. The points represent the mean value over the course of the sampling and the error bars represent the standard deviation over the course of the sampling.

**Hydraulic Conductivity (K)**

The cores collected at each location during each sampling event provided statistically similar K values for each location (Table 1 and fig. 6). The Waterfall Passage had the highest K values, which ranged from $3.04 \times 10^{-7}$ m/s to $3.28 \times 10^{-7}$ m/s, with a standard deviation range of $1.65 \times 10^{-9}$ m/s to $1.63 \times 10^{-8}$ m/s. Middle Main Passage had the lowest K values that ranged from $2.01 \times 10^{-7}$ m/s to $2.17 \times 10^{-7}$ m/s, with a standard deviation range of $2.45 \times 10^{-10}$ m/s to $1.02 \times 10^{-8}$ m/s. The Drum Passage had K values that ranged from $2.31 \times 10^{-7}$ m/s to $2.45 \times 10^{-7}$

m/s, with a standard deviation range of $4.12 \times 10^{-10}$ m/s to $1.49 \times 10^{-8}$ m/s. K values at No Name Passage ranged from $3.03 \times 10^{-7}$ m/s to $3.16 \times 10^{-7}$ m/s, with a standard deviation range of $9.12 \times 10^{-10}$ m/s to $2.10 \times 10^{-8}$ m/s.

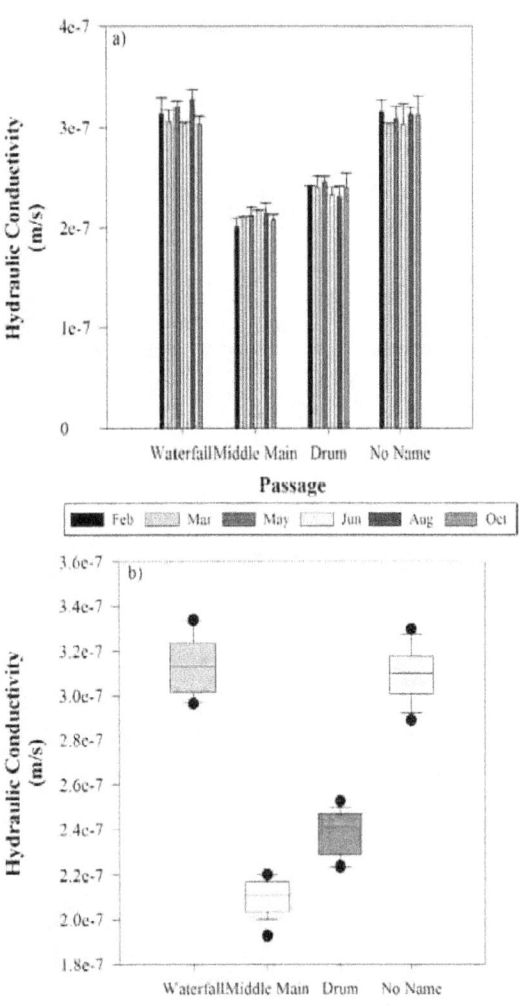

Figure 6: Hydraulic conductivity values for the sediments. a) comparison of the individual sampling events at the four sampling locations. b) Comparison of the collective K values among the four passages. The ends of the boxes represent the 25th and 75th percentiles with the solid line at the median; the error bars depict the 10th and 90th percentiles and the points represent the 5th and 95th percentiles.

The K values from the different sampling events were pooled to provide an overall comparison of the K values among the sites (fig. 6b). There exists a statistically significant variation among the K values among the sites (Table 2). Overall, Waterfall Passage and No Name Passage had the highest K values at $3.13 \times 10^{-7} \pm 1.26 \times 10^{-8}$ m/s and $3.10 \times 10^{-7}$

$\pm 1.27 \times 10^{-8}$ m/s. Middle Main had the lowest K value at $2.13 \times 10^{-7} \pm 7.61 \times 10^{-8}$ m/s and Drum Passage had a slightly higher K value at $2.38 \times 10^{-7} \pm 9.28 \times 10^{-9}$ m/s.

## DISCUSSION

During the nine-month period of sampling, the sediment properties remained stable, resulting in consistent values among the various sediment properties (Table 1 and fig. 3). Exceptions to the stable properties include $\theta$ for the sediments at Waterfall Passage and Middle Main Passage and $O.C.$ at Drum Passage and No Name Passage. A noticeable difference in the $\theta$ occurs in the May sampling event, which followed a precipitation event. Both Waterfall and Middle Main sediments had higher $\theta$ values, which may be a direct result of the wetter conditions. For the variations in $O.C.$ values two possible reasons exist. First, there was a large vertical heterogeneity as dictated by the different beds that make up the sediment profile. If each sample did not capture the exact same beds, then a difference would result even if the individual beds were homogenous. Second, there was a large horizontal heterogeneity, meaning that as the sample points moved along the sediment pile, the heterogeneity caused deviation in the data points and could be interpreted as a temporal shift. Small variation in the $\rho_d$, $n$, particle size distributions, and K values existed, but since the sediment piles were immobile, there is reduced likelihood that these properties were changing.

A previous study reports that organic carbon content of karst sediments are less than 1.5 percent (Peterson and Wicks, 2003). The only site that was in-line with the reported data is the Middle Main Stream Passage. The common trait that the Middle Main Stream Passage has compared to the previous study is that there is a large volume of water constantly moving by the sediment pile. The less than 1.5 percent organic carbon content may be due to the washing effect of this moving water. Organic debris caught in this swift water may not have the opportunity to deposit with the sediment. This is further supported by the lack of the forest-litter rich layer that was found at the other sites. Compared to the Middle Main Stream Passage and previous studies mentioned, the Waterfall

Passage, Drum Passage, and No Name Passage have more than twice the organic content (fig. 3). The Waterfall Passage had a very strong presence of the forest litter rich layer. Some core samples had a simple black layer while other cores were rich in recognizable pine needle remains. It is likely that the waterfall feeds in from a sinkhole. Forest litter at the bottom of the sinkhole was washed in and deposited around the plunge pool. The Drum Passage showed greater variation in organic carbon content, which may be related to the back flooding and presence of water pools. Near the pools, more organic carbon has been deposited. Away from the pools, less organic material is deposited. The most revealing part of the organic carbon content was revealed in the No Name Passage. Here, a lot of insect activity was observed in the form of beetles and crickets. Earth worms crawl through the sediment piles and fresh plant debris, including tree seeds that attempt to germinate in the dark cave environment, is carried in. The rich organic nature of the sediment, almost 5percent, is likely due to the fresh inputs.

The clay armor seems to make the sediment piles a somewhat permanent feature of the cave streams and only very high flow events are capable of altering them. The permanence of the sediment piles was demonstrated in August when high flow conditions only caused plastic deformation of previous sampling points. Although the sediment piles are composed of grains that should become mobile according to the study done by Dogwiler and Wicks (2004), the finer sized particles make the sediment piles behave like larger particles.

The controlling factor on the hydraulic conductivity of the sediment piles is likely the hydraulic conductivity of the silt. Silt tends to have a hydraulic conductivity that ranges from $1 \times 10^{-9}$ m/s to $1 \times 10^{-5}$ m/s. The K values measured for the sediments, on the order of $10^{-7}$ m/s, at each location were within this range and were similar to those measured by Peterson and Wicks (2003). The data presented here further support the claim by Peterson and Wicks (2003) that the cave sediments can be modeled as an extension of the matrix rock because the hydraulic conductivity of the sediment piles are similar to the hydraulic conductivity of

limestone. A limitation to this claim would be if the sediment piles are highly mobile.

Compared to other published work, the sediment piles appear to be less mobile than the sediment within the karst streams. Dogwiler and Wicks (2004) observed movement of the $d_{50}$ particle on a 2.4 month interval; over the nine months, there was no movement of the sediment pile in Berome Moore Cave. While comparison is made between the studies, caution must be exercised. Dogwiler and Wicks (2004) were examining larger bedload particles, the $d_{50}$ particle was 70 mm, located in the thalweg of the stream. The sediment piles are finer grained, $d_{50}$ particle ~0.04 mm, but require similar stream velocities to become entrained as the larger particles. Additionally, the sediment piles are located along the banks of the stream where velocities are lower than in the thalweg. Thus, erosion is limited as a result of the need for higher velocities to entrain the sediment and the lower velocities that occur along the edges of the stream.

Finer sized particles do move through systems. Herman and others (2008) indentified flow thresholds that needed to be exceeded before deposited sediment in the conduits would be entrained and moved through the system. For the systems examined by Herman and others (2008), the entrainment thresholds were reached as a consequence of extremely high flows, greater than two orders of magnitude above baseflow, associated with precipitation from remnant hurricanes. During the nine months, the Berome Moore system did not experience extreme flows remotely close to the levels observed by Herman and others (2008). This work was not able to address the point when the flow threshold for entrainment was exceeded.

## CONCLUSIONS

Over the course of the nine-month sampling period, the sediment piles within Berome Moore Cave did not move. The hydrologic properties, $n$ and K, did not change over the study, confirming the stability of the piles. While the sediment was not entrained and the properties remained stable at the individual sampling locations, the locations behaved differently. Particle size distributions were similar among the sites, but $\rho_d$, n, $\theta$, O.C., K., statistically varied among the sites. This variability appears to be a result of the difference in flow regime among the sampling locations.

Table 1. ANOVA results examining temporal variation of the sediment properties at the four locations.

| Parameter | Waterfall Passage | Main Middle Passage | Drum Passage | No-Name Passage |
|---|---|---|---|---|
| Dry Bulk Density | $F(5, 12) = 2.12, p = .13$ | $F(5, 12) = 2.03, p = .15$ | $F(4, 10) = 0.14, p = .96$ | $F(5, 12) = 0.10, p = .99$ |
| Porosity | $F(5, 12) = 2.44, p = .09$ | $F(5, 12) = 2.02, p = .15$ | $F(4, 10) = 0.15, p = .97$ | $F(5, 12) = 0.11, p = .99$ |
| Volumetric wetness | $F(5, 12) = 3.61, p = .04$ | $F(5, 12) = 5.20, p = .01$ | $F(4, 10) = 1.31, p = .33$ | $F(5, 12) = 0.71, p = .99$ |
| Organic Carbon Content | $F(5, 12) = 0.55, p = .73$ | $F(5, 12) = 0.25, p = .93$ | $F(5, 12) = 22.32, p < .01$ | $F(5, 12) = 6.74, p < .01$ |
| Hydraulic Conductivity[1] | $F(5, 12) = 3.09, p = .06$ | $F(5, 12) = 2.03, p = .17$ | $F(5, 12) = 0.95, p = .49$ | $F(5, 12) = 0.45, p = .80$ |

[1]Data were subjected to a logarithmic transformation

Table 2. ANOVA results examining the spatial variation of the sediment properties among the four locations.

| Parameter | |
|---|---|
| Dry Bulk Density | $F(3, 65) = 53.04, p < 0.01$ |
| Porosity | $F(3, 65) = 52.16, p < 0.01$ |
| Volumetric wetness | $F(3, 65) = 58.61, p < 0.01$ |
| Organic Carbon Content | $F(3, 68) = 68.66, p < 0.01$ |
| Hydraulic Conductivity[1] | $F(3, 68) = 449.69, p < 0.01$ |

[1]Data were subjected to a logarithmic transformation

## REFERENCES

Anthony, D. M., and Granger, D. E., 2004, A Late Tertiary origin for multilevel caves along the western escarpment of the Cumberland Plateau, Tennessee and Kentucky, established by cosmogenic 26Al and 10Be: Journal of Cave and Karst Studies, v. 66, no. 2, p. 46-55.

Blake, G. R., 1986, Particle density, in Klute, A., ed., Methods of soil analysis, Part 1: Madison, WI, American Society of Agronomy, p. 377-382.

Dean, T. J., 1977, Engineering Geology along Interstate-55, in Thacker, J. L., and Satterfield, I. R., eds., Guidebook to the Geology along Interstate-55 in Missouri, Volume Report of Investigation Number 62: Rolla, MO, Missouri Department of Natural Resources Division of Geology and Land Survey,.

Dogwiler, T., and Wicks, C. M., 2004, Sediment entrainment and transport in fluviokarst systems: Journal of Hydrology, v. 295, p. 163-172.

Engel, A. S., Lascu, C., Badescu, A., Sarbu, S., Sasowsky, I. D., and Huff, W., 1997, A study of cave sediment from Movile Cave, southern Dobrogea, Romania: Proceedings of the 12th international congress of Speleology, v. 1, p. 25-28.

Gardner, W. H., 1986, Water content, in Klute, A., ed., Methods of soil analysis, Part 1: Madison, WI, American Society of Agronomy, p. 493-544.

Granger, D. E., Fabel, D., and Palmer, A. N., 2001, Pliocene--Pleistocene incision of the Green River, Kentucky, determined from radioactive decay of cosmogenic 26Al and 10Be in Mammoth Cave sediments: GSA Bulletin, v. 113, no. 7, p. 825–836.

Herman, E. K., Tancredi, J. H., Toran, L., and White, W. B., 2007, Mineralogy of suspended sediment in three karst springs Hydrogeology Journal, v. 15, no. 2, p. 255-266 doi:210.1007/s10040-10006-10108-10042.

Herman, E. K., Toran, L., and White, W. B., 2008, Threshold events in spring discharge: Evidence from sediment and continuous water level measurement: Journal of Hydrology, v. 351, p. 98-106.

Krekeler, M. P. S., Engel, A. S., Engel, S., Mixon, D., and Ragsdale, M., 1997, Sedimentology, clay mineralogy, and geochemistry of cave sediment from Hard Baker Cave, Rockcastle County, Kentucky, USA, Proceedings of the 12th international congress of Speleology; Symposium 7, Physical speleology; Symposium 8, Karst geomorphology, Volume 12, Vol. 1,; [location varies], United States, International Union of Speleology, p. 21-24.

Lee, K., and Fetter, C. W., 1994, Hydrogeology Laboratory Manual, Upper Saddle River, New Jersey, Prentice Hall, 135 p.

Martin, J. A., and Wells, J. S., 1966, Guidebook to Middle Ordovician and Mississippian Strata, St. Louis and St. Charles Counties, Missouri, Missouri Geological Survey and Water Resources, Rolla, MO, Report of Investigation 34, 48 p.

Murray, A. S., Stanton, R., Olley, J. M., and Morton, R., 1993, Determining the origins and history of sedimentation in an underground river system using natural and fallout radionuclides: Journal of Hydrology, v. 146, no. 1-4, p. 341-359.

National Climatic Data Center, 2011, NNDC Climate Data Online, in NOAA, ed., Volume 2011.

Panno, S. V., Weibel, C. P., Wicks, C. M., and Vandike, J. E., 1999, Geology, hydrology, and water quality of the karst regions of southwestern Illinois and southeastern Missouri, Illinois State Geological Survey, Champaign,, ISGS Guidebook 27.

Peterson, E. W., and Wicks, C. M., 2003, Characterization of the physical and hydraulic properties of the sediment in karst aquifers of the Springfield Plateau, Central Missouri, USA: Hydrogeology Journal, v. 11, no. 3, p. 357-367.

Schulte, E. E., and Hopkins, B. G., 1996, Estimation of soil organic matter by weight loss-on-ignition, in Magdoff, F. R., Tabatabia, M. A., and Hanlon, E. A., eds., Soil organic matter: Analysis and interpretation: Madison, WI, Soil Science Society of America, p. 21-31.

Springer, G. S., 2002, Caves and their potential use in paleoflood studies, in Webb, R. H., Baker, V. R., Levish, D. R., and House, P. K., eds., Ancient floods, modern hazards; principles and applications of paleoflood hydrology: Washington, DC, American Geophysical Union, p. 329-343.

White, W. B., 1977, Characterization of Karst Soils by Near Infrared Spectroscopy: NSS Bulletin, v. 39, no. 1, p. 27-31.

White, W. B., 1988, Geomorphology and Hydrology of Karst Terrains, New York, Oxford University Press, 464 p.

# Evaluating the Stormwater Filters at Mammoth Cave, Kentucky

By Ashley West[1], Rickard Toomey[2], Michael W. Bradley[3] and Thomas D. Byl[1,3]

[1]College of Engineering, Tennessee State University, 3500 John A Merritt Blvd, Nashville, TN 37209
[2] Geography/Geology Dept., Western Kentucky University, Bowling Green, KY 42101-1066
[3] U.S. Geological Survey, 640 Grassmere Park, Suite 100, Nashville, TN 37211

## Abstract

Mammoth Cave is home to many unique animals that could be harmed by contaminants carried into the cave system during storm events. This project was conducted to determine if leaf-pack filter-systems attenuated storm runoff at seven parking lots in Mammoth Cave National Park. Grab samples were collected at the inlet and outlet of the filter systems, and analyzed for oil and grease, sediments, turbidity, gasoline compounds, nitrate, ammonia, fecal bacteria, dissolved iron, and chemical oxygen demand (COD). For the first sampling round, the filters had not been serviced for 8 years and did very little to remove contaminants. The contaminant concentrations at the outlet were similar to those at the inlet, with the exception of removing 20 to 70 percent of the oil and grease. After replacing leaf packs and cleaning out debris, the re-conditioned filters did not remove oils and greases and did little to remove copper and ammonia from runoff waters. However, the re-conditioned filters removed up-to 99 percent of the benzene, toluene, ethyl-benzene and xylene, and, up to 90 percent of the turbidity, *E. coli*, COD, and iron from the storm runoff. These results indicate that well-maintained filtration systems are more effective than clogged filters at removing many, but not all, contaminants from parking lot runoff.

# Using Labeled Isotopes to Trace Groundwater Flow Paths in a Northwestern Arkansas Cave

By Katherine J. Knierim[1], Erik D. Pollock[2], and Phillip D. Hays[3]

[1]Environmental Dynamics Program, 113 Ozark Hall, University of Arkansas, Fayetteville, AR 72701

[2]University of Arkansas Stable Isotope Laboratory, 116 Ferritor Hall, University of Arkansas, Fayetteville, AR 72701

[3] U.S. Geological Survey, Arkansas Water Science Center, 700 W. Research Center Blvd MS36, Fayetteville, AR 72701

## Abstract

Tagged water was used to investigate the conditions of carbon and nitrogen cycling over small spatial and temporal scales while mimicking natural conditions (that is, small concentrations and slow addition of tracers) in a small (< 200 meter) and shallow (< 9 meter) cave system in northwestern Arkansas. Approximately 950 liters of isotopically tagged water with labeled glucose ($^{13}C_6H_{12}O_6$) and potassium nitrate ($K^{15}NO_3$) was released to a trench above the cave at a rate of 2 liters per minute to simulate infiltration by precipitation. The signature of the labeled glucose was observed in the carbon dioxide and dissolved inorganic carbon isotopic compositions of soil water samples closest to the trench. The labeled tracers were not detected in the cave stream after 2 months, as evidenced by the isotopic compositions of carbon dioxide, dissolved inorganic carbon, dissolved organic carbon, and nitrate in the cave stream. Although the labeled tracers were not detected in the cave stream, stored soil water did infiltrate into the cave, as evidenced by changes in the pH and conductance of the cave stream and the concentration of inorganic and organic carbon in the cave stream. Most of the heavy isotopic signature was likely respired as carbon dioxide and passed out of the soil zone. The nitrate was also processed in the soil zone, but gaseous nitrogen was not monitored during the tracing experiment.

## INTRODUCTION

Northwestern Arkansas is characterized by karst, leading to water-quality degradation in cave and spring systems because of the large concentration of agricultural operations and population growth in the region (Davis and others, 2000; Gillip, 2007; Peterson and others, 2002). Nationally, Arkansas is ranked second in poultry production and the three top counties for agricultural sales in Arkansas are located in the northwestern region of the state (U.S. Department of Agriculture, 2010). Nitrate ($NO_3^-$) is often the dominant and most mobile form of nitrogen (N) in agricultural settings (Böhlke, 2002) and can contribute to eutrophication of surface water and degradation of karst ecosystems. Concentrations and stable isotopic compositions of $NO_3^-$ ($\delta^{15}N$ and $\delta^{18}O$) and dissolved inorganic/organic carbon ($\delta^{13}C$) have been used to characterize denitrification – an important process for removal of $NO_3^-$ – in natural systems because of the interaction between $NO_3^-$ and carbon (C) species (Aravena

and Robertson, 1998; Einsiedl and others, 2005; Nascimento and others, 1997). Stable isotopes provide an economical and powerful tool for tracing biogeochemical cycling of nutrients in groundwater systems because of the observable and predictable changes in isotopic compositions as compounds are cycled between reservoirs (Kendall, 1998).

This study investigated C and N cycling for a short temporal scale experiment (2 months) while mimicking natural conditions (that is, small concentrations and slow addition of tracers) in a small (< 200 meter (m) long) and shallow (< 9 m deep) cave system in northwestern Arkansas. We used isotopically tagged water with labeled glucose ($^{13}C_6H_{12}O_6$) and potassium nitrate ($K^{15}NO_3$) as tracers to monitor C and N cycling from the soil zone into the cave to better understand how nutrients and organic substrate may be utilized in shallow karst settings.

## STUDY SITE

Jack's Cave is located in Madison County, Arkansas in the Ozark Plateaus (Ozarks). The cave is in the Sneeds Dolomite Member of the Middle Ordovician Everton Formation and is capped by the Kings River Sandstone Member of the Everton Formation. Meteoric water infiltrates into a very gravelly silt loam (National Cooperative Soil Survey, 2008), which is developed from colluvium deposited on the Kings River Sandstone Member. Gravel is predominantly remnant chert of the Early Mississippian Boone Formation. After infiltrating through the silt loam, water enters the cave through fractures in the sandstone cap rock, collects in a main room of Jack's Cave (location C1, Fig. 1), and flows southwest along the cave passage towards C2 and C3. The cave ends in a sediment-filled constriction; the exit point(s) for the cave stream were not identified. The cave extends an additional 55 m beyond C3.

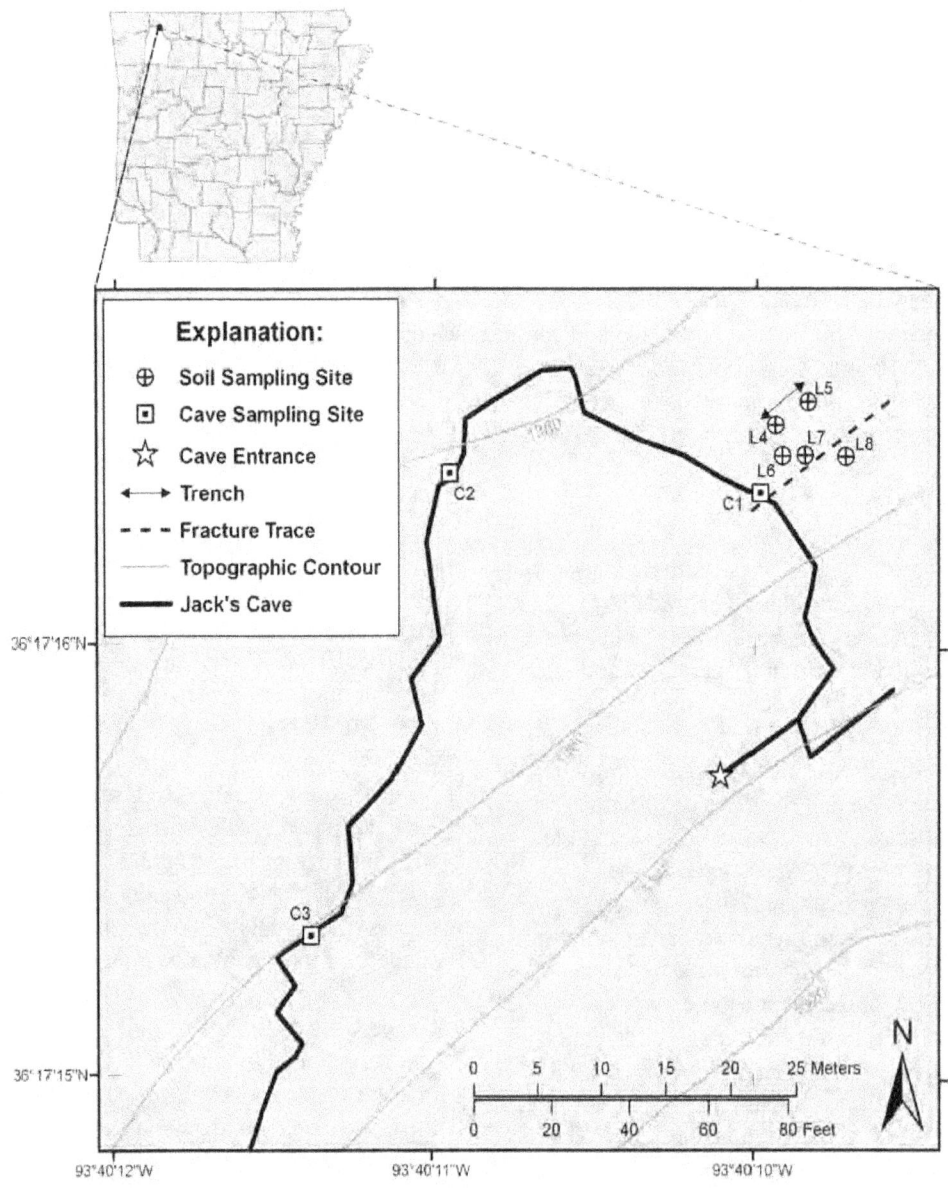

**Figure 1.** Site map showing the locations of soil sampling sites on the surface and cave sampling sites along the trace of Jack's Cave. Water and gas samples were collected from the soil sampling sites (L4 through L8). Water and gas samples were collected from the cave at C1, and only water samples were collected at C2 and C3. Topographic contours are elevation in feet above mean sea level.

## METHODS

The position of the main room in Jack's Cave and a fracture trace trending along the cave were located on the surface by surveying from the cave map to define the boundaries for the experiment. A trench (4.3 m long, 0.3 m wide, 0.2 m deep) was dug upslope from the fracture trace and five suction lysimeters were installed for collecting soil-water samples downslope from the trench at varying depths over a 6-m by 9-m area (Fig. 1). The shallow trench was designed to temporarily pond the tagged water and ensure infiltration into the soil. Water introduced to the trench mimicked groundwater flow paths where meteoric water infiltrates into the soil and travels along fractures into the cave. Groundwater flow paths between the soil surface and the cave were tested by releasing approximately 900 liters (L) of untagged, surface water into the trench. Prior to release of water at the surface, no water was infiltrating along fractures into Jack's Cave because of dry conditions at the site. The water arrived in the main room of the cave approximately 23 minutes following release at the surface, as evidenced by water flowing through fractures in the sandstone cap rock. Water flowing along fractures from the land surface is considered the source of water flowing in the cave stream; therefore, cave water samples were collected from the cave stream where water dripping from fractures immediately pooled in the cave (C1).

Water and gas samples for C and N isotopic analyses were collected between September and November 2009. Background sampling at all sampling sites was conducted prior to the release of labeled water. Water samples from the cave stream (C1, C2, and C3) and soil lysimeters (L4 through L8) were collected in 40-milliliter (mL) precombusted vials and filtered through 0.45-micrometer ($\mu$m) filters. The concentration and isotopic composition ($\delta^{13}C$) of dissolved inorganic carbon (DIC) and dissolved organic carbon (DOC) were analyzed at the Colorado Plateau Stable Isotope Laboratory using continuous flow isotope ratio mass spectrometry (IRMS) (St-Jean, 2003). Nitrate concentration ([$NO_3^-$]) and isotopic composition ($\delta^{18}O$ and $\delta^{15}N$) were analyzed at the University of Arkansas Stable Isotope Laboratory using the bacterial denitrifier method for analysis on a Gas Bench II interfaced to an IRMS (Casciotti and others, 2002; Sigman and others, 2001). Gas samples from the main room of Jack's Cave (C1) and soil atmosphere pumped from the soil lysimeters (L4 through L8) were collected in precombusted 100-mL serum vials purged with helium. Gas samples were analyzed for [$CO_2$] and isotopic composition of $CO_2$ ($\delta^{13}C$-$CO_2$) at the University of Arkansas Stable Isotope Laboratory using gas chromatography-combustion IRMS. All isotope ratios are reported using delta notation ($\delta$) in parts per thousand, or per mil (‰). The $\delta$ notation represents the ratio of the heavier to lighter stable isotope ($^{13}C/^{12}C$, $^{15}N/^{14}N$, or $^{18}O/^{16}O$) relative to a standard (Vienna Peedee Belemnite, air, or Standard Mean Ocean Water, respectively) (Clark and Fritz, 1997).

The labeled tracers used in this experiment were applied at small concentrations to avoid stimulating the biologic system. Approximately 950 L of water tagged with 0.2713 grams (g) of $^{13}C_6H_{12}O_6$ (99 atom %) and 1.134 milligrams (mg) of $K^{15}NO_3$ (99 atom %) was released to the trench from a tank beginning on 9/12/09 at 15:00 at a rate of approximately 2 L/min to simulate infiltration by precipitation (approximately 2 mm/hr). Water temporarily ponded in the trench and on the soil surface, and then infiltrated into the soil. Periodic sampling was conducted over the next 2 days at all sites following the release. Approximately 2 months following the addition of the tracer, a final sampling event was conducted at all sites. Water samples were kept at 4°C until analysis; water temperature, conductivity, and pH were measured on site.

## RESULTS AND DISCUSSION

Background samples could not be collected at every sampling site because of dry conditions at the study location prior to addition of the tagged water. In the soil, there was insufficient water for DIC analysis. The isotopic composition of DOC ($\delta^{13}C$-DOC) averaged -25.0‰ and the isotopic composition of $NO_3^-$ ($\delta^{15}N$-$NO_3$) averaged 6.3‰ at L5 and L6. Soil

$\delta^{13}$C-CO$_2$ values averaged -19.2‰ for all sites. Background values in the cave are only provided for site C1; $\delta^{13}$C-DIC was -14.2‰, $\delta^{13}$C-DOC was -23.3‰, and $\delta^{15}$N-NO$_3$ was 5.4‰. During the summer and fall, $\delta^{13}$C-CO$_2$ values in the atmosphere of Jack's Cave averaged -20.4‰ (Knierim, 2009).

Water samples collected from the trench immediately following release of the tagged water from the tank were enriched in $^{15}$N of NO$_3^-$ and $^{13}$C of DIC/DOC because of addition of the labeled glucose and potassium nitrate. Trench water had a $\delta^{15}$N-NO$_3$ value of 135.8‰ and $\delta^{13}$C values of 208.7‰ for DIC and 1,007.8‰ for DOC.

In the soil, $\delta^{13}$C-DIC values ranged between -0.4‰ and -24.3‰ (n = 10) and varied by lysimeter; the heaviest values occurred closest to the trench. At sites L4 and L5 (Fig. 1), the heaviest $\delta^{13}$C-DIC values occurred on 9/13/09 at 13:30 (-3.9‰) and on 9/15/09 at 17:00 (-0.44‰), respectively (Fig. 2). The $\delta^{13}$C-DOC values showed a smaller range between lysimeters (-23.7‰ to -27.4‰, n = 16), but isotopic composition varied by location. In the cave, $\delta^{13}$C-DIC values ranged between -11.1‰ and -15.0‰ (n = 7) and $\delta^{13}$C-DOC between -21.9‰ and -26.1‰ (n=7). Relative minimum and maximum concentrations of DIC and DOC, respectively, were observed at C1 on 9/13/09 at 12:30 (Fig. 2, DIC data only).

Nitrate isotopic composition showed a greater range in $\delta^{15}$N and $\delta^{18}$O values in the soil (n=18) than in the cave (n=18). The $\delta^{15}$N-NO$_3$ values were between -7.7‰ and 12.8‰ in the soil and 0.7‰ and 6.4‰ in the cave (Fig. 3). Nitrate concentrations as N ([NO$_3^-$]) ranged from 0.80 to 2.36 milligrams per liter (mg/L) in the soil and 0.35 to 3.15 mg/L in the cave. Similar to DIC concentration, a relative minimum in [NO$_3^-$] (0.83 mg/L) was observed on 9/13/09 at 12:30 at C1.

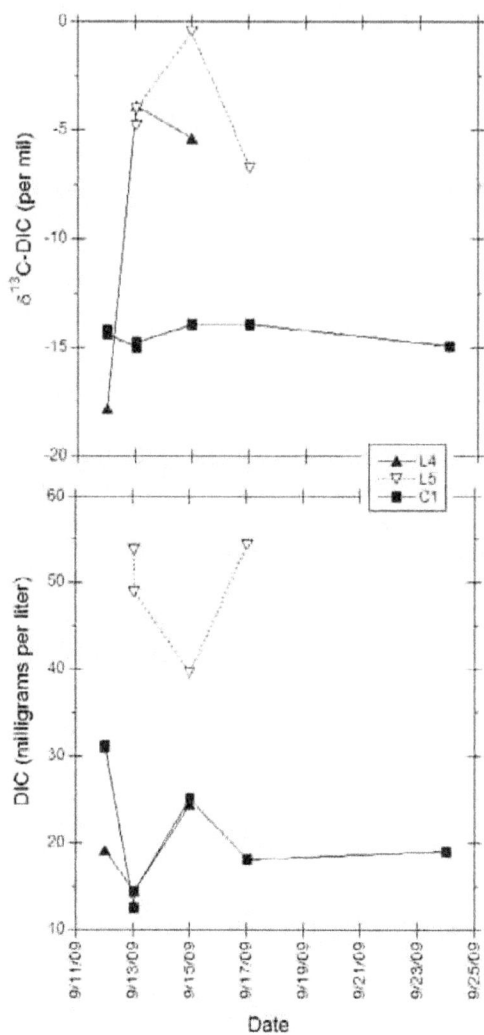

**Figure 2.** Concentration and isotopic composition of dissolved inorganic carbon (DIC) in the soil (L4 and L5) and in the cave stream in the main room of Jack's Cave (C1) over time. Note how the $\delta^{13}$C-DIC values in the soil are heavier than in the cave due to the heavy signature from the labeled glucose.

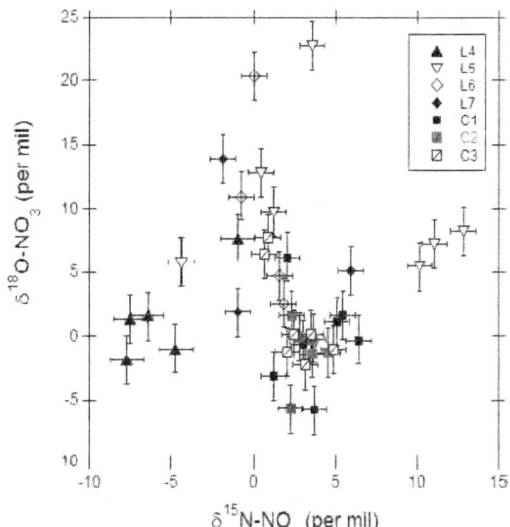

**Figure 3.** Isotopic composition of nitrate ($NO_3^-$) in the soil (L4 through L7) and the cave (C1 through C3).

Concentration of $CO_2$ ([$CO_2$]) in L4, L5, and L6 was greater than 20,000 parts per million (ppm) (instrument range). In the cave, [$CO_2$] averaged 2,950 ppm in September (n=7) and decreased to 750 ppm in November. During the warmer months of the summer and fall, [$CO_2$] was found to increase in Jack's Cave because respired $CO_2$ from the soil zone moved into the cave through gaseous and aqueous phases (Knierim, 2009). The $\delta^{13}C$-$CO_2$ values ranged between 1.7‰ and -22.4‰ in the soil (n=13) and -20.0‰ and -22.7‰ in the cave (n=5). In L4 and L5, $\delta^{13}C$-$CO_2$ values varied over time and in L4 the heaviest value (1.8‰) occurred on 9/13/09.

The isotopic signature of the labeled tracer was observed in the $CO_2$ and DIC isotopic compositions in L4 and L5 because the signature of the labeled glucose was converted to inorganic C species through respiration. L4 was closest to the trench (approximately 1 m) and the heaviest $\delta^{13}C$-$CO_2$ values were observed 18.5 hours after releasing the tagged water into the trench. The isotopic compositions of $CO_2$ and DIC began to approach background values during the experiment, but had not returned to the typical range of values 5 days after adding the tracer. From a previous study conducted at Jack's Cave, soil $\delta^{13}C$-$CO_2$ averaged -21.3‰ over a one-year time period (Knierim, 2009). Most of the heavy isotopic signature was likely

respired as $CO_2$ and fluxed out of the soil zone because heavy $\delta^{13}C$-DOC values were not observed during the study.

The tracer was not detected in the two lysimeters further from the trench (L6 and L7). During the experiment the volumes of water collected from L6 and L7 were much smaller than L4 and L5, likely because the tagged water did not reach the more distant lysimeters located between 3 to 4 m from the trench, and any water collected from L6 and L7 had a meteoric source. The lysimeter installed farthest from the trench (L8) did not yield water during the experiment; L8 was purposefully installed on the downslope side of the infiltration fracture. Any infiltrating water (either meteoric or tagged) could have flowed into the fracture before reaching L8, or the lysimeter installed at L8 may not have developed a proper seal with the surrounding soil.

The isotopic compositions of $NO_3^-$ were not outside the range of typical compositions for $NO_3^-$ in soil or precipitation (Clark and Fritz, 1997); therefore, the $^{15}N$ signal from the labeled water could not be confirmed. Although the heavy N signature was not observed during the experiment, processing of $NO_3^-$ in the soil zone did occur, as evidenced by the variation in the concentration and isotopic composition of $NO_3^-$ in L4. Mixing between two $NO_3^-$ sources (small concentration and heavier composition with large concentration and lighter composition) also accounted for some of the variation, but the effects of mixing compared to processing could not be separated.

No labeled tracers were detected in the cave stream or atmosphere, as evidenced by the isotopic compositions of $CO_2$, DIC, DOC, and $NO_3^-$ in the cave, which were all within the range of values observed previously in Jack's Cave (Knierim, 2009) or the literature for C and N species (Kendall, 1998; Spötl and others, 2005). The isotopic compositions of $CO_2$ were typical of values for Jack's Cave during the warmer periods of summer and fall (Knierim, 2009). Although the $\delta^{13}C$-DIC values were heavier than $\delta^{13}C$-$CO_2$ values, the composition of DIC was likely controlled by interaction with carbonate bedrock which would provide a heavier, inorganic source of C. Additionally, the heaviest $\delta^{13}$C-DIC value that occurred on 11/8/09 (-11.1‰) coincided with the occurrence of the smallest $CO_2$ concentration (750 ppm); the heavier composition is because of surface atmosphere being drawn into the cave and interaction between gaseous and aqueous C species. Atmospheric $[CO_2]$ is approximately 380 ppm and has a $\delta^{13}$C-$CO_2$ value of -7.7‰ (Faure and Mensing, 2005), which would provide a source of isotopically heavier and smaller $[CO_2]$.

Although the tagged water was not detected in the cave environment, stored soil water did infiltrate into the cave. On 9/13/09 at 12:30, the concentration of $NO_3^-$ and DIC in the cave (C1) reached relative minima, DOC concentration peaked (2.8 mg/L), and the heaviest $\delta^{13}$C-DOC value (-21.9‰) was observed. The changes in the concentration and isotopic compositions of C and N species in the cave stream were not outside of the range of background values and, therefore, not caused by tagged water entering the cave. Following release of the tagged water, precipitation fell on the morning of 9/13/09, although the rainfall was not recorded at nearby weather stations. The decrease in pH and specific conductance observed in the cave on 9/13/09 at 12:30 provided further evidence that stored soil water was moved into the cave through fractures in the sandstone cap rock following infiltration of labeled water and precipitation. The stored soil water was relatively more acidic, lower in $NO_3^-$ and DIC, and higher in recalcitrant DOC than the cave water. Although the labeled tracers were not detected in the cave, the pulse of stored soil water provided evidence that allochthonous C infiltrates into the cave environment during precipitation events. Many caves are oligotrophic, and the external source of C could provide an important organic substrate for organisms in the cave environment. The decrease in $NO_3^-$ concentration because of the pulse of stored soil water was delayed at the downstream sites (C2 and C3) compared to the upstream site (C1). The small water flow in the cave and abundance of clastic sediments may provide micro-environments suitable for denitrifcation, although $NO_3^-$ processing cannot be confirmed in the cave because the labeled tracers did not reach the cave stream. However, DOC did enter the cave from the soil zone and could serve as an electron donor during denitrification.

## SUMMARY

Glucose ($^{13}C_6H_{12}O_6$) was likely utilized in the soil and the isotopic signature was rapidly converted to inorganic C species through respiration. Most of the heavy isotopic signature was likely respired as $CO_2$ and fluxed out of the soil zone. The $K^{15}NO_3$ was also utilized in the soil zone, but $N_2$ was not monitored during the tracing experiment and the heavy isotopic signature was not observed in soil or cave $NO_3^-$. Monitoring soil respiration gases ($N_2$, $CO_2$, and likely $N_2O$) is important for understanding biological processing during isotopic tracing experiments.

Although the labeled tracer was not detected in the cave environment, stored soil water did infiltrate into the cave through fractures in the sandstone caprock. The stored soil water was relatively more acidic, lower in $NO_3^-$ and DIC, and higher in recalcitrant DOC than the cave water. The pulse of stored soil water provided evidence that allochthonous C infiltrates into the cave environment during precipitation events.

## ACKNOWLEDGMENTS

Support for this project was provided by the Cave Research Foundation.

## REFERENCES

Aravena, R., and Robertson, W.D., 1998, Use of multiple isotope tracers to evaluate denitrification in ground water: Study of nitrate from a large-flux septic system plume: Ground Water, v. 36, p. 975-982.

Böhlke, J.K., 2002, Groundwater recharge and agricultural contamination: Hydrogeology Journal, v. 10, p. 153-179.

Casciotti, K. L., Sigman, D. M., Hastings, M. G., Bohlke, J. K., Hilker, A., 2002, Measurement of the oxygen isotopic composition of nitrate in seawater and freshwater using the denitrifier method: Analytical Chemistry, v. 74, p. 4905-4912.

Clark, I., and Fritz, P., 1997, Environmental Isotopes in Hydrogeology: Boca Raton, N.Y., Lewis Publishers.

Davis, R. K., Brahana, J. V., and Johnston, J. S., 2000, Ground water in northwest Arkansas: Minimizing nutrient contamination from non-point sources in karst terrane: Arkansas Water Resources Center, Publication Number MSC-288, http://www.uark.edu/depts/awrc/pdf_files/MSC/MSC-288.pdf.

Einsiedl, F., Maloszewski, P., and Stichler, W., 2005, Estimation of denitrification potential in a karst aquifer using the $^{15}N$ and $^{18}O$ isotopes of $NO_3^-$: Biogeochemistry, v. 72, p. 67-86.

Faure, G., and Mensing, T.M., 2005, Carbon, *in* Isotopes: Principles and Applications: Hoboken, N.J., John Wiley and Sons, Inc., p. 753-802.

Gillip, J. A., 2007, The effects of land-use change on water quality and speleogenesis in Ozark cave systems – a paired cave study of Civil War and Copperhead Caves, northwestern Arkansas [M.S. thesis]: Fayetteville, University of Arkansas, 77 p.

Kendall, C., 1998, Tracing nitrogen sources and cycling in catchments, *in* (Eds.) C. Kendall and J. J. McDonnell, Isotope Tracers in Catchment Hydrology: Amsterdam, Elsevier, p. 519-576.

Knierim, K. J., 2009, Seasonal variation of carbon and nutrient transfer in a northwestern Arkansas cave [M.S. thesis]: Fayetteville, University of Arkansas, 141 p.

Peterson, E.W., Davis, R.K., Brahana, J.V., and Orndorff, H.A., 2002, Movement of nitrate through regolith covered karst terrane, northwest Arkansas: Journal of Hydrology, v. 256, p. 35-47.

Nascimento, C., Atekwana, E.A., and Krishnamurthy, R.V., 1997, Concentrations and isotope ratios of dissolved inorganic carbon in denitrifying environments: Geophysical Research Letters, v. 24, p. 1511-1514.

National Cooperative Soil Survey, 2008, United States Department of Agriculture, Custom Soil Resource Report for Madison County, Arkansas, 17 p.

Sigman, D. M., Casciotti, K. L., Andreani, M., Barford, C., Galanter, M., Bohlke, J. K., 2001, A bacterial method for the nitrogen isotopic analysis of nitrate in seawater and freshwater: Analytical Chemistry, v. 73, p. 4145-4153.

Spötl, C., Fairchild, I.J., and Tooth, A.F., 2005, Cave air control on dripwater geochemistry, Obir Caves (Austria): Implications for speleothem deposition in dynamically ventilated caves: Geochimica et Cosmochimica Acta, v. 69, p. 2451-2468.

St-Jean, G., 2003, Automated quantitative and isotopic ($^{13}C$) analysis of dissolved inorganic carbon and dissolved organic carbon in continuous-flow using a total organic carbon analyser: Rapid Communications in Mass Spectrometry, v. 17, p. 419-428.

U.S. Department of Agriculture, 2010, State Facts Sheets: Arkansas, http://www.ers.usda.gov/StateFacts/AR.htm, [accessed February 26, 2010].

# Interpreting a Spring Chemograph to Characterize Groundwater Recharge in an Urban, Karst Terrain

By Victor Roland[1], Carlton Cobb[1], Lonnie Sharpe[1], Patrice Armstrong[2], Dafeng Hui[2], and Thomas D. Byl[3]

[1]College of Engineering, Tennessee State University, 3500 John A Merritt Blvd, Nashville, TN 37209
[2]Biology Dept., Tennessee State University, 3500 John A Merritt Blvd, Nashville, TN 37209
[3] U.S. Geological Survey, 640 Grassmere Park, Suite 100, Nashville, TN 37211

## Abstract

Karst aquifers in urban locations are considered particularly vulnerable to contamination for three reasons. First, karst features and hydraulic processes tend to promote rapid movement of surface waters into the sub-surface with little or no filtration. Second, urban settings tend to have a higher density of contaminant sources. Third, urban settings have a greater amount of impervious surfaces that were frequently designed to direct surface runoff to sinkholes or losing streams with no filtration. The objective of this project was to better understand the vulnerability of Nashville's shallow aquifer to contamination by using tracers to estimate residence time in the shallow karst aquifer. Two years of near continuous temperature and conductivity monitoring was done at Tumbling Rock Spring; located on Tennessee State University's campus. Additional synoptic sampling was done to augment the electronic monitoring. The data include $dO_2$, pH, $NO_3$, $SO_4$, iron, discharge, and turbidity. Attention was also paid to the weather to ascertain patterns in the spring chemographs associated with rain events. The data show that groundwater discharging from the spring maintained a temperature of $17.5°$ centigrade (C) plus or minus 1 degree C year round. Specific conductance generally dropped during the drier summer months and then rose during the wet winter season. This pattern was punctuated by sharp peaks and valleys associated with rain events. Based on tracer studies using sodium chloride, up to 10 percent of the spring's flow has an estimated aquifer residence time of less than 1 month.

# Investigation of Nitrate Processing in the Interflow Zone of Mantled Karst, Northwestern Arkansas

By Jozef Laincz

Environmental Dynamics, University of Arkansas, 113 Ozark Hall, Fayetteville, AR 72701

## Abstract

Anthropogenic nitrate contamination of groundwater is a common problem in vulnerable terrains dominated by karst topography. Elucidation of in-situ nitrate dynamics is important to the design of sustainable land-management practices in these terrains. A field-scale study was conducted at a manure-amended, mantled karst site in the Ozark Highlands to characterize multiple potential sources and processes affecting nitrate in the interflow zone. Increased water residence time and water-matrix contact may favor nitrate attenuation processes, such as denitrification, in the interflow zone, which is situated between the soil and focused-flow (bedrock) zones. Groundwater samples were collected along the hydrologic gradient in and below the study plot and analyzed for reactive species concentration (nitrate), conservative species concentration (chloride), nitrate isotopic compositions ($\delta^{15}N$ and $\delta^{18}O$), and dissolved organic carbon concentration and bioavailability. $\delta^{15}N$ and $\delta^{18}O$ indicated a mixed soil organic matter/manure origin for nitrate. Mass-balance calculations indicated that although mixing was the primary process decreasing nitrate concentration along flowpaths through the interflow zone, up to 33 percent of nitrate moving through the interflow zone may have been removed through microbial processing. The magnitude of this processing varied spatially and temporally. Dissolved organic carbon bioavailability was elevated and its concentration decreased in the interflow zone relative to the focused-flow zone, suggesting the availability and utilization of the substrate required for nitrate processing in this zone. Overall, the observations suggest that the interflow zone may be important for nitrate attenuation in karst systems.

## INTRODUCTION

High nitrate ($NO_3^-$) concentrations in water are detrimental to man and the environment. High intake of $NO_3^-$ may result in formation of potentially carcinogenic compounds in the human gastric system (Tenovuo, 1986) as well as low oxygen levels in blood of infants, a potentially fatal condition and the reason for the U.S. Environmental Protection Agency to regulate $NO_3^-$ in drinking water by establishing maximum contaminant level of 10 mg/L $NO_3^-$ as nitrogen (N) (Fan and Steinberg, 1995). Large amounts of $NO_3^-$ discharging from agricultural watersheds have been implicated in the development of hypoxic zones around the world including in the Gulf of Mexico (Rabalais and others, 1996; Goolsby and Battaglin, 2001).

Intensive animal production in Northwest Arkansas generates large volumes of manure rich in $NO_3^-$ precursors, organic N compounds. Counties in Northwest Arkansas rank among the top 2 percent of counties in the nation which together generate 12 percent of the nation's total manure nitrogen. With the mass of manure generated, these counties do not have sufficient crop and pasture land to assimilate the N generated (Gollehon and others, 2001). Moreover, these large volumes of manures are typically applied to pastures with soils which often are, as is the case in Northwest Arkansas, extremely thin with limited nutrient bioremediation potential. To accentuate problems further, Northwest Arkansas is underlain by karst; a carbonate rock terrain made up of interconnected and hierarchically enlarged fractures, conduits and large voids which can readily transport manure-derived contaminants into deeper subsurface with little chance for attenuation. Unsurprisingly, numerous studies reported $NO_3^-$ contamination in springs and wells throughout Northwest Arkansas (Adamski, 1997; Steele and McCalister, 1990; Davis and others, 2000; Laubhan, 2006) where demand for water is rapidly increasing, with the region being the sixth most dynamically growing

Metropolitan Statistical Area in the country (U.S. Census Bureau, 2001).

To protect groundwater in this vulnerable landscape, effective manure management practices need to be designed taking into account the capacity of the local mantled karst environment to attenuate $NO_3^-$. That, in turn, requires elucidation of in-situ $NO_3^-$ attenuation processes among which primarily are vegetative uptake, dilution with waters low in $NO_3^-$, and denitrification, i.e. microbial reduction of $NO_3^-$ to N gases (NO, $N_2O$, $N_2$). Denitrification is the most important $NO_3^-$ attenuation process in as much as it removes $NO_3^-$ from the watershed in the form of N gases (Martin and others, 1999). These gases represent a longer term sink for N relative to N immobilization processes that lead to generation of organic N which may later be subjected to remobilization. Denitrification has been successfully measured using a range of methods in a wide range of terrestrial, marine, and freshwater environments (see review by Groffman and others, 2006). The occurrence of denitrification generally requires a set of conditions including $NO_3^-/NO_2^-$ availability, presence of denitrifiers, low $O_2$ concentrations, sufficiently long residence time, and a supply of bioavailable organic matter (Seitzinger and others, 2006).

Karst environments are considered unfavorable to denitrification because of high flow velocities and low nutrient supply. However, many karst terrains, including Northwest Arkansas, are mantled by regolith containing impermeable layers such as relict insoluble chert ledges and fragipan. These along with the surface of the underlying limestone bedrock divert infiltrating groundwater laterally (Al-Rashidy, 1999; Little, 2001), creating a hydrologic compartment known as the interflow zone. The interflow zone delays the movement of water into the deeper, focused-flow compartment of the system and facilitates contact between water and soil or rock where $NO_3^-$ utilizing microbes reside. In addition, the interflow zone may contain surface-derived bioavailable dissolved organic matter as an energy source for denitrifiers. Thus the interflow zone may be an important zone of $NO_3^-$ attenuation. This study aimed to investigate the potential of this zone for $NO_3^-$ attenuation

including denitrification and other biogeochemical as well as physical processes affecting the fate of $NO_3^-$, using a mass balance approach involving reactive ($NO_3^-$) and conservative ($Cl^-$) species, $NO_3^-$ $\delta^{15}N$ and $\delta^{18}O$ isotopic composition, and dissolved organic carbon concentration (DOC) and bioavailability. Previous research of these topics has been sparse.

## METHODS

### Site

The study was conducted at the Savoy Experimental Watershed (SEW), near the town of Savoy, Arkansas. The area is typical of the mantled karst setting of the Springfield Plateau of the Ozarks where regolith covers the underlying chert-rich limestone. Topography is ridge and valley with elevation in the watershed ranging from 317 to 376 m. Land cover consists of hardwood forest (57%) and pasture (43%). The study focused on and around an instrumented plot (Fig. 1) in Basin 1, which is drained by an ephemeral stream that flows towards the southwest and discharges into the Illinois River. Average annual rainfall for the area is 1,119 mm with mean January and July air temperatures of 1.1 and 25.9°C, respectively (Owenby and Ezell, 1992).

### Sampling

The fate of $NO_3^-$ was studied across a hydraulic gradient; from an infiltration area within and in the vicinity of an instrumented plot located on a ridge-top pasture (15% slope) in Basin 1, through the subsurface interflow zone, to the main drainage points of the basin – Langle and Copperhead Spring; special attention was given to the interflow zone. Hydrogeology of the system was established by previous research focusing on various aspects such as runoff and infiltration mechanisms (Sauer and others, 2000; Sauer and Logsdon, 2002, Leh, 2006), regolith and bedrock geophysics (Ernenwein and Kvamme, 2004), water and solute movement in the regolith mantle (Al-Qinna, 2003; Brahana and others, 2005; Brahana and others, 2006; Laincz, 2007) as well as saturated-flow characteristics and controls in the underlying bedrock (Brahana, 1995; Brahana, 1997;

Brahana, 1999; Whitsett, 2002; Ting, 2005). Based on this research, runoff events are rare and rainfall tends to quickly infiltrate into the regolith mantle where water moves via diffuse and macropore flow in both vertical and lateral directions; lateral movement – interflow – occurs owing to permeability contrasts presented by chert ledges and the bedrock surface. Macropores including root channels, worm holes, and pores created by the loose contact between chert fragments and soil matrix are abundant and greatly accelerate water and solute flow through the mantle. At the study site, interflow paths begin in the upland area underneath a thin (30-75 cm) soil horizon relatively rich in organics. These emerge on the side slope down-gradient as springs. During wet antecedent conditions the time of travel of a solute from the upland plot area to these springs (through the interflow zone) can range from 10 to 40 hours. Down-gradient of the emergence points, the flow forms a losing stream, with the water entering focused-flow dominated bedrock to emerge after approximately 500 meters at Langle and Copperhead Springs. The time of travel for this focused flow for a conservative tracer can be approximately 11.5 hours with a peak concentration arriving 15-19 hours after tracer injection (Whitsett, 2002; Ting, 2002).

Three distinct hydrologic components or zones of the system were characterized (Fig. 1): diffuse flow, interflow, and focused flow. Diffuse flow (soil water) was sampled by 8 suction-cup lysimeters installed 75 cm deep in the ground inside the plot; these samples required compositing owing to low volume yield of individual lysimeters. The interflow zone was sampled at an upper and lower point on the flowpath in order to evaluate $NO_3^-$ evolution within this zone. The upper point (referred to as 'interflow in') was an interceptor trench (Smettem and others 1991) and constructed at the down-slope boundary of the plot. The second point (interflow out) were three seeps (J2, J2b, J3) on the slope side, downgradient from the plot and the trench, representing the terminus of the interflow zone flowpaths. Finally, focused-flow sampling points included two main springs draining Basin 1, Copperhead and Langle, and one smaller spring adjacent to the plot area, J1. Dye tracing (Laincz, 2007) confirmed that J1 has no hydrologic connection to the plot area,

and distinct physical and chemical parameters of its water suggest that it is dominated by focused flow. In addition, surface runoff was collected for three events. Discharge and field parameters including pH, conductivity, temperature, and dissolved oxygen were measured. Sampling was conducted for seven events between July 2004 and June 2005, three of which were conducted under storm flow and four under base flow conditions. The plot area was amended with chicken litter in June and October 2004 at a rate of 5 tons per acre to simulate common agricultural practices in terms of $NO_3^-$ loading and to increase $NO_3^-$ signal.

## Analysis

Samples were analyzed for major anions and cations using ion chromatography (EPA method 300.0) and ICP (EPA method 200.7), respectively. Stable isotopes of $NO_3^-$ ($\delta^{15}N$ and $\delta^{18}O$) were using the Denitrifier Method (Sigman and others, 2001; Casciotti and others, 2002). The method precision and accuracy for $\delta^{15}N$ were 0.35‰ and 0.43‰, respectively, and for $\delta^{18}O$ they were 0.66‰ and 0.78‰,

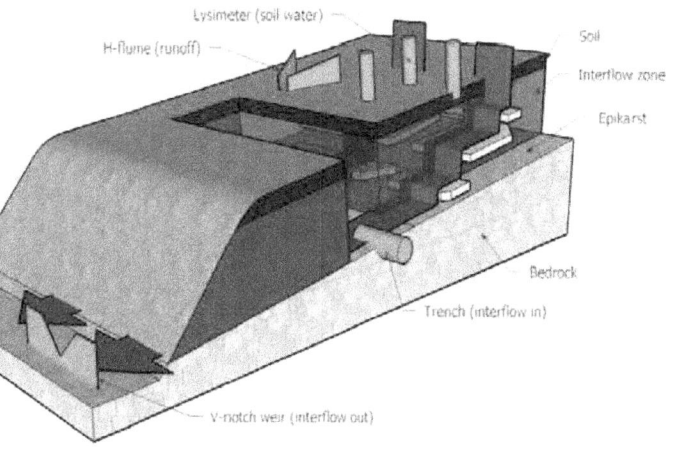

Figure 1. Study plot with hydrogeologic components and sampling instruments. Langle and Copperhead springs are located about 500 yards from the site.

respectively. DOC concentration was determined using high-temperature combustion with a Shimadzu TOC5050 (Benner and Strom, 1993). Bioavailability of DOC was measured as groundwater community respiration normalized to DOC concentration and bacterial abundance, which was determined using epifluorescent microscopy (Porter and Feig, 1980). Details pertaining to bioavailability measurement as well as to all of the above mentioned sampling and analytical procedures can be found in Winston (2006).

## RESULTS AND DISCUSSION

### DOC concentration and DOM bioavailability

DOC comprised 80% of total organic carbon in SEW water samples, providing the necessary substrate for microbial processes. DOC concentration ranged from 0.14 to 22.44 mg/L with overall median of 1.43 mg/L. The median for high-flow events was slightly higher (3.03 mg/L) compared to low-flow events (1.10 mg/L), and DOC concentration of high-flow events also exhibited greater variability. Mean DOC concentrations for the July and December 2004 events were elevated relative to the other events sampled. DOC concentrations tended to decrease downgradient (Fig. 2) suggesting possible utilization for microbial metabolic processes.

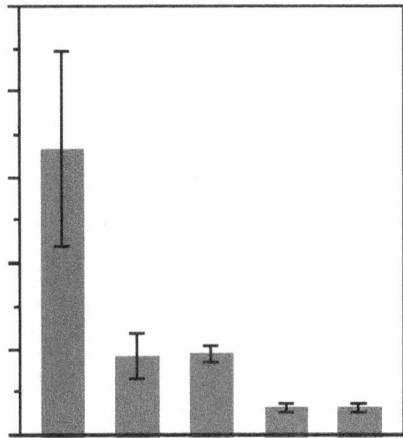

Figure 2. Mean (± 1 Std. Error) DOC concentration across the hydrologic gradient.

DOM bioavailability was determined for only 12 samples all of which were collected from the higher discharge sites – J1, J2, Copperhead

Spring and Langle Spring – due to difficulty obtaining sufficient representative sample volumes from the rest of the sampling sites. In addition, no one sampling event had all of the four sites sampled, and the July 2004 event was not sampled for bioavailability at all. Nevertheless, the limited data indicate seasonal and spatial variation of DOM quality. DOM bioavailability in the interflow and focused-flow zones was lower during fall events relative to spring events (Fig. 3), suggesting a relatively more refractory DOC pool during fall and influx of labile organic carbon from growing plants and microbial exudates during spring. Spatially, comparing the individual sampling points, bioavailability was elevated in J2 and Langle Spring relative to J1 and Copperhead Spring. The sites representing the interflow zone exhibited about 2.6 times greater bioavailability than the focused-flow zone sites (Fig. 4), providing the former a needed substrate for $NO_3^-$ microbial processing in this zone.

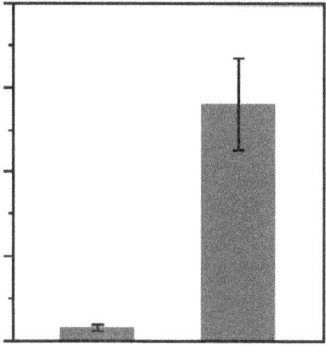

Figure 3. Mean (± 1 Std. Error) relative bioavailability of DOC by season.

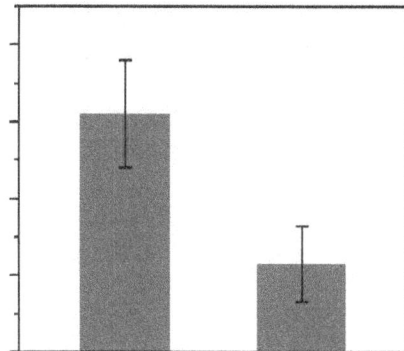

Figure 4. Mean (± 1 Std. Error) relative bioavailability of DOC by flow type. Focused flow is a mean of average bioavailabilities of seep J1, Langle and Copperhead Springs. Interflow (out) represents seep J2.

## Nitrate concentration and isotopic composition

$NO_3^-$ concentration ranged from 0.64 to 48 mg/L throughout the study with a median of 3.3 mg/L. Average $NO_3^-$ concentrations were not significantly different among events. Along the gradient, $NO_3^-$ concentration averages for lysimeters, trench, and Copperhead spring were 10.2, 36.6, and 5.3 mg/L, respectively, which was higher relative to the interflow zone seeps and L angle spring where concentrations were within a relatively narrow range of 2.8 to 3.4 mg/L.

$\delta^{15}N$-$NO_3^-$ ranged from 0.35 to 9.66‰ and had a mean of 4.85‰. Seasonally, $\delta^{15}N$ values appeared to slightly increase during spring and fall; averages for spring and fall events were elevated relative to the overall mean by about 1.19 and 0.42‰, respectively. Across the flow path, the highest average value was in the diffuse-flow zone (8.08‰) and the lowest in the runoff samples (3.02‰). $\delta^{18}O$-$NO_3^-$ values ranged from -0.89 to 19.67‰ with a mean of 5.13‰. Similar to $\delta^{15}N$, $\delta^{18}O$ values also tended to increase during spring and fall. Average values for spring and fall events were elevated relative to the overall mean by about 3.24 and 0.19‰. Across the flow path, runoff had the highest average value (8.01‰) while the 'interflow in' (trench) samples had the lowest (3.76‰). The measured $\delta^{15}N$ and $\delta^{18}O$ values are similar to those obtained for ground water draining karst (e.g. Panno and others, 2001; Einsiedl and others, 2005). Ninety percent of both $\delta^{15}N$ and $\delta^{18}O$ values fell between approximately 2 and 8‰, a range indicative of a mixture of nitrate derived from soil organic matter and animal manure (Clark and Fritz, 1997).

Denitrification is known to cause the $\delta^{15}N$ and $\delta^{18}O$ values of the residual $NO_3^-$ to increase as a result of kinetic isotopic fractionation (Böttcher and others, 1990). In addition to this isotopic enrichment of the remaining $NO_3^-$ pool, the ratio of the enrichment of oxygen to N tends to be close to 1:2 (Kendall and McDonnell, 1998) imparting an additional signal for recognition of denitrification. Nitrate $\delta^{15}N$ and $\delta^{18}O$ of the samples from the interflow zone-draining seeps and the adjacent small spring J1

positively correlate indicating simultaneous enrichment in both isotopes likely due to denitrification. While the slope deviates from 0.5 and the correlation is only moderately strong (r = 0.4), it may not be reasonable to expect a much stronger fit with the model considering the geochemical and physical heterogeneity of the system,

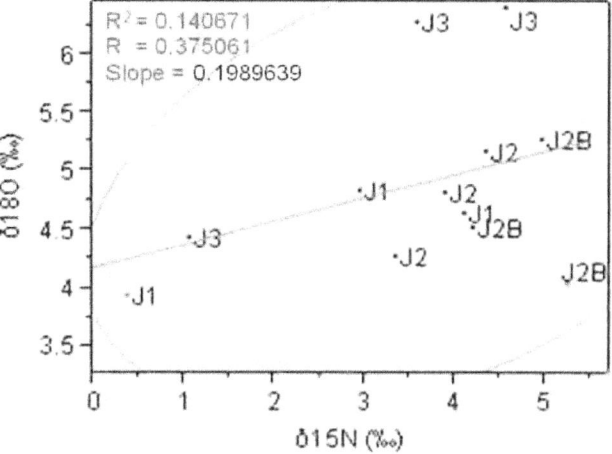

Figure 5. Co-linearity between δ15N and δ18O of seep and J1 samples.

## Nitrate processing assessment through mass balance

Three post-storm sampling events (July 3, December 1 and 7 2004) were conducted under hydrologic conditions wet enough to induce flow in the trench, and these events thus enabled assessing $NO_3^-$ evolution between the trench as an upper interflow zone point and interflow zone exit points, the seeps. Concentrations of solutes in the seeps were lower as compared to the trench. For example, average $Cl^-$ concentration at the seeps was 75% lower than the average concentration at the trench. In addition, for a given trench-seep flowpath, the magnitude of the decrease was unequal between conservative and reactive species; $NO_3^-$ decrease was on average 16% greater than $Cl^-$ decrease. While the decrease in $Cl^-$ indicated dilution or mixing along the interflow zone flowpaths, the greater decrease in $NO_3^-$ concentration indicated the existence of an additional mechanism of $NO_3^-$ removal – microbial processing.

The amount of microbial processing was evaluated for the three sampling events by modeling $NO_3^-$ concentration, $NO_3^-$ $\delta^{15}N$ and $\delta^{18}O$, and DOC concentration expected at the seeps as a result of mixing (dilution) and comparing these model values to measured values, which represent the combined effects of mixing and microbial processes, such as microbial assimilation and denitrification. Microbial processing would have decreased $NO_3^-$ concentration beyond the decrease resulting from mixing and would also have isotopically enriched the remaining $NO_3^-$ pool (Kendall and Mcdonell, 1998).

The evidence of dilution along the trench-seep flowpaths led us to adopt a flow model that assumed two types of water mixing along these flow paths, producing a mixture discharging out of the seeps; one type infiltrating in and carrying the chemical and isotopic signature of the up-gradient plot, where manure amendment had provided a strong solute source, and another type infiltrating in and carrying the chemical and isotopic signature of the surrounding unamended off-plot area. Based on dye-tracing evidence, the interceptor trench immediately down-gradient from the plot was considered representative of the plot water, and spring J1 with lack of hydrologic connection to the plot was regarded as representative of off-plot water type.

Nitrate and DOC concentrations expected in the mixtures of plot and off-plot waters were calculated using a common two-component mass-balance equation (Faure and Mensing, 2005; Fry, 2006):

$$X_M = X_P * f_P + X_{OP} * f_{OP}$$

where $X_M$, $X_P$, and $X_{OP}$ are the concentrations of DOC or $NO_3^-$ in a seep mixture, plot and off-plot water components, respectively, and $f$ represents the mass fractions or mixing proportions of these components. The mass fractions were calculated by solving the equation above for $f_P$ using conservative species ($Cl^-$) as concentrations. The off-plot mass fraction $f_{OP}$ was then obtained by subtracting $f_P$ from 1.

Table 1 lists the mass fractions of plot and off-plot components calculated for each seep

mixture, $NO_3^-$ concentrations expected in the seep mixtures from mixing, corresponding measured concentrations, and the difference between calculated and measured. Average calculated $NO_3^-$ concentration in the seep mixture was 6.6 mg/L. Average measured concentration, however, was about 3.1 mg/L, and so the average 3.47 mg of "missing" $NO_3^-$ in a liter of the seep mixture was possibly denitrified and/or microbially immobilized. Since plant roots at the site do not reach into the interflow zone, assuming that microbial processing is indeed the process responsible for the missing $NO_3^-$ is reasonable. Performing similar calculations with DOC, 0.23 mg/L on average of DOC loss in the interflow zone mixtures could not be accounted for by plot/off-plot mixing and was likely consumed by some DOC-utilizing microbial process. Although the calculated amount of missing DOC appear modest, it may be underestimated. Other potential inputs of DOC percolating into the interflow zone flow-paths and not accounted for in the mixing equation may have increased the final measured DOC concentration and thus reduced the calculated difference ascribable to processes other than mixing.

## CONCLUSIONS

Mass-balance calculations based on two-component mixing indicated that most of the $NO_3^-$ moving through the interflow zone was removed by dilution. Between 1 and 33%, however, depending on flow-path, was removed by additional processes which likely included microbial processing since uptake by vegetation does not occur in the interflow zone. The $\delta^{15}N$ and $\delta^{18}O$ values of this remnant $NO_3^-$ present in the seep samples and the adjacent site J1 exhibited moderately strong positive correlation (r = 0.4), indicating denitrification. Nitrate $\delta^{15}N$ and $\delta^{18}O$ also proved useful in determining correctly the source of nitrate – a mixture of organic matter and animal manure. DOC concentration generally decreased along the flowpath and mass balance computation showed that for the majority of the seep samples dilution could not account for this decrease entirely and, as a result, a different process such as utilization by microbes might have also occurred. Both DOC concentration and DOM bioavailability were elevated in the interflow zone relative to focused-flow paths, providing the former with increased potential for $NO_3^-$ attenuation within this karst system. Temporally, DOC as well as $NO_3^-$ increased during high-flow conditions, creating favorable conditions for microbial processing of $NO_3^-$. In the karst watershed as a whole, DOM bioavailability appeared to be influenced by seasonality, with spring having greater DOM bioavailability than fall, suggesting an influx of labile organic matter during spring. Whereas all measured parameters varied considerably in both time and space, the overall biogeochemical evidence from this study implies that the interflow zone is a zone of significant $NO_3^-$ attenuation by dilution and denitrification. It seems that more research would be warranted to establish more definitive patterns of how these processes change with environmental factors, flow path, and land use.

Table 1. $NO_3^-$ concentration measured and predicted by the binary mixing model, and the difference between the two ascribable to microbial processing.

| Sampling event | Seep | Endmember fractions | | Nitrate, mg/L | | | Nitrate Processing, % |
|---|---|---|---|---|---|---|---|
| | | Plot (Trench) | Off-plot (J1) | Modeled | Measured | Mod.-Meas. | |
| 3-Jul-04 | J2 | 0.11 | 0.89 | 3.84 | 2 | 1.84 | 8 |
| 3-Jul-04 | J2b | 0.39 | 0.61 | 9.75 | 2.42 | 7.33 | 33 |
| 3-Jul-04 | J3 | 0.09 | 0.91 | 3.50 | 1.4 | 2.10 | 9 |
| 1-Dec-04 | J2 | 0.01 | 0.99 | 3.70 | 3.26 | 0.44 | 1 |
| 1-Dec-04 | J2b | 0.03 | 0.97 | 4.30 | 3.42 | 0.88 | 2 |
| 1-Dec-04 | J3 | 0.02 | 0.98 | 4.00 | 2.63 | 1.37 | 3 |
| 7-Dec-04 | J2 | 0.15 | 0.85 | 7.96 | 3.42 | 4.54 | 12 |
| 7-Dec-04 | J2b | 0.34 | 0.66 | 15.06 | 7.96 | 7.10 | 18 |
| 7-Dec-04 | J3 | 0.12 | 0.88 | 7.01 | 1.4 | 5.61 | 14 |

## ACKNOWLEDGMENTS

I am grateful to Dr. Van Brahana and Katherine Knierim, Univ. of Arkansas, for reviewing this manuscript. I also thank my advisor, Dr. Phillip Hays, for his guidance throughout this project.

## REFERENCES

Adamski, J. (1997). Nutrients and pesticides in ground water of the Ozark Plateaus in Arkansas, Kansas, Missouri, and Oklahoma (U.S. Geological Survey Water-Resources Investigations Report No. 96-4313)

Al-Qinna, M. I. (2003). Measuring and modeling soil water and solute transport with emphasis on physical mechanisms in karst topography. Unpublished Ph.D. Dissertation, University of Arkansas, Fayetteville.

Al-Rashidy, S. M. (1999). Hydrogeologic controls of groundwater in the shallow mantled karst aquifer, Copperhead Spring, Savoy Experimental Watershed, Northwest Arkansas. Unpublished M.S. Thesis, University of Arkansas, Fayetteville.

Benner, R. and Strom, M. (1993) A critical evaluation of the analytical blank associated with DOC measurements by high-temperature catalytic oxidation. Marine Chemistry, 41, 1-3, 153-160.

Brahana, J. V. (1995). Controlling influences on ground-water flow and transport in the shallow karst aquifer of Northeastern Oklahoma and Northwestern Arkansas. Proceedings of the Arkansas Water Resources Center 1994 Research Conference, Fayetteville, Arkansas. 25.

Brahana, J. V. (1997). Rationale and methodology for approximating spring-basin boundaries in the mantled karst terrane of the Springfield Plateau, Northwestern Arkansas. In B. F. Beck, and J. B. Stephenson (Eds.), The engineering geology and hydrogeology of karst terranes (pp. 77-82). United States (USA): A.A. Balkema, Rotterdam.

Brahana, J. V., Hays, P. D., Kresse, T. M., Sauer, T. J., and Stanton, G. P. (1999). The Savoy Experimental Watershed; early lessons for hydrogeologic modeling from a well-characterized karst research site. In A. N. Palmer, M. V. Palmer and I. D. Sasowsky (Eds.), Karst modeling: Special publication no. 5 (pp. 247-254). Charlottesville, VA: Karst Water Institute.

Brahana, J. V., Killingbeck, J. J., Stielstra, C., Leh, Mansoor Delali Kwasi, Murdoch, J. F., and Chaubey, I. (2006). Elucidating flow characteristics of epikarst springs using long-term records that encompass extreme hydrogeologic stresses. Geological Society of America Abstracts with Programs, Philadelphia, Pennsylvania. , 38(7) 196.

Brahana, J. V., Ting, T. E., Al-Qinna, M., Murdoch, J. F., Davis, R. K., Laincz, J., (2005). Quantification of hydrologic budget parameters for the vadose zone and epikarst in mantled karst. U. S. Geological Survey Karst Interest Group Proceedings, Rapid City, South Dakota. , SIR 2005-5160 144-152.

Casciotti, K. L., Sigman, D. M., Hastings, G. M., Böhlke, J. K., and Hilkert, A. (2002). Measurement of the oxygen isotopic composition of nitrate in seawater and freshwater using the denitrifier method. Analytical Chemistry, 74, 4905-4912.

Clark, I.D. and Fritz, P. (1997). Environmental isotopes in hydrogeology. New York, Lewis Publishers.

Davis, R. K., Brahana, J. V., and Johnston, J. S. (2000). Groundwater in northwest arkansas: Minimizing nutrient contamination from non-point sources in karst terrane. (Arkansas Water Resources Center Publication No. MSC-288). Fayetteville, Arkansas: Arkansas Water Resources Center.

Drury, C. F., McKenney, D. J., and Findlay, W. I. (1991). Relationships between denitrification, microbial biomass and indigenous soil properties. Soil Biology and Biochemistry, 23(8), 751-755.

Einsiedl, F., Maloszewski, P., and Stichler, W. (2005). Estimation of denitrification potential in a karst aquifer using the 15N and 18O isotopes of NO 3−. Biogeochemistry 72, 1, 67-86.

Ernenwein, E. G., and Kvamme, K. L. (2004). Geophysical investigations for subsurface fracture detection in the Savoy Experimental Watershed, Arkansas. University of Arkansas, Fayetteville: Department of Biological and Agricultural Engineering.

Fan, A. M., and Steinberg, V. E. (1996). Health implications of nitrate and nitrite in drinking water: An update on methemoglobinemia occurrence and reproductive and developmental toxicity. Regulatory Toxicology and Pharmacology, 23(1), 35-43.

Faure, G., and Mensing, T. M. (2005). Isotopes: Principles and applications (3rd ed.). Hoboken, New Jersey: John Wiley and Sons.

Gollehon, N., Caswell, M., Ribaudo, M., Kellogg, R., Lander, C., and Letson, D. (2001). Confined animal production and manure nutrients. USDA Agriculture Information Bulletin, (AIB771), 1-40.

Goolsby, D. A., and Battaglin, W. A. (2001). Long-term changes in concentrations and flux of nitrogen in the Mississippi River Basin, USA. Hydrological Processes, 15(7), 1209-1226.

Kendall, C., and McDonnell, J. J. (Eds.). (1998). Isotope tracers in catchment hydrology. Amsterdam: Elsevier Science B.V.

Laincz, J. (2007). Qualitative dye tracer test at the Savoy Experimental Watershed plot (Unpublished data).

Laubhan, A. C. (2007). A hydrogeologic and water-quality evaluation of the Springfield Aquifer in the vicinity of north-central Washington County, Arkansas. Unpublished M.S. Thesis, University of Arkansas, Fayetteville.

Leh, Mansoor Delali Kwasi. (2006). Quantification of rainfall-runoff mechanisms in a pasture dominated watershed. Unpublished M.S. Thesis, University of Arkansas, Fayetteville.

Little, P. R. (2001). Characterization and development of a conceptual model of groundwater flow and transport in basin 2, Savoy Experimental Watershed, Northwest Arkansas. Unpublished M.S. Thesis, University of Arkansas, Fayetteville.

Martin, T. L., Kaushik, N. K., Trevors, J. T., and Whiteley, H. R. (1999). Review: Denitrification in temperate climate riparian zones. Water, Air, and Soil Pollution, 111(1), 171-186.

Owenby, J. R., and Ezell, D. S. (1992). Monthly station normals of temperature, precipitation, and heating and cooling degree days, 1961-1990, Arkansas. Climatography of the United States No. 81. Asheville, NC: U.S. Department of Commerce, National Climatic Data Center.

Panno, S. V., Hackley, K. C., Hwang, H. H. and Kelly, W.R., 2001. Determination of the sources of nitrate contamination in karst springs using isotopic and chemical indicators. Chemical Geology, 179, 1-4, 113-128.

Porter K. G. and Feig, Y. S. (1980). The Use of DAPI for Identifying and Counting Aquatic Microflora. Limnology and Oceanography, 25, 5, 943-948.

Rabalais, N. N., Turner, R. E., Justic, D., Dortch, Q., Wiseman, W. J., and Sen Gupta, B. K. (1996). Nutrient changes in the Mississippi River and system responses on the adjacent continental shelf. Estuaries and Coasts, 19(2), 386-407.

Sauer, T. J., Daniel, T. C., Nichols, D. J., West, C. P., Moore, P. A.,Jr., and Wheeler, G. L. (2000). Runoff water quality from poultry litter-treated pasture and forest sites. Journal of Environmental Quality, 29, 515-521.

Sauer, T. J., and Logsdon, S. D. (2002). Hydraulic and physical properties of stony soils in a small watershed. Soil Science Society of America Journal, 66(6), 1947-1956.

Seitzinger, S., Harrison, J. A., Böhlke, J. K., Bouwman, A. F., Lowrance, R., Peterson, B., (2006). Denitrification across landscapes and waterscapes: A synthesis. Ecological Applications, 16(6), 2064-2090.

Sigman, D. M., Casciotti, K. L., Andreani, M., Barford, C., Galanter, M., and Böhlke, J. K. (2001). A bacterial method for the nitrogen isotopic analysis of nitrate in seawater and freshwater. Analytical Chemistry, 73, 4145-4153.

Smettem, K. R. J., Chittleborough, D. J., Richards, B. G., and Leaney, F. W. (1991). The influence of macropores on runoff generation from a hillslope soil with a contrasting textural class. Journal of Hydrology, 122(1-4), 235-252.

Tenovuo, J. (1986). The biochemistry of nitrates, nitrites, nitrosamines and other potential carcinogens in human saliva. Journal of Oral Pathology, 15(6), 303-307.

Ting, T. (2005). Assessing bacterial transport, storage and viability in mantled karst of Northwest Arkansas using clay and Escherichia coli labeled with lanthanide series metals. Unpublished Ph.D.

Dissertation, University of Arkansas, Fayetteville.

U.S. Census Bureau. (2001). Metropolitan areas ranked by percent population change: 1990 to 2000. Retrieved January 1, 2008, from http://www.census.gov/population/cen2000/phc-t3/tab05.pdf

Whitsett, K. S. (2002). Sediment and bacterial tracing in mantled karst at Savoy Experimental Watershed, Northwest Arkansas. Unpublished M.S. Thesis, University of Arkansas, Fayetteville.

Winston, B. A. (2006). The biogeochemical cycling of nitrogen in a mantled karst watershed. Unpublished M.S. Thesis, University of Arkansas, Fayetteville.

# KARST MAPPING, DYE TRACING, AND GEOGRAPHIC INFORMATION SYSTEMS

## Using a Combination of Geographic Information System Techniques and Field Methods to Analyze Karst Terrain in Selected Red River Sub-Watersheds, Tennessee and Kentucky

By David E. Ladd

U.S. Geological Survey, 640 Grassmere Park, Suite 100, Nashville, TN, 37211

**Abstract**

Karst features such as closed depressions and their catchments present challenges to natural-resources management and topographic analysis. Diversion and collection of surface-water runoff by these features can complicate analyses of stream flow, groundwater recharge and flow, and contaminant transport. In karst areas, some component of surface-water runoff is diverted to closed depressions that recharge groundwater at various rates, thus affecting the amount of direct runoff reaching streams and providing potential pathways for contaminant entry to a groundwater system.

The U.S. Geological Survey (USGS) Tennessee Water Science Center (WSC), in cooperation with the Tennessee Department of Environment and Conservation (TDEC), is applying Geographic Information System (GIS) terrain analysis along with field data-collection methods to identify karst features, characterize their surface-water/groundwater connectivity, and determine the fate of contaminants that enter these features in support of water-resources protection in selected sub-watersheds of the Red River in Tennessee and Kentucky.

The analysis focuses on quantifying the number, locations, catchment areas, storage capacities, and drainage rates of closed depressions in this watershed. The results are incorporated into analytical, statistical, and conceptual models to explore the role of closed depressions and their internal drainage characteristics as controls on basin-scale hydrologic response and contaminant delivery to streams and springs.

Calculations of depression water storage volume and catchment area provide estimates of the rainfall required to fill a closed depression. If the amount of rainfall runoff diverted into a depression during a storm event exceeds the depression drainage rate, flooding may occur. A flooded depression that can no longer accept water may reject storm flow and re-divert it to runoff. Depressions with low storage capacity will flood and overflow quickly if poor connection to a groundwater system exists. Depressions with these characteristics are less likely to transmit significant amounts of contaminated runoff to groundwater during storm events. In contrast, depressions with high storage capacity are less likely to overflow regardless of their ability to transmit runoff as recharge to a groundwater system. These depressions have the potential to collect more runoff than low-storage capacity depressions, and they may recharge a groundwater system relatively rapidly if they are hydraulically well connected to the subsurface. If depression drainage rates can be determined, techniques to define which depressions are most likely to quickly transmit the highest quantities of runoff to a groundwater system could be developed. Such techniques could be used to categorize depressions based on their potential to contribute contaminants to groundwater and surface water, and to focus more-detailed efforts such as dye-tracer tests on depressions that have the highest contaminant-transport potential. These methods then could be applied in karst areas at a regional scale to help define surface-water/groundwater interaction at lower costs than typically labor intensive field-only studies.

# Parellelism in Karst Development Suggested by Quantitative Dye Tracing Results from Springfield, Missouri

By Douglas R. Gouzie and Katherine Tomlin
Department of Geography, Geology, and Planning, Missouri State University, 901 S. National Ave, Springfield, MO 65897

**Abstract**

Recent and continuing urban development on the southern edge of the city of Springfield, Missouri has drawn public attention to typical karst processes such as sinkhole formation and surface stream capture by underlying conduit piracy. Four dye traces were performed in the Ward Branch watershed to better understand the subsurface karst drainage system in this rapidly urbanizing area – including a trace from a sinkhole that opened in 2009 in the stormwater detention basin of a McDonald's restaurant. Quantitative dye tracing involved using automatic samplers in springs to collect discharging groundwater and direct-water fluorometry of these samples to analyze for the presence and concentration of dye at each of these springs. This extensive data collection using hourly sample data allowed construction of well-defined dye-trace break-through curves. The results of these dye traces indicated that dye injected into each of four sinkholes travelled to springs in two distinct discharge areas located approximately one-quarter mile apart on opposite sides of US 160. The average flow velocities calculated during this study ranged from 0.11 to 0.27 feet per second for the springs in the western area, and from 0.22 to 0.35 feet per second for the springs in the eastern area. Follow-up work, which is ongoing in this study area, includes use of a dye-injection pump to improve discharge determinations along the flowpaths during a dye trace and surface geophysics to better delineate likely conduits controlling groundwater flow. Preliminary results from the dye-injection system suggest the presence of a conduit connecting the groundwater system between the eastern and western springs (within, or essentially beneath, the rapidly developing US 160 corridor) and that this conduit may be transporting as much as five cubic feet per second ($ft^3/s$) of flow. This volume is similar to the typical 4 to 5 $ft^3/s$ in the surface portions of Ward Branch flowing in the study area.

# Geologic Controls on Karst Landscapes in the Buffalo National River Area of Northern Arkansas: Insights Gained from Comparison of Geologic Mapping, Topography, Dye Tracers and Karst Inventories

By Mark R. Hudson[1], Kenzie J. Turner[1], Chuck Bitting[2], James E. Kaufmann[3], Timothy M. Kresse[4] and David N. Mott[5]
[1]U.S. Geological Survey, Box 25046, Mail Stop 980, Denver, CO 80225
[2]National Park Service, 402 N. Walnut, Suite 136, Harrison, AR 72601
[3]U.S. Geological Survey, 1400 Independence Road, Mail Stop 543, Rolla, MO 65401
[4]U.S. Geological Survey, 401 Hardin Road, Little Rock, AR 72211
[5]U.S. Geological Survey, 2617 E. Lincolnway, Suite B, Cheyenne, WY 82001

## Abstract

The Ozark Plateaus host extensive karst landscapes developed on soluble bedrock. Geologic mapping by the USGS at 1:24,000 scale in the western Buffalo River region of northern Arkansas highlights the geologic controls on karst features and provides a scientific basis for resource management at Buffalo National River (BNR), a 214 kilometer (km) long, river-corridor park. The western Buffalo River flows eastward through plateau surfaces of the Boston Mountains and Springfield Plateau where it has eroded a valley 130 to 400 meter (m) deep. Stratigraphically, the watershed exposes a 500 m aggregate thickness of carbonate, sandstone, and shale formations of Pennsylvanian, Mississippian, and Ordovician age. As potential karst hosts, limestone and dolostone intervals are significant in five of the eight major map formations. Noteworthy are (1) the 120 m thick cherty limestone of the Mississippian Boone Formation that has widespread surface exposures, forming the Springfield Plateau aquifer, and (2) dolostone and lesser limestone intermixed with sandstone in the underlying Ordovician Everton Formation, forming the upper part of Ozark aquifer. Structurally, most rocks in the Buffalo River region dip gently (less than 5°) but are broken by a series of faults and folds that formed during late Paleozoic development of the southern Ozark Dome. These structures produce vertical relief of rock units of as much as 300 m across the region. Multiple joint sets are pervasive in bedrock; most-common strike directions are north to south, northeast to southwest, and west-northwest to east-southeast.

In BNR, comparison of a cave inventory to the geology demonstrates that most caves either lie within limestone of the Boone Formation (55 percent), or limestone or dolostone intervals within the Everton Formation (38 percent). Large sinkholes are preferentially concentrated near the contact of Boone Formation with overlying Batesville Sandstone. Most springs within the watershed discharge near the base of the Boone Formation, particularly in its basal Saint Joe Limestone Member. The largest of these springs in the Springfield Plateau aquifer are localized in structural lows formed by faults and folds; dye-tracer studies demonstrate that springs in three structural lows gather interbasin recharge from outside the Buffalo River watershed. A secondary concentration of springs, including the largest in BNR, discharge from the lower part of the Ordovician Everton Formation. Development of this lower karst aquifer in the Everton Formation was facilitated by a thickening of carbonate-rich facies compared to more sand-rich facies of the formation farther west. In contrast to relations for the Boone Formation, structural highs localize recharge and discharge of this lower Everton karst aquifer. In summary, both stratigraphic and structural characteristic have had important influence on karst development in the Buffalo River watershed.

# Determining the Relation between Water Loss from Beaver Creek and Kamas Fish Hatchery Springs, Samak, Utah

By Lawrence E. Spangler

U.S. Geological Survey, 2329 Orton Circle, Salt Lake City, Utah 84119

## Abstract

On October 13, 2010, fluorescein dye was injected into a sinkhole that developed alongside Beaver Creek about 4,000 feet upstream of the Kamas State Fish Hatchery in Samak, Utah. Activated charcoal packets used for adsorption of the dye were placed in the north and south hatchery springs as well as in several other springs in the vicinity of the hatchery and at selected locations along Beaver Creek. An automatic water sampler also was installed in the south spring to collect water samples on a 6-hour interval for determining groundwater time of travel from the sinkhole to the spring. The injected dye was first detected in a water sample collected at 2200 on October 16, or approximately 3 days and 8 hours after injection. The concentration of dye in the spring water peaked in the sample collected at 1000 on October 17. The dye also was detected on activated charcoal samples that were placed in both hatchery springs as well as those placed in other groundwater discharge points on the hatchery property. In addition, dye was detected on charcoal samples placed in the outflow from a private fish hatchery on the south side of Beaver Creek, in the outflow from springs discharging in the lower part of the Left-hand Canyon drainage, and in the outflow from a spring located about 4,600 feet west of the hatchery. A hydrologic connection between the sinkhole and the hatchery springs as verified by dye tracing also was indicated by measured changes in specific conductance and metered discharge of water from the south spring during the dye test, as well as by visual changes in turbidity. Results of the dye tracing indicate that only part of the water diverted into the sinkhole discharges from the hatchery springs and that divergent flow paths exist to other springs. Groundwater flow from the sinkhole to the hatchery springs may be in part, along preferential flow paths within the unconsolidated valley-fill deposits.

## INTRODUCTION

In early September 2010, residents in the community of Samak, Utah, about 50 miles east of Salt Lake City, noticed that flow in Beaver Creek had diminished substantially in the vicinity of the Kamas Fish Hatchery, one of 10 State-operated spring-fed hatcheries. Several days later, it was discovered that a sinkhole had developed along the south side of the creek approximately 4,000 feet upstream of the hatchery and was taking most of the flow of the creek, reported to have been as much as 1,000 gallons per minute (gpm). Personnel at the Kamas hatchery also noticed that flow of the two principal springs that supply the hatchery had increased by about 60 to 80 gpm and had become increasingly turbid. These relations indicated that at least some of the water losing into the sinkhole was likely discharging from the hatchery springs. More importantly, because fish in Beaver Creek upstream of the hatchery tested positive for the spore that causes whirling disease, a potential existed that water carrying these spores could move along fractures or other

high permeability flow paths from the Beaver Creek sinkhole to the hatchery springs. As a result, the U.S. Geological Survey (USGS) in cooperation with the Utah Division of Wildlife Resources conducted a dye-tracing study to determine if a hydraulic connection existed between the sinkhole and one or both of the hatchery springs.

During 2004 and 2006, the USGS also conducted investigations at the Kamas Fish Hatchery to identify sources of water to the north and south hatchery springs (Spangler, unpublished data) and to assess the potential for spores that can transmit whirling disease to move from contaminated surface-water sources along underground flow paths to the hatchery springs (Spangler and others, 2005). During this investigation, the area of focus was on drainages along the flank of the adjacent mountains, where streamflow losses were known to occur. Results of this investigation indicated that streamflow losses in Left-hand Canyon and a tributary drainage discharged at Left Fork Spring, at the confluence of these drainages (fig. 1), but did not discharge from either of the two springs at the hatchery. Increases in turbidity in the hatchery springs during the annual snowmelt runoff period, however, indicated that unidentified, discrete (focused) sources of recharge to the springs likely existed.

## METHODS

Dye-tracing techniques were used to determine if a hydrologic connection existed between the sinkhole along Beaver Creek and the springs at the hatchery. For this study, fluorescein (uranine) dye was selected because of its relatively conservative nature in the environment, detectability at low concentrations and over long distances, ease of analytical detection, and very low toxicity. Activated charcoal contained in nylon screen packets was used for adsorption of the dye and represent an integration of dye concentration during the period they are deployed. In addition, an automatic water sampler at the hatchery's south spring collected water samples on a 6-hour interval to refine groundwater time of travel between the sinkhole and the spring. Extraction of dye from the activated charcoal was done in the Utah Water Science Center laboratory by using a 5-percent solution of potassium hydroxide and isopropyl alcohol, a common eluent for extraction of fluorescein dye. Water samples also were analyzed in the Center laboratory using a Turner Designs Model 10 filter fluorometer, and selected samples were sent to the Edwards Aquifer Authority laboratory in San Antonio, Texas, for confirmation on a Perkin-Elmer scanning spectrofluorometer. Prior to injection of the dye, activated charcoal packets were placed in the north and south hatchery springs, other selected springs in the vicinity of the hatchery, and at selected locations along Beaver Creek. A total of 18 locations were monitored during the test (table 1). The automatic sampler began collecting water samples one day prior to injection of the dye to establish a baseline (background) of natural fluorescence in the water. The activated charcoal packets were collected and replaced initially after 5 days, then every week thereafter over the next month. The bottles in the automatic sampler were collected and replaced every 6 days (4 samples per day on

a 6-hour interval). In addition, water samples were collected from two domestic-use wells located near the sinkhole. Landowners of these wells collected the samples on a daily basis.

Discharge of the hatchery springs (combined outflow) was measured at a concrete weir on the south side of the hatchery building that encloses the fish runs. Discharge of the south spring was obtained from a continuous flow meter located inside a degassing unit (to reduce nitrogen concentrations) that also allows access to the spring vault. Discharge recorded by this meter was observed to vary over about 20 gpm. The meter at the north spring was not functioning during the tracer test but discharge was obtained by subtracting the metered flow of the south spring from the measured combined flow of both springs. Discharge of inflow to the sinkhole from Beaver Creek was measured at the time of the dye injection. Discharge also was measured at other selected locations in the vicinity of the hatchery (table 1). These measurements were used for evaluating the relation between the amount of flow going into the sinkhole compared with discharge from the hatchery and other springs in the area. Flow was measured using a standard pygmy current meter and wading rod.

Specific-conductance and temperature measurements were taken intermittently at the hatchery springs, in Beaver Creek, and at other selected springs and outflows in the vicinity of the hatchery (table 1). These data were used to assess similarities between the water in Beaver Creek (and going into the sinkhole) and that discharging from the hatchery and other springs in the area, and to observe changes in these parameters during the dye test.

In addition, samples of water from Beaver Creek at the sinkhole and the south hatchery spring were collected and sent to the Utah State Health Laboratory for analysis of total and fecal coliform bacteria. This was done to determine the concentration of bacteria in the creek water in comparison to that in the spring water to evaluate the potential for a particle-sized organism such as the parasitic spore that causes whirling disease to be transported underground from the sinkhole to the hatchery springs.

Table 1. Field measurements and results of tracer test for monitored sites in the Kamas Fish Hatchery study area, Samak, Utah.

[Map ID, refer to figure 1; gpm, gallons per minute; microS/cm, microSiemens per centimeter @ 25 degrees Celsius (C); NMT, no measurement taken; NM, not monitored; NMP, no measurement possible; E, estimate based on visual observation or rough calculation; --, not applicable; <, less than]

| Site name | Map ID | Date | Discharge (gpm) | Specific Conductance (microS/cm) | Water Temperature (degrees C) | Detector In (Date-Time) | Detector Out (Date-Time) | [1]Result |
|---|---|---|---|---|---|---|---|---|
| Beaver Creek sinkhole inflow | | 10/13/2010 | 386 | 134 | 8.8 | -- | -- | -- |
| | | 10/18/2010 | NMT | 138 | 11 | -- | -- | -- |
| | | 10/24/2010 | NMT | 135 | 6.8 | -- | -- | -- |
| | | 10/30/2010 | E 300 | NMT | NMT | -- | -- | -- |
| | | 11/13/2010 | 269 | NMT | NMT | -- | -- | -- |
| Beaver Creek at sinkhole | | 10/13/2010 | 1,480 | NMT | NMT | -- | -- | -- |
| | | 10/24/2010 | 525 | NMT | NMT | -- | -- | -- |
| Stevens well | 1 | NMT | NMT | NMT | NMT | [2]10/12/10 1900 | [3]10/30/10 1150 | Negative |
| Worthen well | 2 | NMT | NMT | NMT | NMT | [2]10/13/10 1130 | [3]10/30/10 1215 | Negative |
| Private hatchery spring source | 3 | NMT | NMT | NMT | NMT | NM | NM | NM |
| North hatchery spring (source vault) | 4 | 10/18/2010 | NMP | 432 | 11.9 | 10/12/10 1635 | 10/18/10 1515 | Very positive |
| | | 10/24/2010 | NMP | 429 | 11.8 | -- | -- | -- |
| | | 11/13/2010 | NMP | 440 | 12.1 | -- | -- | -- |
| North hatchery spring (degasser vault) | 5 | NMT | NMP | NMT | NMT | 10/12/10 1625 | 10/18/10 1435 | Very positive |
| South hatchery spring (source vault) | 6 | 10/18/2010 | NMP | 443 | 11.6 | 10/12/10 1645 | 10/18/10 1525 | Very positive |
| | | 10/24/2010 | NMP | 438 | 11.5 | -- | -- | -- |
| | | 11/13/2010 | NMP | 454 | 11.6 | -- | -- | -- |
| South hatchery spring (degasser vault) | 7 | 10/12/2010 | 745 | NMT | NMT | 10/12/10 1610 | 10/18/10 1405 | Very positive |
| | | 10/18/2010 | 770 | NMT | NMT | -- | -- | -- |
| | | 10/24/2010 | 795 | NMT | NMT | -- | -- | -- |
| | | 10/30/2010 | 780 | NMT | NMT | -- | -- | -- |
| | | 11/13/2010 | 745 | NMT | NMT | -- | -- | -- |
| Groundwater inflow (vault) | 8 | 10/12/2010 | E 100 | NMT | NMT | 10/12/10 1700 | 10/18/10 1455 | Very positive |
| Combined hatchery springs outflow | 9 | 10/13/2010 | 1,500 | 476 | 12.2 | 10/12/10 1655 | 10/18/10 1450 | Very positive |
| | | 11/19/2010 | 1,400 | NMT | NMT | -- | -- | -- |
| Beaver Creek above confluence | 10 | NMT | NMT | NMT | NMT | 10/12/10 1720 | 10/18/10 1650 | Negative |
| Private hatchery spring outflow | 11 | 10/13/2010 | NMT | 452 | 13.6 | 10/12/10 1555 | 10/18/10 1720 | Positive |
| | | 10/24/2010 | 255 | 424 | 10.6 | -- | -- | -- |
| | | 11/13/2010 | NMT | 431 | 8.3 | -- | -- | -- |
| Left Fork spring | 12 | 10/20/2010 | E <100 | NMT | NMT | 10/20/10 2010 | 11/13/10 1750 | Negative |
| Thompson springs outflow | 13 | 11/19/2010 | 191 | NMT | NMT | 10/13/10 1240 | 10/18/10 1735 | Positive |
| Left-Hand Canyon outflow | 14 | 11/19/2010 | E 180 | NMT | NMT | 10/18/10 1840 | 10/30/10 1055 | Positive |
| Spring near Beaver Creek Inn | 15 | 10/24/2010 | E 100 | 394 | 9.7 | 10/13/10 1255 | 10/24/10 1645 | Negative |
| Beaver Creek at Country Lane | 16 | NMT | NMT | NMT | NMT | 10/13/10 1810 | 10/18/10 1745 | Very positive |
| Gazebo spring | 17 | 10/20/2010 | E <100 | NMT | NMT | 10/20/10 1925 | 10/30/10 1045 | Weakly positive |
| Beaver Creek at Kamas | 18 | NMT | NMT | NMT | NMT | 10/13/10 1835 | 10/24/10 1800 | Weakly positive |

[1]Result - Based on relative qualitative visual observation of eluted dye sample
[2]First water sample collected
[3]Last water sample collected

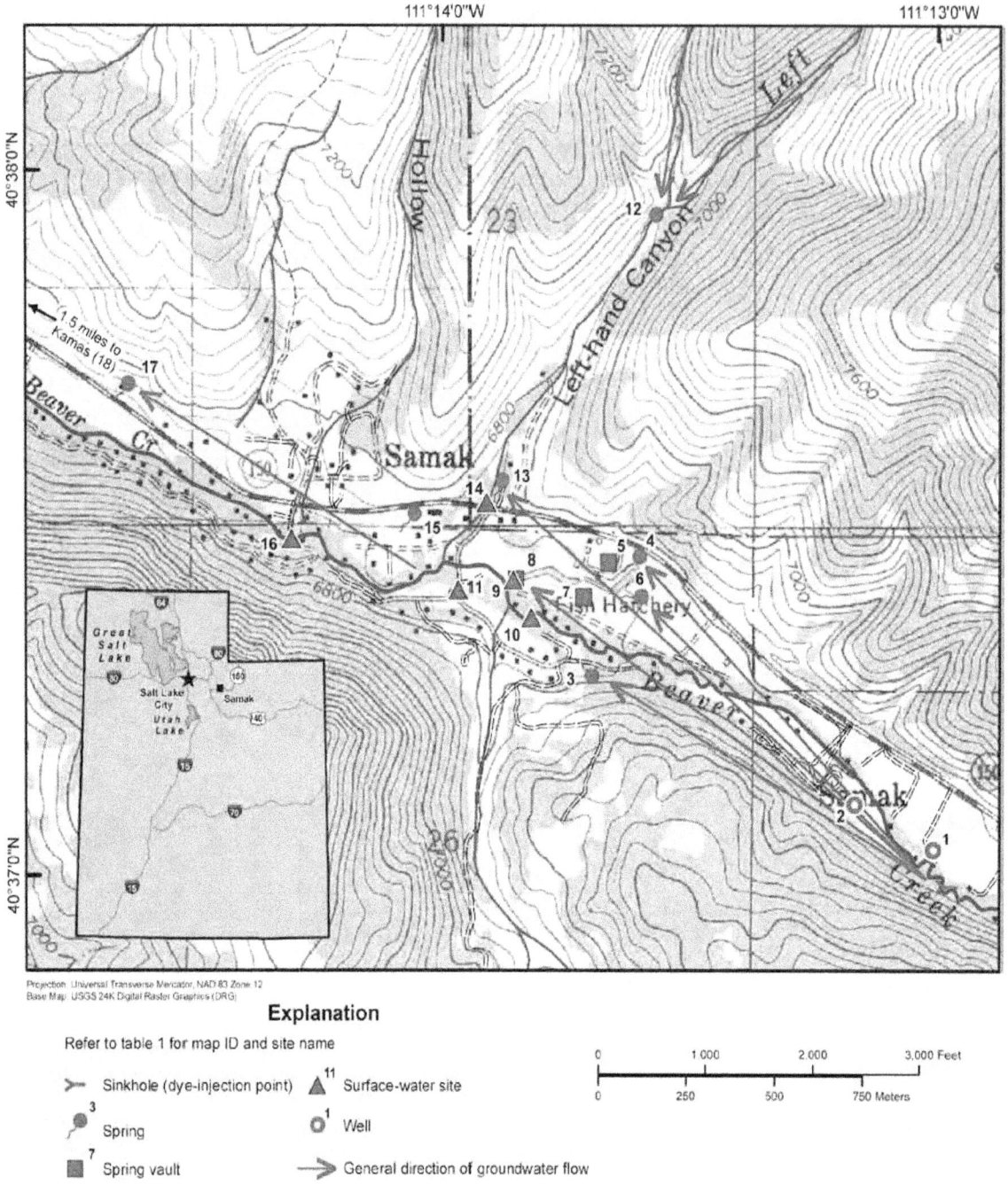

Figure 1. Location of monitoring sites and general direction of groundwater flow in the vicinity of the Kamas Fish Hatchery, Samak, Utah.

## RESULTS AND DISCUSSION

On October 12, 2010, at approximately 1745, water was diverted from Beaver Creek into the sinkhole by breaching a rock berm that had been constructed around the sinkhole. Because no flow had been going into the

sinkhole for an extended period of time, water was diverted into the sinkhole prior to the injection to "prime" the flow path. The rate of flow into the hole was adjusted to approximately 200 gpm so that no pooling would occur. The following day, October 13, from about 1320 to 1350, 300 grams of fluorescein dye were injected into the flow of water going into the sinkhole (fig. 2). Measured flow into the sinkhole at this time was 386 gpm, with no pooling. Streamflow loss was through large cobbles and boulders; no bedrock was exposed in the hole. Flow in Beaver Creek just above the sinkhole was measured to be 1,480 gpm and specific conductance of water in the creek was 134 microsiemens per centimeter (microS/cm) at 8.8 degrees Celsius.

Figure 2. Injection of fluorescein dye into the Beaver Creek sinkhole.

On October 18, 5 days after dye injection, activated charcoal detectors were pulled from the north and south hatchery springs and from other springs and sites monitored during the test (fig. 1 and table 1) and extracted for the presence of dye. Detectors from both of the hatchery springs and their associated vaults (sites 4 – 7) were very positive for the dye, and dye was reported to be visual in the spring vaults (sites 5 and 7) the previous day, October 17 (Ted Hallows, Division of Wildlife Resources, oral commun.). Dye also was recovered from detectors placed in a vault into which groundwater is collected from other springs on the hatchery property (site 8). In the vicinity of the hatchery, dye was recovered from detectors placed in the outflow (site 11) from a private hatchery spring on the south side of Beaver Creek (site 3), from detectors placed in the outflow from several springs discharging along the valley margin in the lower part of the Left-hand Canyon drainage (site 13), about 1,000 feet northwest of the hatchery, and from detectors placed in the outflow from the main drainage

(site 14). In addition, dye was detected in a spring located on the north side of the highway about 4,600 feet west of the hatchery (site 17).

Dye was not detected downstream in Beaver Creek above the confluence with outflow from the hatchery (site 10) and was not present on detectors collected from a spring located near the Beaver Creek Inn (site 15). Left Fork Spring (site 12) also was monitored during the test although the altitude of the spring is about 40 feet higher than the altitude of the sinkhole; samples from this site were negative. Water samples collected daily from the domestic-use wells (fig. 1, sites 1 and 2) also tested negative for the dye.

Although dye injected into the sinkhole was detected on activated charcoal samples collected from both hatchery springs about 5 days after injection, samples analyzed from the automatic water sampler at the south hatchery spring degasser unit (site 7) showed that dye was present in the sample collected at 2200 on October 16 or about 3 days and 8 hours after injection. No dye was detected in the sample collected 6 hours prior, at 1600; thus, dye arrival at the south spring occurred sometime between 1600 and 2200. The concentration of dye in the spring water peaked in the sample collected at 1000 on October 17, or almost 4 days after injection, and dye was barely detectable in water from the spring in the last sample taken at 0600 on October 30, indicating an overall transit time of the dye slug of about 17 days. Measured combined flow of both springs at the hatchery on October 13, the day of injection, was about 1,500 gpm, and metered discharge of the south spring was about 745 gpm. By difference, flow from the north spring at this time was estimated to be about 755 gpm. During the tracer study, the metered flow of the south spring increased from about 745 gpm on October 12, prior to diversion of water into the sinkhole, to 775 gpm on October 16, just prior to first arrival of the dye, to almost 790 gpm on October 18, at about the time peak dye concentration occurred at the spring, or an overall increase of about 45 gpm. Response of the south spring to the diversion of water into the sinkhole in relation to the dye breakthrough curve is shown in figure 3.

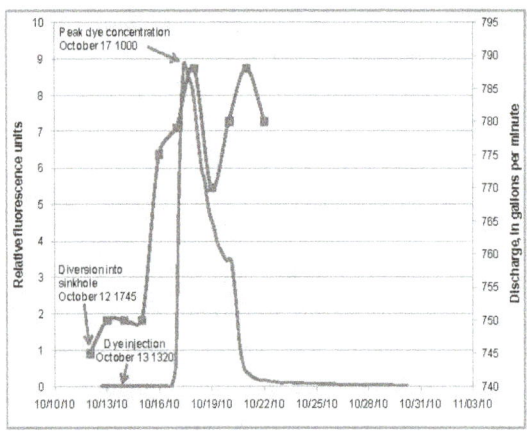

Figure 3. Comparison of discharge and dye breakthrough at the south hatchery spring.

The specific conductance of water from the north and south springs also changed in response to the diversion of water into the sinkhole during the tracer study, decreasing from 432 and 443 microS/cm on October 18 to 429 and 438 microS/cm on October 24, respectively. On November 13, when observed flow into the sinkhole had dropped substantially and flow at the south hatchery spring had decreased to 745 gpm, specific conductance of the north and south spring waters increased to 440 and 454 microS/cm, respectively. Although these changes are small in relation to the substantial difference in specific conductance between water from the creek and the spring, taken in conjunction with changes in other parameters and similarity to historical measurements documenting changes in specific conductance during snowmelt runoff, they are considered significant.

A bacteria sample collected from Beaver Creek at the sinkhole showed a total coliform count of more than 200 cells per 100 ml of sample and a fecal coliform count of 53 cells per 100 ml of sample. The bacteria sample collected from the south spring source vault (site 6) showed a total coliform count of about 16 cells per 100 ml of sample and a fecal coliform count of 1 cell per 100 ml of sample. Although substantially reduced in numbers and considerably smaller in size (1 to 3 microns) than the spores that cause whirling disease (about 30 microns), the presence of these bacteria in the spring water indicates nonetheless that the permeability of and rate of groundwater

movement through the aquifer are high enough for transporting particles or suspended materials (turbidity) to the springs from focused (point) surface-water sources such as the sinkhole.

Mississippian-age carbonate rock units crop out on the flank of the adjacent mountains and dip generally to the southwest beneath the valley of Beaver Creek in the vicinity of the hatchery. Well log data indicate, however, that the overlying unconsolidated deposits are as much as 150 feet thick along the Beaver Creek drainage (Hurlow, 2002). The log of the sampled domestic well only about 300 feet northwest of the sinkhole (site 1) indicates no bedrock at a depth of 200 feet. The upper 30 feet of these unconsolidated deposits consist primarily of alluvial-fan and debris-fan gravels, sands, and boulders, which can have relatively high hydraulic-conductivity values that are similar to those in fractured (non-cavernous) carbonate rocks (Heath, 1989). In addition, on the basis of the transit time of the dye from the sinkhole to peak concentration at the south hatchery spring, and assuming a linear flow path of about 3,800 feet, average groundwater velocity would have been on the order of 1,000 feet per day, a value that is comparatively slow for conduit flow through karst aquifers, where average travel times of about 5,800 feet per day have been documented from dye-tracer tests (Worthington, 2007).

## CONCLUSIONS

Results of the tracer study have shown not only that a hydrologic connection exists between the sinkhole and the hatchery springs but that groundwater also discharges at a number of other springs in the vicinity of and downgradient from the hatchery, both north and south of Beaver Creek. This distributary network of discharge points indicates diverging or possibly sub-parallel groundwater flow paths, rather than a pattern of converging flow paths to one principal spring such as Left Fork Spring and typical of many karst flow systems. The sinkhole along Beaver Creek probably lies on the margin or boundary of the hatchery springs groundwater basin that defines a weak divide from which water can flow in multiple directions. Variations in flow of the hatchery springs in comparison with the amount of water diverted into the sinkhole, relative changes in specific conductance of water from the hatchery springs when compared with the low specific conductance of water diverted into the sinkhole, and most significantly, the detection of dye at springs other than the hatchery springs indicate that only part of the total amount of water diverted into the sinkhole during the tracer study discharged from the hatchery springs and that most of the water discharging from the springs originates from other sources. The amount of flow discharging from other springs in the vicinity of the hatchery that also tested positive for dye more than accounts for the volume of water diverted into the sinkhole from the creek during the study. Given the distribution of discharge points, groundwater travel time, and geology of the study area, groundwater movement between the sinkhole and the hatchery and other springs may be at least in part, along preferential flow paths within the unconsolidated valley-fill deposits, in conjunction with flow along fractures in the underlying carbonate bedrock where near the surface.

## ACKNOWLEDGMENTS

The author would like to acknowledge the assistance of Murray Stevens and Taigon Worthen with regard to the collection of water samples from their wells during the dye study. Ted Hallows, Kamas Fish Hatchery manager also provided assistance throughout the dye study. The analysis of dye samples by Steven Johnson with the Edwards Aquifer Authority is greatly appreciated. Finally, reviews of the manuscript by Bert Stolp, U.S. Geological Survey, and Dan Aubrey and Ed Fall, Utah Division of Water Resources, helped to improve readability and technical accuracy.

## REFERENCES

Heath, R.C., 1989, Basic ground-water hydrology: U.S. Geological Survey Water-Supply Paper 2220, 84 p.

Hurlow, H.A., 2002, The geology of the Kamas-Coalville region, Summit County, Utah, and its relation to ground-water conditions: Utah Geological Survey Water Resource Bulletin 29, plates 1-5.

Spangler, L.E., Tong, M., and Johnson, W., 2005, A multi-tracer approach for evaluating the transport of whirling disease to Mammoth Creek fish hatchery springs, southwestern Utah, *in* U.S. Geological Survey Karst Interest Group Proceedings, Rapid City, South Dakota, September 12-15, 2005: U.S. Geological Survey Scientific Investigations Report 2005-5160, p. 116-121.

Worthington, S.R.H., 2007, Ground-water residence times in unconfined carbonate aquifers: Journal of Cave and Karst Studies, v. 69, no. 1, p. 94–102.

# Using Geographic Information System to Identify Cave Levels and Discern the Speleogenesis of the Carter Caves Karst Area, Kentucky

By Eric Peterson[1], Toby Dogwiler[2], and Lara Harlan[1]

[1]Department of Geography-Geology, Illinois State University, Normal, IL 61790
[2]Southeastern Minnesota Water Resources Center, Department of Geoscience, Winona State University, Winona, MN 55987

## Abstract

Cave level delineation often yields important insight into the speleogenetic history of a karst system. Various workers in the Mammoth Cave System (MCS) and in the caves of the Cumberland Plateau Karst (CPK) have linked cave level development in those karst systems with the Pleistocene evolution of the Ohio River. This research has shown that speleogenesis was closely related to the base level changes driven by changes in global climate. The Carter Caves Karst (CCK) in northeastern Kentucky has been poorly studied relative to the MCS to the west and the CPK karst to the east. Previously, no attempt had been made to delineate speleogenetic levels in the CCK and relate them to the evolution of the Ohio River.

In an attempt to understand cave level development in CCK we compiled cave entrance elevations and locations. The CCK system is a fluviokarst typical of many karst systems formed in the Paleozoic carbonates of the temperate mid-continent of North America. The CCK discharges into Tygarts Creek, which ultimately flows north to join the Ohio River. The lithostratigraphic context of the karst is the Mississippian Age carbonates of the Slade Formation. Karst development is influenced by both bedding and structural controls. We hypothesize that cave level development is controlled by base level changes in the Ohio River, similar to the relationships documented in MCS and the karst of the Cumberland Plateau

The location and elevation of cave entrances in the CCK was analyzed using a GIS and digital elevation models (DEMs). Our analysis segregated the cave entrances into four distinct elevation bands that we are interpreting as distinct cave levels. The four cave levels have mean elevations (relative to sea level) of 228 m (L1), 242 m (L2), 261 m (L3), and 276 m (L4). The highest level—L4—has an average elevation 72 m above the modern surface stream channel. The lowest level—L1—is an average of 24 m above the modern base level stream, Tygarts Creek. The simplest model for interpreting the cave levels is as a response to an incremental incision of the surface streams in the area and concomitant adjustment of the water table elevation. The number of levels we have identified in the CCK area is consistent with the number delineated in the MCS and CPK. We suggest that this points toward the climatically-driven evolution of the Ohio River drainage as controlling the speleogenesis of the CCK area.

## INTRODUCTION

Surface rivers play an integral role in the formation of many karst systems. In fluviokarst settings, the formation of phreatic cave passages is thought to occur at, or just below, the water table; hence as the rivers incise, lower levels of conduits are formed at increasingly lower elevation (Ford and Williams, 2007; Palmer, 1987). As a karst system evolves, subsurface drainage at the current base level becomes more efficient at draining the watershed and surface systems and upper cave levels go dry (Kaufmann, 2009). Just as the development of terraces represent periods of river stability in surficial fluvial settings, the formation of cave levels across a region provide an archive recording periods of base level stability (Kaufmann, 2009; Palmer, 1987).

The Ohio River and its tributaries provide ample evidence of this phenomenon. The entrenchment of the Ohio River and its tributaries produced multiple levels in the Mammoth Cave System (MCS) (Granger and others, 2001; Palmer, 1989) and in the Cumberland Plateau region (CPK) (Anthony and Granger, 2004). In the Mammoth Cave region and the Cumberland Plateau, investigators have demonstrated that alternating periods of climate-

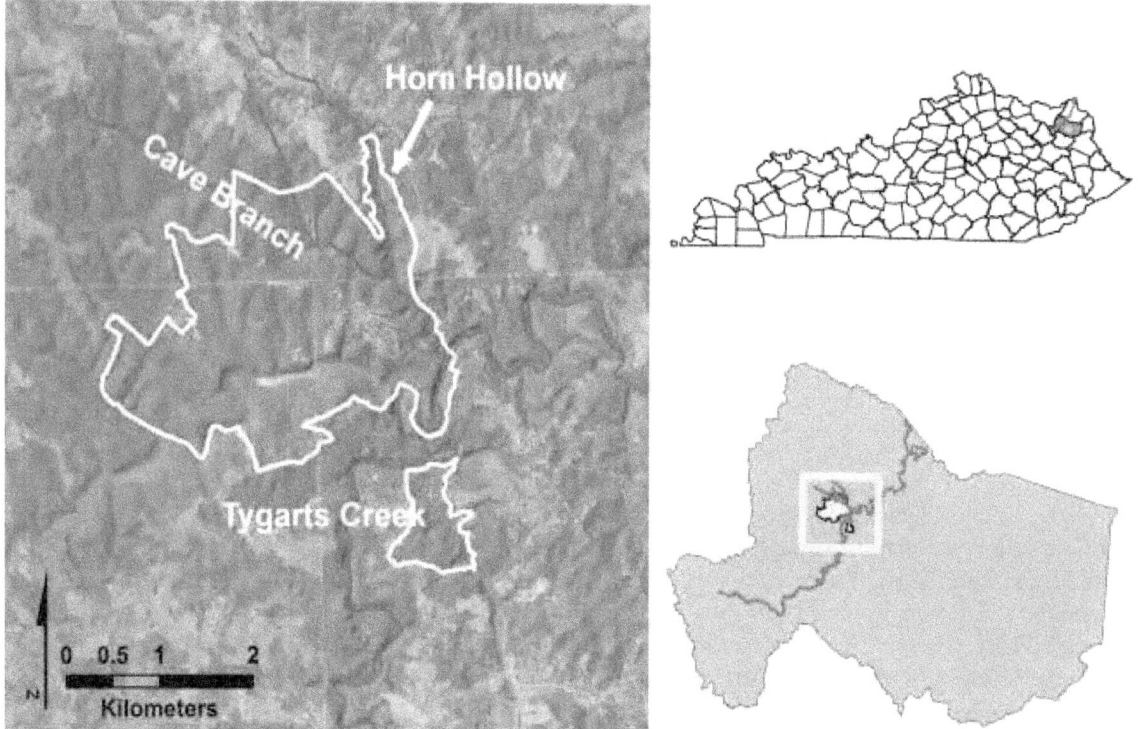

Figure 1. Location of the study area, highlighting Carter Caves State Resort Park. In the aerial image on the left, the park boundary is outlined in white. Rivers are represented as blue lines. The location of Carter County is shown in the map in the upper right. The location of the state park within Carter County is shown in the lower right figure.

driven incision and aggradation of the Ohio River have strongly influenced the evolution of the those karst systems over the past 3-5 million years (Anthony and Granger, 2004; Granger and others, 2001). In both cases, pre-existing cave systems experienced significant vertical development and modification during the Plio-Pleistocene due to changes in erosional base level, which forced alternating periods of incision and aggradation in river valleys south of the glacial margin (Teller and Goldthwait, 1991). Isostatic responses of the continental crust to the waxing and waning of ice sheets and the resulting transmission of glacially-derived sediment packages down the Ohio River valley also played a role in the rates of stream erosion and deposition (Granger and others, 2001; Potter, 1955).

The MCS has four main cave levels[1] (Granger and others, 2001) which generally correlate with similar levels in the CPK (Anthony and Granger, 2004). These cave levels

all ultimately coincide with base level changes in the Ohio River and its predecessor drainages. The oldest and highest cave levels formed in the Pliocene during a period of extremely low rates of river incision and landscape denudation. This led to long-term stabilization of water table levels and the development of extensive and large conduit systems (e.g., Collins Avenue in the MCS). With the onset of Pleistocene glaciation and the evolution of the modern Ohio River drainage, base level stability ended and a sequence of rapid incision and aggradation led to the development of several new cave levels. We hypothesize that the Carter Caves Karst (CCK) in Carter County, northeastern Kentucky, developed due to a similar speleogenetic response to the reorganization of the Ohio River drainage.

## Geologic Context

The CCK is located about 40 km south of the Ohio River in northeastern Kentucky (Fig. 1) and has a stratigraphic and geologic setting similar to that of the MCS. The system has been described by several authors including (e.g., McGrain, 1966; Tierney, 1985). More recently, Engel and Engel (2009) provide a thorough and

[1] In MCS and CPK the cave levels are lettered with level 'A' representing the upper-most level in the system. In CCK we designated our lowest level 'L1' and increased the number (i.e., 'L2', 'L3', 'L4') with each higher elevation.

up-to-date discussion of the local geologic setting, including an updated synthesis of the area's stratigraphy. Unlike the larger karst systems in western Kentucky or West Virginia, karstification in the Carter Caves area is constrained by a relatively thin sequence of karstifiable carbonates (Engel and Engel, 2009). The carbonate sequence is Mississippian (latest Osagean to late Chesterian times) in age and is sandwiched between Mississippian and Middle Pennsylvanian siliciclastics (Ettensohn and others, 1984).

Northeastern Kentucky has experienced a complex drainage evolution through the Plio-Pleistocene as the Teays River system was abandoned and reorganized into the Ohio River drainage (Andrews, 2006; Rhodehamel and Carlston, 1963; Teller and Goldthwait, 1991; Ver Steeg, 1946). Prior to the onset of Pleistocene glaciation, northeastern Kentucky was part of the Teays River basin (Janssen, 1953). Currently, the CCK area is highly dissected and characterized by deeply-incised stream valleys that are graded to Tygarts Creek, the regional baselevel. Tygarts Creek flows north through Carter and Greenup Counties toward its confluence with the Ohio River. Locally, Tygarts Creek has a very low gradient of 0.0007 m/m. Tygarts Creek is currently incised through the carbonate sequence into the shales of the underlying Borden Formation (Engel and Engel, 2009; Tierney, 1985). Thus, the lower stratigraphic limit of surficial karst development has been reached. Within the study area, Horn Hollow Creek drains to Cave Branch which in turn flows to Tygarts Creek. These tributary valleys are steeply graded (average of 0.053 m/m) and are underlain by the karst-forming carbonates. The tributary streams are characterized by numerous small waterfalls, sinking streams, and numerous resurgent springs. In Horn Hollow the stream is diverted into the subsurface by several caves (during floods, some flow is diverted to normally dry surface channels).

The CCK is a fluviokarst system and is comprised of the surface and subterranean drainage associated with Tygarts Creek and its tributaries. The karst system includes a number of watersheds tributary to Tygarts Creek. Engel and Engel (2009) attribute the consistent distribution of caves with the same stratigraphic units throughout the various watersheds as evidence of simultaneous karstification. Engel and Engel (2009) also note two morphologically distinct cave passage types. The first are large trunk passages, that are stratigraphically high and whose development is controlled both by structure and stratigraphy. The second passage type are smaller passages, that are stratigraphically lower in the bedrock section and are characterized by simple passage segments with morphologies indicative of incision-driven water table lowering. The development of these passages was strongly controlled by bedrock fractures. Engel and Engel (2009) interpret these passage types as representing at least two distinct periods of karstification.

## Methodology

Our objective is to assess the feasibility of applying the incision driven model of speleogenesis derveloped for the MCS(Granger and others, 2001) and the Cumberland Plateau region(Anthony and Granger, 2004) to the CCK. More specifically, we posit that if cave levels can be distinguished in the CCK system, it may be possible to link them to the evolution of the Ohio River drainage, and thus devise a model for the timing and style of speleogenesis in CCK. The evolution of those systems has been worked out using sophisticated and expensive geochemical analysis, particularly of speleothems and cave sediments. Herein, we attempt to extrapolate the results of those studies in combination with GIS-based analysis of remotely-sensed data and direct observations into a robust model for the speleogenetic history of the CCK. Such an approach may be attractive to workers deciphering systems that lack the geoarchives (i.e., well-understood and wide spread sediments) that exist in these other systems or the technical and financial resources required for the requisite analytical methodologies employed in those systems. Studies such as ours may also provide a framework and an impetus for employing sophisticated analytical methods such as those used in MCS and CPK.

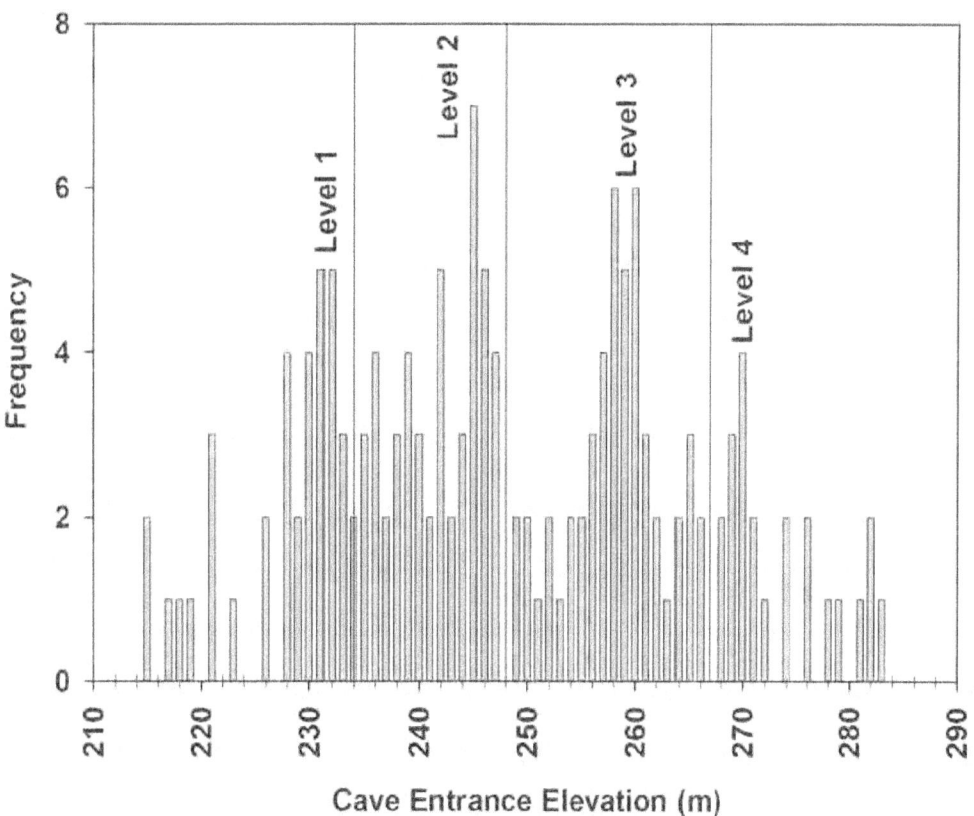

**Figure 2. Histogram of all the cave entrance elevations in CCK. Cave levels were delineated based on where there was a high frequency of caves at one elevation with breaks on either side or a reverse in the frequency trend.**

We employed a GIS (ESRI ArcMap™ 9.2) to visualize and analyze various data from the karst system. Although the use of GIS to study karst has grown in the last decade, many early applications of GIS to karst focused on GIS as a database and management tool for information (e.g., Florea and others, 2002; Gao and others, 2006; Ohms and Reece, 2002). More recently, the use of GIS has increasingly included sophisticated data analysis, geoprocessing, and modeling as a central aspect of the research endeavor. For example, GIS has been used to indentify sinkholes, faults, and fractures (Angel and others, 2004; Florea, 2005; Seale and others, 2008), to model depressions (Yilmaz, 2007), to create virtual field trips through caves (McNeil and others, 2002), to model karst hazards (McNeil and others, 2002), to delineate karst watersheds (Choi and Engel, 2003; Glennon and Groves, 2002), and to identify critical source areas of contaminants (Dockter and Dogwiler, 2010).

## GIS Data Sources

Digital topography, hydrography, a digital elevation model, and orthophotos layers for the study area were obtained from the Kentucky Geological Survey[2] and the United States Geological Survey[3] web sites. The DEM for the study area has a spatial resolution of 30m. DEM accuracy is expressed in terms of root mean square error (RMSE) and is assessed based on ground control points, the National Map Accuracy Standards (NMAS), and the National Standard for Spatial Data Accuracy (NSSDA). The NMAS is the RMSE that bounds 90 percent of the values, while the NSSDA is the RMSE that bounds 95 percent of the values. Based on these various methods of assessment the CCK DEM accuracy ranges from 3.74 m to 7.34 m. The hydrography data were derived from the USGS National Hydrography Dataset (NHD), which was created from 1:24,000 Digital Line

---

[2] http://www.uky.edu/KGS/gis/kgs_gis.htm
[3] http://seamless.usgs.gov/

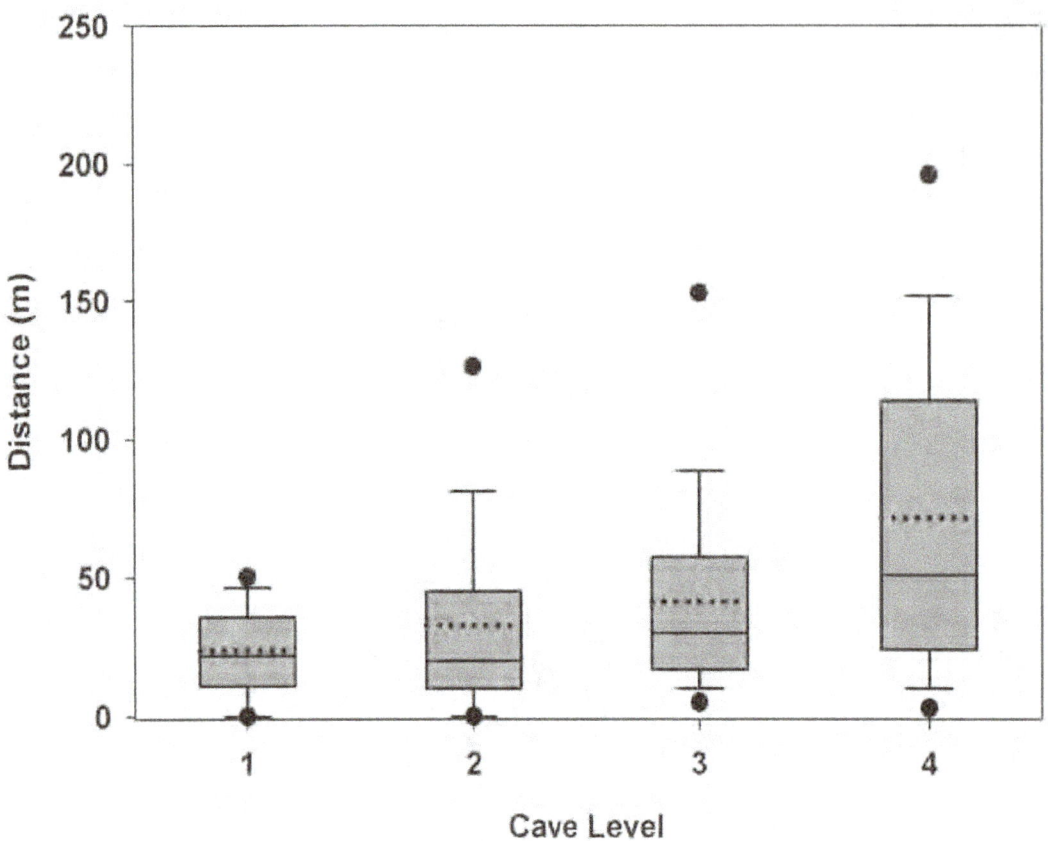

**Figure 3.** Box plots of the distance cave entrances are from streams. The ends of the boxes represent the 25th and 75th percentiles with the solid line at the median; the error bars depict the 10th and 90th percentiles, and the dots represent the 5th and 95th percentile. The dotted line represents the mean.

**Table 1.** Summary of cave level elevations, lateral distance to the stream valley axis, and selected example caves from each level. Harlan (2009, p. 32-34) provides a detailed description and statistical summary of each of the cave levels. Engel and Engel (2009) provide additional description and context regarding the specific caves.

[All units are in meters]

| Cave Level | # of Entrances | Mean / Mode Elevation | Elevation Range | Mean Distance to Stream ± Std. Dev. | Example Caves |
|---|---|---|---|---|---|
| L4 | 25 | 276 / 270 | 268 – 283 | 72 ±56 | X Cave, Coon-in-the-Crack Cave |
| L3 | 49 | 261 / 260 | 249 – 266 | 41 ± 37 | Saltpetre Cave, Rat Cave |
| L2 | 47 | 242 / 245 | 234 – 247 | 33 ± 37 | Cool James Cave |
| L1 | 36 | 228 / 231 | 215 – 234 | 24 ± 15 | Laurel Cave, Lake Cave |

Graphs (DLG). The NHD accuracy is reported as 98.5 percent.

ESRI's ArcCatalog™ was used to build a geodatabase for data collected in the field. Cave locations and descriptions were obtained from the Wittenberg University Speleological Society. These data represent a thorough and systematic reconnaissance of the study area and surrounding areas and represents most—if not all—of the discoverable karst features in the area. The Wittenberg data lacked elevations of

the documented features. All elevation data used in our analysis were obtained from the DEM and compared to elevations directly determined for selected features and fixed reference locations in the field using a combination of a differential GPS, an electronic altimeter, and an analog altimeter (Gorecki, 2008). Based upon temporally repeated measurements of fixed reference locations within the study area we estimate the electronic altimeter error at ±3.2 to 4.3 m and the analog altimeter error at ±0.6 to 0.9 m. The accuracy of the post-processed digital GPS data was limited at many locations by the steep topography and dense tree canopy. Most positions had a vertical accuracy of 1 to 5 m (one standard deviation error).

Cave levels were determined based on the elevations of the cave entrances. The karst features geodatabase was filtered to remove vertically oriented features, such as vadose pits and sinks. Although the pits and sinks generally represent a vertical connection between the surface and an underlying cave, the depth from the surface to the cave was not known. Thus, the pits and sinks were not incorporated into the cave level determination. The elevation of cave entrances was determined based on the elevation of the corresponding DEM cell. The distribution of cave entrance elevations was analyzed statistically in ArcMap (described more thoroughly in Harlan, 2009). The ArcGIS™ algorithm '*Natural Breaks Classifier*' was used to generate a histogram showing the frequency of cave entrances by elevation (Fig. 2). Cave levels were delineated based on where there was a high frequency of caves at one elevation with breaks on either side or a reverse in the frequency trend.

In addition to identifying the cave entrance elevations, we determined the shortest distance of each cave entrance to the channel of the current surface stream valley. This was accomplished by deriving a stream line network based on the DEM and then using the '*Euclidean Distance*' tool in ArcGIS™ to determine the distance of each DEM cell from the nearest stream line. The resulting raster of values was queried for each cave entrance using the '*Extraction*' tool to yield the distance values which were added as an attribute to the karst features geodatabase.

## RESULTS

### Error Analysis

Forty-three field-collected elevations for cave entrances and the fixed reference locations were compared to elevations obtained from the DEM. The DEM provides slightly higher elevation values. The mean error between the field-collected and DEM-derived elevations is -0.48 m, with a 95percent confidence interval of 1.25 m. However, the DEM elevations are statistically similar to the field-collected

elevations [$t(43) = -0.19$, $p = 0.85$]. The RMSE between the field-collected elevations and the DEM elevations is 3.96 m. This error is only slightly higher than the 3.74 m RMSE of the DEM, but is below the NMAS and the NSSDA values of 6.15 m and 7.34 m, respectively. Both the mean error and the RMSE are within the error associated with electronic altimeter and DGPS; only the mean error is within the error of the analog altimeter. Overall, the data indicate that the DEM provides acceptable estimates of elevations.

### Cave Levels

The location of 157cave entrances were analyzed in this study. Based on our analysis of the distribution of cave entrances by elevation we have delineated four cave levels (Figure 2, Table 1). We have denoted these levels as L1 (mean elevation of 228 m above sea level), L2 ($\bar{x} = 242$ m), L3 ($\bar{x} = 261$ m), and L4 ($\bar{x} = 276$ m). The mean lateral (horizontal) distances of entrances in each cave level to the nearest stream valley axis are shown in table 1. In general, L4 cave entrances were furthest from the streams, as would be expected if the cross-sectional (normal to flow) valley shape is approximated by the classic 'V-shaped' valley of fluvial origin. Because the slope of valley walls varies significantly, there is also a reasonable expectation that this metric will show significant overlap in the lateral distances ranges. Nonetheless, analysis of variance (ANOVA) suggests that the cave levelsare a significant predictor of the lateral cave entrance to valley axis dimension [$F(3, 156) = 8.78$, $p = <0.001$].

## DISCUSSION

Figure 2 shows our delineation of cave levels in the CCK system. The histogram contains many breaks and admittedly lends itself to a number of interpretations. However, an ongoing graduate project at Illinois State University is working to refine the delineation of cave levels presented here, and the preliminary results (Jacoby and Peterson, 2010) support our current interpretation. The development of larger trunk passages in levels L4 and L3 seem to be controlled by subtle changes in dip of the bedrock (Engel and Engel, 2009) and likely pre-date the onset of glaciation to the north of the study area the led to the reorganization of the Ohio and Teays River Drainage systems. Several of the upper level passages in the study area occur at similar vertical elevations and may be truncated remnants of a formerly integrated cave system. The lower cave levels in the CCK system are controlled by a combination of stratigraphic and structural influences and correlate strongly to modern-day surface stream patterns (Engel and Engel, 2009).

## The Horn Hollow Valley

We will focus our discussion of cave hydrology and geomorphology in CCK on the Horn Hollow Valley portion of the system (Figure 1). Horn Hollow has numerous caves, sinking streams, springs, sinks, and pits that are all well-studied and documented in the literature (Angel, 2010; Dogwiler and Wicks, 2004; Engel and Engel, 2009; Hobbs and Pender, 1985; McGrain, 1966; Ochsenbein, 1974; Tierney, 1985) . The upper section of Horn Hollow is largely under-drained by an active cave system (variously referred to as Boundary Cave or Upper Horn Hollow Cave). The surface stream channel in this section of the valley is poorly maintained and ill-defined indicating a paucity of flow events large enough to inundate the active cave system and flow across the surface. In several places this section of surface stream is occupied by large blocks of limestone displaying anastomoses, scallops, and other dissolutional features associated with caves. It is likely that at least some of this portion of the surface stream is a former L3 cave that has been hydrologically abandoned and subsequently unroofed.

Several caves higher up in the stratigraphic section occur in the valley flanks. Some, such as Fudge Ripple Cave, are fairly near the contact with the siliciclastic units that overlie the carbonate sequence. Fudge Ripple Cave and another cave—Volcano Cave—appear to be examples of phreatic passages that have been overprinted with a vadose signature formed as waters have cut through the passage floors seeking pathways to lowering water tables. Stratigraphically, and in terms of elevation, these caves represent L3 and L4. In numerous places pits and sinks dot the hillslopes along the valley walls. Currently, the hydrologic function of these caves and pits is to direct water vertically down toward the modern phreatic zone.

Dye tracing and water chemistry data (Angel, 2010), confirm that Bowel Spring (L2), in the central part of the Horn Hollow Valley, is a resurgence point for water flowing from Volcano Cave and Fudge Ripple Cave through Boundary Cave—hydrologically spanning several cave levels. From Bowel Spring the flow alternates from the surface to the subsurface through Cobble Crawl Cave, Horn Hollow Cave, New Cave, and H2O caves. Thus, in the lower part of Horn Hollow Valley, it is possible to explore several of the active L2 and L1 caves. H2O Cave and New Cave are phreatic tubes that meander along bedding planes and drain significant amounts of water during large flow events. H2O Cave (L1) emerges from Horn Hollow as a waterfall along the contact between the St. Louis limestone and the Borden Shale. As such, H2O Cave is formed at the carbonate/siliciclastic contact that forms the lower stratigraphic boundary of cave development in the region.

## Comparison to Regional Karst Systems

In Mammoth Cave, Palmer (1987) and Granger and others (2001) identify four levels centered around 150 m, 167 m, 180 m, and 200 m. The number of levels within the MCS corresponds well to the CCK area, but there is an absolute difference of ~80 m between the levels of the two systems. We assume this difference in absolute elevations is a function of regional dips. However, the relative elevation differences between individual levels in each area are roughly comparable. Additionally, the three lowest levels in the MCS are also in the Ste. Genevieve Limestone, which is correlative between the Mammoth Cave and CCK areas.

In the MCS the upper levels (Level A, 200m and Level B, 180 m) formed in the Pliocene and early Pleistocene due to slow valley deepening and aggradation, while the lower levels (Level C, 167 m and Level D, 150) developed during the Pleistocene glacial intervals during periods of base level stability (Palmer, 1987). Using cosmogenic[26]Al and [10]Be dating, Granger and others (2001) determined that Levels A and B were both formed prior to 3.25 Ma and constrain the formation of Levels C and D as prior to 1.39 Ma and 1.24 Ma, respectively.

The CPK also has four levels (Anthony and Granger, 2004). Cosmogenic[26]Al and [10]Be analysis demonstrates that the upper-most level (Level 1) was formed between 5.7 and 3.5 Ma, the second level was formed between 3.5 and 2.0 Ma, the third level was formed between 2.0 and 1.5 Ma, and the fourth level was formed after 1.5 Ma. Thus, levels one and two formed in the Pliocene and levels three and four formed in the Pleistocene.

MCS, CPK, and CCK are geographically close (within 300 km of one another), contain many of the same stratigraphic units, and are ultimately controlled by the base flow of the Ohio River. Thus, it is reasonable to hypothesize that CCK may share a similar history of cave development. However, unlike the Green River and the Cumberland Rivers which flowed west into the Old Ohio River in pre-Glacial times (Granger and others, 2001; Teller, 1973), northeastern Kentucky was part of the southern branch of the Teays River drainage that flowed from eastern North Carolina toward northwestern Ohio and Indiana (Hansen, 1995; Janssen, 1953). Whereas, the Green and Cumberland joined their master streams south of the glacial margin, the Teays drainage downstream of Kentucky was overrun by advancing ice sheets and flow was impounded south of the glacial margin (Andrews, 2006; Teller, 1973). It is difficult to ascertain precisely what effect these events had on karst development in CCK and how its progression may have differed from the other two karst systems.

Nonetheless, we believe that enough similarities in cave level sequences, bedrock geology, and relative elevations exist between the three systems to pose some preliminary hypotheses regarding the development of the CCK system. Certainly, these hypotheses would benefit from future geochronology studies of CCK sediments and additional geomorphic field work in the study area. The L4 and L3 trunk passages in CCK, such as Saltpetre Cave and the upper level of Laurel Cave likely correlate with the upper cave levels in MCS and CPK and represent Pliocene or early Pleistocene karst development. These passages contain fine- to coarse-grained silt and sand deposits that Engel and Engel (2009) suggest are fluvial in origin. These sediments may be suitable for cosmogenic or paleomagnetic analysis.

The Ohio River initially occupied its current course approximately 1.4 Ma and drove a rapid incision event that is attributed to the formation of MCS level D (Granger and others, 2001). After 1.24 Ma, the incision and aggradation history of the Ohio River becomes more complicated and Granger and others (2001) attribute the relative instability of the river level to the lack of well-defined levels below level D. During this time period in northeastern Kentucky, it is possible that Tygarts Creek incised at times well into the siliciclastics underlying the carbonates—leaving the CCK hydrologically abandoned. Whereas, L4 and L3 cave entrance elevations are tightly distributed across narrow distributions, cave entrance elevations in L2 and L1 are more broadly distributed. Thus, the "noise" in the L2 and L1 distributions may represent the complex base level evolution of the Ohio River drainage over the last 1.24 Ma.

## CONCLUSIONS

The number of levels within the CCK shows that the area has experienced changes in the elevation of the water table. We posit that the upper-level trunk passages in the CCK may represent the remnants of a more extensive karst system that developed in the Plio-Pleistocene during a period of relatively slow landscape denudation prior to the abandonment of the Teays River network and the development of the Ohio River drainage. The lower level caves in the CCK system likely formed during periods of base level stability during the wax and wane of the Pleistocene ice sheets.

We propose that accepted models for the Plio-Pleistocene development of the Mammoth Cave and Cumberland Plateau karst systems are appropriate starting points for deciphering the history of cave level development in the CCK area. Additional geomorphic analysis of the system, including geochronologic analysis of the cave sediments, could provide important insight into the demise of the Teays drainage and development of the modern Ohio River.

## ACKNOWLEDGMENTS

The authors would like to thank the Wittenberg University Speleological Society for providing the karst features database, the

Kentucky Department of Parks for access to CCSRP, and Kimberly Gorecki Boland who helped to collect the altimeter and GPS control point elevations.

# REFERENCES

Andrews, W. M., Jr., 2006, Geologic Controls on Plio-Pleistocene Drainage Evolution of The Kentucky River in Central Kentucky: Kentucky Geological Survey, Series XII, Thesis 4, 11125: Kentucky Geological Survey, University of Kentucky, p. 213.

Angel, J. C., 2010, Characterization and comparison of the geochemistry and evolutionary controls of Horn Hollow Valley karst waters to other karst systems [Master's thesis]: Illinois State Unviersity.

Angel, J. C., Nelson, D. O., and Panno, S. V., 2004, Comparison of a new GIS-based technique and a manual method for determining sinkhole density: An example from Illinois' sinkhole plain: Journal of Cave and Karst Studies, v. 66, no. 1, p. 9-17.

Anthony, D. M., and Granger, D. E., 2004, A Late Tertiary Origin for Multilevel Caves along the Western Escarpment of the Cumberland Plateau, Tennessee and Kentucky, Established by Cosmogenic 26AL and 10BE: Journal of Cave and Karst Studies, v. 66, no. 2, p. 46-55.

Choi, J.-Y., and Engel, B. A., 2003, Real-time watershed delineation system using Web-GIS: Journal of Computing in Civil Engineering, v. 17, no. 3, p. 189-196.

Dockter, D. L., and Dogwiler, T., 2010, Use and verification of DEM-based terrain analysis to identify source regions of erosion and saturation in southeastern Minnesota watersheds: GSA Abstracts with Programs, v. 42, no. 5, p. 601.

Dogwiler, T., and Wicks, C. M., 2004, Sediment entrainment and transport in fluviokarst systems: Journal of Hydrology, v. 295, no. 1-4, p. 163-172.

Engel, A. S., and Engel, S. A., 2009, A field guide for the karst of Carter Caves State Resort Park and the surrounding area, northeastern Kentucky, Field Guide to Cave and Karst Lands of the United States, Karst Waters Institute Special Publication 15, Karst Waters Institute.

Ettensohn, F. R., Rice, C. R., Dever, G. R., Jr., and Chesnut, D. R., 1984, Slade and Paragon Formations: New stratigraphic nomenclature for Mississippian rocks along the Cumberland Escarpment in Kentucky, U.S. Geological Survey Bulletin 1605-B, U.S. Government Printing Office, 37 p.

Florea, L., 2005, Using state-wide GIS data to identify the coincidence between sinkholes and geologic structure: Journal of Cave and Karst Studies, v. 67, no. 2, p. 120-124.

Florea, L. J., Paylor, R. L., Simpson, L., and Gulley, J., 2002, Karst GIS advances in Kentucky: Journal of Cave and Karst Studies, v. 64, no. 1, p. 58-62.

Ford, D. C., and Williams, P., 2007, Karst hydrogeology and geomorphology: Chichester, John Wiley & Sons Ltd, 562 p.

Gao, Y., Tipping, R., and Alexander Jr., E. C., 2006, Applications of GIS and database technologies to manage a karst feature database.: Journal of Cave and Karst Studies, v. 68, no. 3, p. 144-152.

Glennon, A., and Groves, C., 2002, An examination of perennial stream drainage patterns within the Mammoth Cave watershed, Kentucky: Journal of Cave and Karst Studies, v. 64, no. 1, p. 82-91.

Gorecki, K., 2008, On The Level: Shedding Light on the Dark Mysteries of Eastern Kentucky Cave Evolution [Undergraduate thesis]: Winona State University, 11 p.

Granger, D. E., Fabel, D., and Palmer, A. N., 2001, Pliocene--Pleistocene incision of the Green River, Kentucky, determined from radioactive decay of cosmogenic 26Al and 10Be in Mammoth Cave sediments: Geological Society of America Bulletin, v. 113, no. 7, p. 825-836.

Hansen, M. C., 1995, The Teays River, Ohio Geology GeoFacts No. 10: Columbus, OH, Ohio Department of Natural Resources, Division of Geological Survey, 2 p.

Harlan, L. E., 2009, Expanding the conceptual model for the Carter caves system, Carter County, Kentucky using GIS [Master's thesis]: Illinois State University, 68 p.

Hobbs, H. H., III, and Pender, M. M., 1985, The Horn Hollow Cave system, Carter County, Kentucky: Pholeos, v. 5, no. 2, p. 17-22.

Jacoby, B., and Peterson, E. W., 2010, Uncovering the Speleogenesis of Carter Caves, Kentucky Using GIS: Geological Society of America Abstracts with Programs, v. 42, no. 5, p. 450.

Janssen, R. E., 1953, The Teays River, ancient precursor of the East: The Scientific Monthly, v. 77, no. 6, p. 306-314.

Kaufmann, G., 2009, Modelling karst geomorphology on different times scales: Geomorphology, v. 106, p. 62-77.

McGrain, P., 1966, Geology of Carter and Cascade Caves Area, Series X, Special Publication 12: Lexington, KY, Kentucky Geological Survey, University of Kentucky, p. 32.

McNeil, B. E., Jasper, J. D., Luchsinger, D. A., and Rainsmier, M. V., 2002, Implementation and application of GIS at Timpanogos Cave National Monument, Utah: Journal of Cave and Karst Studies, v. 64, no. 1, p. 34-37.

Ochsenbein, G. D., 1974, Origin of caves in Carter Caves State Park, Carter County, Kentucky [Master's thesis]: Bowling Green State University, 64 p.

Ohms, R., and Reece, M., 2002, Using GIS to Manage Two Large Cave Systems, Wind and Jewel Caves, South Dakota: Journal of Caves and Karst Studies, v. 64, no. 1, p. 4-8.

Palmer, A. N., 1987, Cave levels and their interpretation: The NSS Bulletin, v. 49, no. 2, p. 50-66.

Palmer, A. N., 1989, Geomorphic history of the Mammoth Cave system, in White, W. B., and White, E. L., eds., Karst hydrology: Concepts from the Mammoth Cave Area: New York, Van Nostrand Reinhold, p. 317-337.

Potter, P. E., 1955, The Petrology and Origin of the Lafayette Gravel: Part 1. Mineralogy and Petrology: Journal of Geology, v. 63, no. 1, p. 1-38.

Rhodehamel, E. C., and Carlston, C. W., 1963, Geologic History of the Teays Valley in West Virginia: Geological Society of America Bulletin, v. 74, no. 3, p. 251-274.

Seale, L. D., Florea, L. J., Vacher, H. L., and Brinkmann, R., 2008, Using ALSM to map sinkholes in the urbanized covered karst of Pinellas County, Florida—1, methodological considerations: Environmental Geology, v. 54, no. 5, p. 995-1005.

Teller, J. T., 1973, Preglacial (Teays) and Early Glacial Drainage in the Cincinnati Area, Ohio, Kentucky, and Indiana: Geological Society of America Bulletin, v. 84, no. 11, p. 3677-3688.

Teller, J. T., and Goldthwait, R. P., 1991, The Old Kentucky River: A major tributary to the Teays River, in Melhorn, W. N., and Kempton, J. P., eds., Geology and hydrogeology of the Teays-Mahomet bedrock valley systems: Boulder, CO, The Geological Society of America, p. 29-42.

Tierney, J., 1985, Caves of northeastern Kentucky (with speical emphasis on Carter Caves State Park), in Dougherty, P. H., ed., Caves and Karst of Kentucky: Lexington, KY, Kentucky

Geological Survery, University of Kentucky, p. 78-85.

Ver Steeg, K., 1946, The Teays River: The Ohio Journal of Science, v. XLVI, no. 6, p. 297-307.

Yilmaz, I., 2007, GIS based susceptibility mapping of karst depression in gypsum: A case study from Sivas basin (Turkey): Engineering Geology, v. 90, no. 1-2, p. 89-103.

# Surface Denudation of the Gypsum Plain, West Texas and Southeastern New Mexico

By Melinda G. Shaw, Kevin W. Stafford, and B.P. Tate
Department of Geology, Stephen F. Austin State University, 1901 N. Raguet, Nacogdoches, TX 75962

## Abstract

The Castile Formation crops out in Eddy County, New Mexico and Culberson County, Texas, covering more than 1500 square kilometers. The region contains more than ten thousand reported surficial karst features, including closed depressions, sinking streams and caves, with surface karren dominating exposed rock surfaces. A standard gypsum tablet study was conducted over the region for two years to delineate the rate of surface denudation across Castile outcrops of the Gypsum Plain. Fifty five sites were monitored with triplicate gypsum tabs that were deployed for four month intervals. Rates of surface denudation were calculated based on mass loss of gypsum tabs and compared with weather data collected at three control locations within the study area. Surface denudation rates exceeded 50 centimeters per thousand years (cm/ka) in some areas with an average denudation rate of more than 30 cm/ka. Greatest rates of denudation occurred during late summer and fall, in association with the monsoon season, while the lowest rates occurred during late spring and early summer. The Castile Formation is bounded by the Guadalupe Mountains to the northwest, Delaware Mountains and Apache Mountains to the west and southwest, Glass Mountains to the southeast and Rustler Hills to the east; orographic effects of these features contribute to a complex pattern of surface denudation associated with shifting seasonal climate patterns. True denudation within the gypsum plain varies from models developed from standard tablet studies due to variations in surficial deposits; however, standard tablet studies provide a quantitative measurement of the rate at which the gypsum plain is evolving.

## INTRODUCTION

Most gypsum and anhydrite rocks originate as evaporitic formations within somewhat restricted marine basins or epicontinental sea environments. The Castile Formation evaporites are exposed or shallowly buried over approximately 1800 $km^2$ of west Texas and southeastern New Mexico. These rocks crop out in Culberson County, Texas and Eddy County, New Mexico along the western edge of the Delaware Basin and are commonly referred to as the Gypsum Plain (Hill, 1996) (Figure 1). The region is located in the semi-arid southwest on the northern edge of the Chihuahuan Desert; annual precipitation averages 26.7 cm with the greatest rainfall occurring as monsoonal storms in the late summer. Annual temperature averages 17.3°C with an average annual minimum and maximum of 9.2°C and 25.2°C respectively (Sares, 1984).

The Gypsum Plain within the Delaware Basin is an area of low relief 12 – 40 km wide and 90 km long and is characterized by shallow, eastward-trending valleys and broad, low divides. Topographic features of the plain include scattered conical and circular hills rising 15 – 20 meters above the plain, the "Castile Buttes", linear solution subsidence troughs trending eastward to northeastward, and numerous karst features such as sinkholes, caves and dissolution pits (Stafford and others, 2008b). Gypsum is one of the most soluble rocks at or near the earth's surface and is one of the few earth materials that can transition between its hydrated state (gypsum, $CaSO_4 \bullet 2H_2O$) and its dehydrated state (anhydrite, $CaSO_4$) with relative ease.

The extent of the exposed Castile Formation is bounded by the Guadalupe Mountains to the northwest, Delaware Mountains and Apache Mountains to the west and southwest, Glass Mountains to the southeast and Rustler Hills to the east (Figure 1). The thickness of the formation varies from 0 – 540 meters, depending on both depositional and dissolutional factors (Hill, 1996). Surface exposures are influenced and modified by several major streams; Hay Hollow and the Delaware River in the northern section, Salt Creek in the central section and

Cottonwood Creek in the southern section (Figure 2).

Most karst systems develop within carbonate, sulfate or halide rock assemblages, with carbonate rocks (limestone and dolostone) being the most familiar and well documented. In the United States, studies of karst systems associated with sulfate rocks (gypsum and anhydrite - collectively referred to as gypsum karst), have primarily focused on geohazards

found within these sulfate systems (e.g. Johnson and Neal, 2003; Klimchouk, 2007). The high solubility of calcium sulfate rocks enables development of extensive karst features much more rapidly than in carbonate rocks and can be correlated to the evolution of karst surface features such as sinkholes, arroyos, and caves. Karst development within the Gypsum Plain is diverse and widespread with over 10,000 surficial karst features documented for the region (Stafford and others, 2008a).

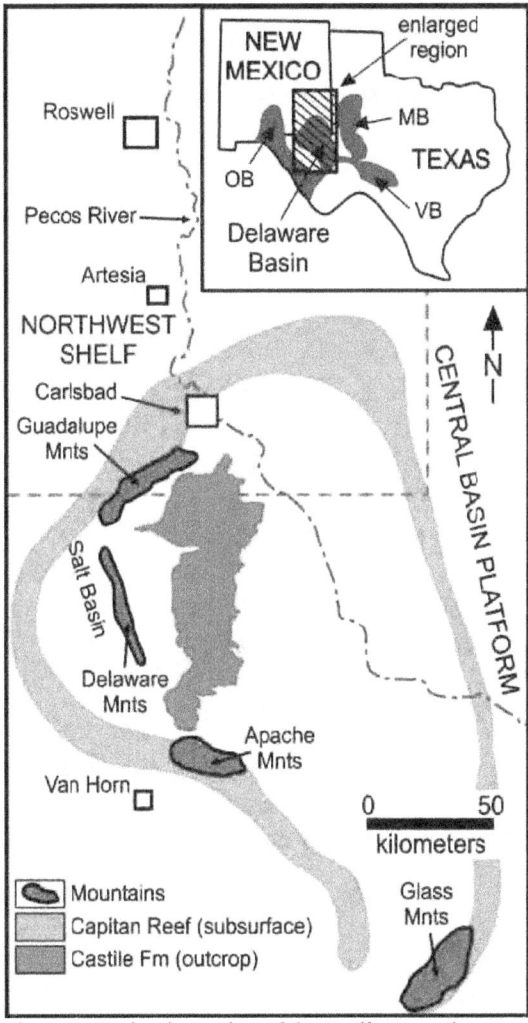

Figure 1. Regional overview of the Castile Formation outcrop area in relation to the Delaware Basin, defined by the boundary of the Capitan Reef. Inset shows location of expanded region and the Delaware Basin in relation to other major seas of the Permian, including: OB – Orogrande Basin; VB – Val Verde Basin; and MB – Midland Basin (adapted from Stafford, 2008c).

and engineering potential. Internationally, gypsum karst has been the focus of extensive speleological studies and worldwide some of the most exotic and interesting karst features are

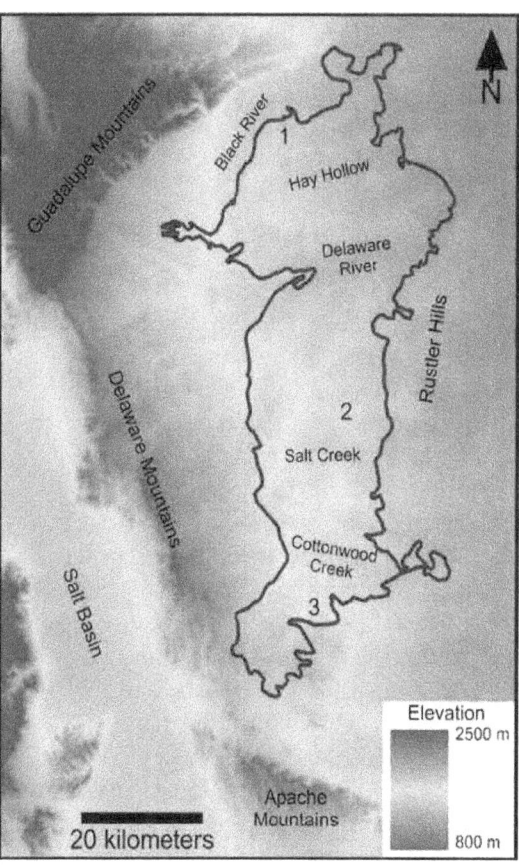

Figure 2. Digital Elevation Model of general study location with outline of Castile Formation. Map shows location of weather stations (numbered), major streams within the study area and geomorphological features adjacent to the Gypsum Plain.

## GEOLOGIC SETTING

The Permian Basin formed as an area of rapid Mississippian-Pennsylvanian subsidence in the foreland of the Ouachita foldbelt. During this time, the basin was situated close to the western margin of Pangaea and lay about 10° north of the paleo-equator (Scotese and Langford, 1995). The region was part of the southern flank of the North American craton with predominantly carbonate sedimentation. During the late Paleozoic, as a result of Pennsylvanian Ouachita-Marathon orogeny, the Permian basin region was subdivided into a series of smaller basins and platforms including the Delaware (northwestern), Midland (northeastern), Val Verde (southeastern) and Marfa (southern) basins (Figure 1) (Scholle, 2004). As basin partitioning commenced, topographical relief between the basins and adjacent platforms was accentuated by differential sedimentation and regional subsidence. Original platform and shelf areas, structural highs, accumulated much higher rates of sediment deposition than low-lying basins. By the end of the Permian, basins that began with relatively shallow depths eventually evolved to deep basins with water depths in excess of 500 meters (Leslie and others, 1996).

Basin filling began in the Pennsylvanian and by the mid Permian, the north and east parts of the Permian Basin had been infilled with clastic sediments. Connection with marine environments to the west became restricted or closed along the eastern parts of the Permian Basin and saline brines began to form (Kirkland and others, 2000). As seas continued to regress, circulation of briny waters in the Delaware Basin became more restricted and the depression was filled by the Castile Formation (Anderson and others, 1972). The Castile Formation originally consisted of approximately 5% calcite, 45% anhydrite and 50% halite, but migration of Tertiary groundwater along the western edge of the basin dissolved and removed most halite, resulting in intrastratal brecciation (Anderson and others, 1978). These evaporitic sediments consist of banded layers of gypsum, anhydrite and calcite with small lenses of halite and were deposited within a steep-walled basin with restricted circulation, coupled to the Permian Sea by a narrow channel known as the Hovey Channel. As seas regressed, circulation of briny waters became more restricted and although there may have been some drop in basinal water levels, most of the Castile was deposited in deep water below the wave base (Anderson and others, 1972). Evidence to support this conclusion includes the absence of shallow water sedimentary structures in most intervals and the presence of fine scale laminations consisting of alternating anhydrite

and darker organic matter and calcite. These laminations have been interpreted as annual varve sequences, although some studies have questioned this characterization (Anderson and others, 1972). Regardless of origin or characterization, darker laminae probably represent a periodic influx of fresher water and the development of plankton blooms; anhydrite layers probably represent more restricted, evaporitic condition.

Sedimentation and precipitation of evaporite layers continued over a period of 260 Ka and coincided closely with termination of reef growth around the Delaware Basin margin. Closure of the Delaware Basin was most probably controlled by tectonic uplift primarily on the west side of the basin. This uplift continued throughout Ochoan time, causing truncation of lower Castile members and non-deposition of upper members on the western side of the basin (Horak, 1985).

## METHODS

A standard gypsum tablet study (e.g. Nance, 1993; Klimchouk and others, 1996) was conducted over the region for two years to delineate the rate of surface denudation across Castile outcrops of the Gypsum Plain. Fifty-five sites were monitored with triplicate gypsum tabs that were deployed for four month intervals. Data was collected during the time period November, 2005 – November, 2007 to determine the rate of surface denudation of the Gypsum Plain within the Delaware Basin. Fifty-five sample locations were selected based on geomorphic attributes and land access. Blocks of homogeneous gypsum were extracted from a quarry site within the southern part study area, cored and sliced into discs 5 cm in diameter and 1 cm thick (Figure 3). Each of the discs were abraded with coarse sand to remove rough edges and rinsed to remove dust. Discs were placed in an oven at 35°C for 48 hours to remove surface and interstitial water, then weighed and tagged for field placement.

Figure 3. Example of typical gypsum tablet used in study.

At each sample location, elevation and coordinates were taken using a portable GPS unit. Three tabs were placed directly on the ground surface at each sample location in areas to minimize mechanical erosion approximately 10 meters apart and left for four months. Upon return, new tabs were placed and weathered tabs were retrieved. Weathered tabs were cleaned and dried in an oven at 35°C for 48 hours, then weighed and measured. Surface denudation was calculated using the loss of tab volume in centimeters per thousand years (cm/kyr).

The four month intervals for data collection were divided based upon distinct weather patterns within the region: November to March, Winter Season; March to July, Summer Season; and July to November, Monsoon Season. Three weather stations were installed within the study area to collect precipitation, temperature, wind direction and wind speed to determine the effect of those parameters on both dissolution and denudation rates of the exposed Castile Formation and mechanical erosion associated with the seasonal and temporal distribution of weather events. Station 1 was installed in the northwestern section near the Guadalupe Mountains, Station 2 in the central eastern area, and Station 3 in the far southeastern part of the study area (Figure 2).

## RESULTS AND DISCUSSION

Surface denudation rates generated from the dataset shows seasonal variation based on the

availability of moisture. Within each of the seasons, higher denudation rates are found in proximity to major stream systems: Hay Hollow and Delaware River in the north, Salt Creek in the central, and Cottonwood Creek in the south. Increased atmospheric moisture proximal to streams results in daily condensation from diurnal temperature variance inducing greater dissolution than regions distal to surficial water supplies. Among the seasons, dissolution rates are higher in the Monsoon Season, correlating with increased precipitation across the study area. Composite figures representing average denudation rates for each of the seasons and an annual average are found in Figure 4.

Precipitation patterns within the study area vary seasonally and spatially: Winter Season is generally dry; Monsoon Season moist; and in general, higher denudation rates are generally reported in the north and can be directly correlated with an increase in moisture and precipitation. General trends also show higher surface denudation rates in the Summer and Monsoon Season in proximity to topographic features such as the Guadalupe Mountains and Delaware Mountains in the northern part of the study area. The exception to this would be the Winter Season, when freezing temperatures could temper liquid precipitation events in the north. Cumulative precipitation during the study period shows the variance of those trends both seasonally and spatially (Figure 5).

As a result of semi-arid climate and little relative humidity, diurnal temperature fluctuations within the study area tend to be greater than those in humid climates (Sares, 1984). The northern section of the study area experienced greater fluctuations in diurnal temperatures; these fluctuations, coupled with readily available moisture in stream channels and springs, produced more effective evaporation / condensation cycles and accelerated denudation (Figure 6). The annual average ranges of temperatures throughout the study area were 9.9°C and 27.9°C, respectively. Dominant wind and weather patterns generally travel from west to east, with a strong

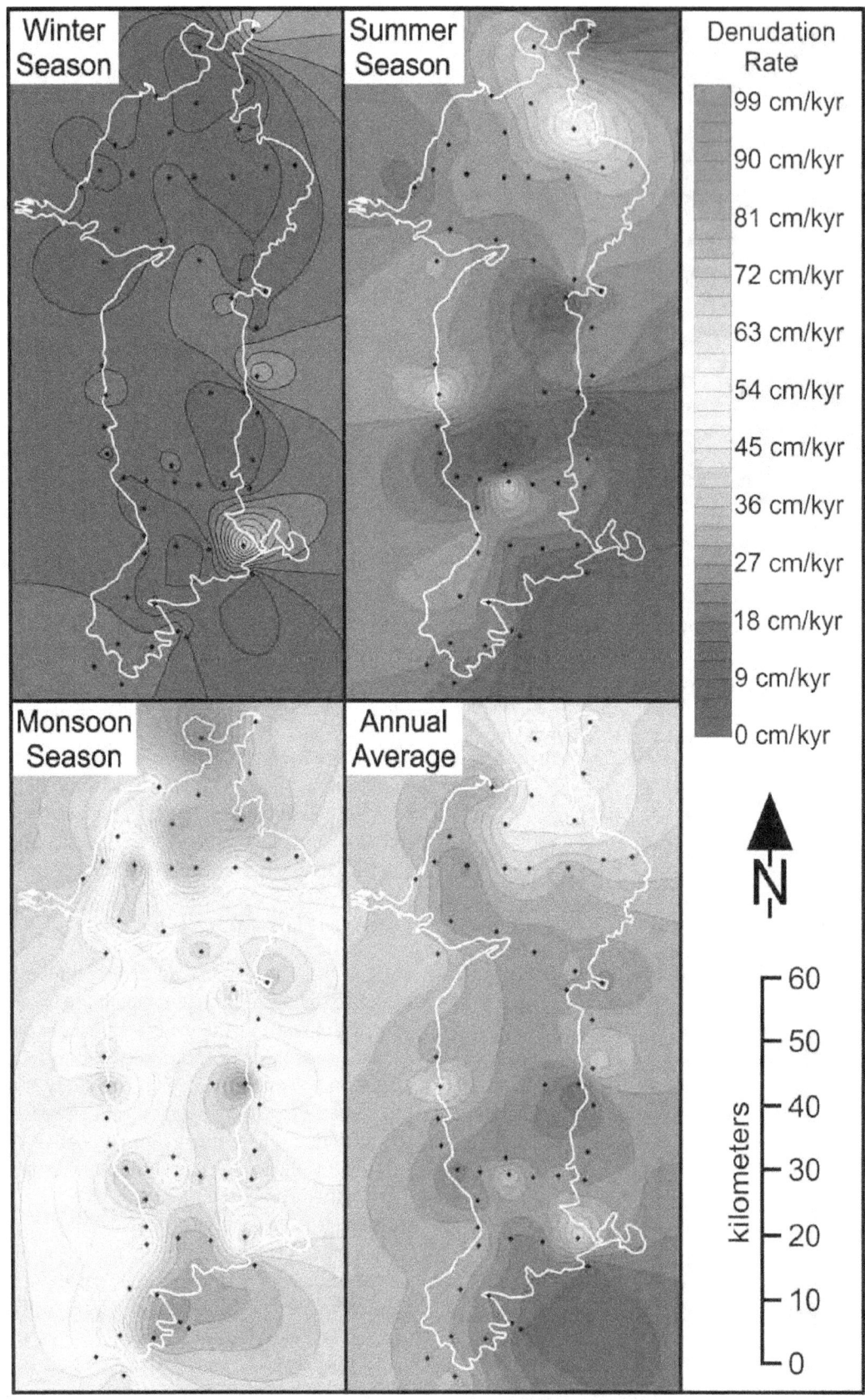

Figure 4. Map of rates of denudation Winter (November – March), Summer (March – July) and Monsoon (July – November) for each season and an annual average. Castile Formation is outlined in white and study sites for tablet placement in black. Note: Contour intervals are 3cm/kyr.

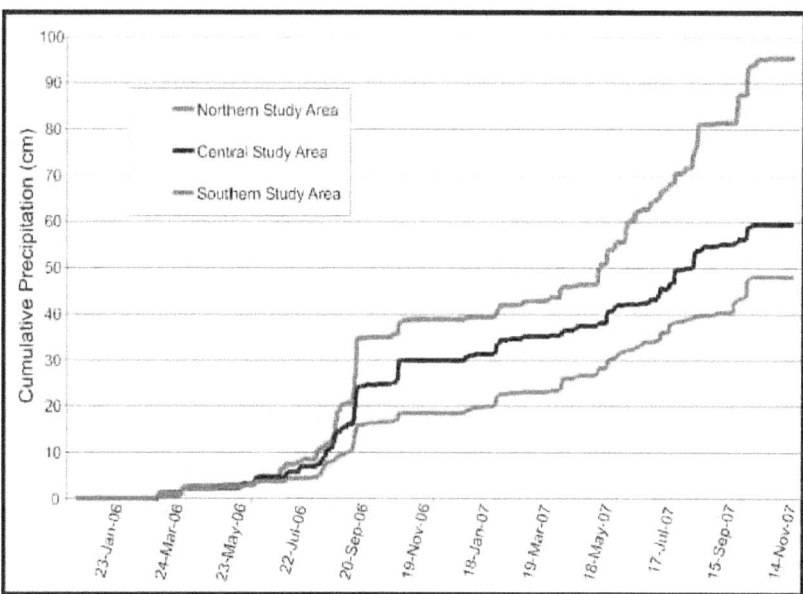

Figure 5. Cumulative precipitation (cm) through the duration of the study period showing increased precipitation rates moving northward across the Gypsum Plain.

Table 1. Average precipitation rates (cm) within seasons and sections of the study area.

| Season | Precipitation Rates (cm) | | | |
|---|---|---|---|---|
| | North | Central | South | Average |
| Winter | 2.7 | 3.4 | 3.3 | 3.1 |
| Summer | 17.5 | 10.0 | 8.9 | 12.1 |
| Monsoon | 27.6 | 16.4 | 11.9 | 18.6 |
| Section Total | 47.7 | 29.7 | 24.1 | 33.8 |

of 2.9°C to an average high of 17.5°C. Predominant wind patterns were generally from the west during the Winter Season, with minimal input from the Gulf Stream (Figure 7). In the northern section of the study area, warm air masses from the Chihuahuan Desert were channeled between the passes in the Guadalupe and Delaware Mountains, along the front of the Guadalupe Mountains. This pattern was repeated in the southern section, with warm desert air channeled between the Delaware and Apache Mountains. Wind patterns in the central section were dominated by cooler air masses moving over the Guadalupe Mountains from the northwest. The highest denudation rates are in proximity to drainage arteries and associated with moisture from the various regional springs that supply water to the atmosphere and surface. Overall, seasonal denudation rates are lowest within the winter, ranging from 5-40 cm, and correlating with the low rates of precipitation throughout the study area.

**Summer Season**

Precipitation rates within the Summer Season, March to July, show a four-fold increase with an annual average precipitation of 12.1 cm; 17.5cm in the north, 10 cm in the central region and 8.9 cm in the south (Table 1). Temperatures during this season ranged from an average low of 7.7°C to an average high of 27.8°C. Wind patterns shifted with each section reporting a strong influence from the southeast (Figure 7). Warm air still flowed up from the desert, channeled between mountain passes in the northern and southern sections, with the central section still dominated by northwestern air masses channeled over the Guadalupe Mountains. The highest denudation rates are again focused along drainage arteries and associated with moisture from the various regional springs that supply moisture to the atmosphere and surface. Seasonal denudation

southeastern component present during the Summer and Monsoon Season (Figure 7). Orographic barriers of mountain ranges on the northwestern and western edge of the Castile Formation and Delaware Basin play a role in channeling wind, storm systems and surface flow (Figure 2). In addition, although the topographic expression within the plain is minimal, a correlation can be made between elevation and denudation; points of higher elevation tend to denude at a greater rate than points of lower elevations.

**Winter Season**

During the study period, annual average precipitation in the Winter Season was 3.1 cm; 2.7 cm in the north, 3.4 cm in the central region and 3.3 cm in the south (Table 1). Temperatures during this season ranged from an average low

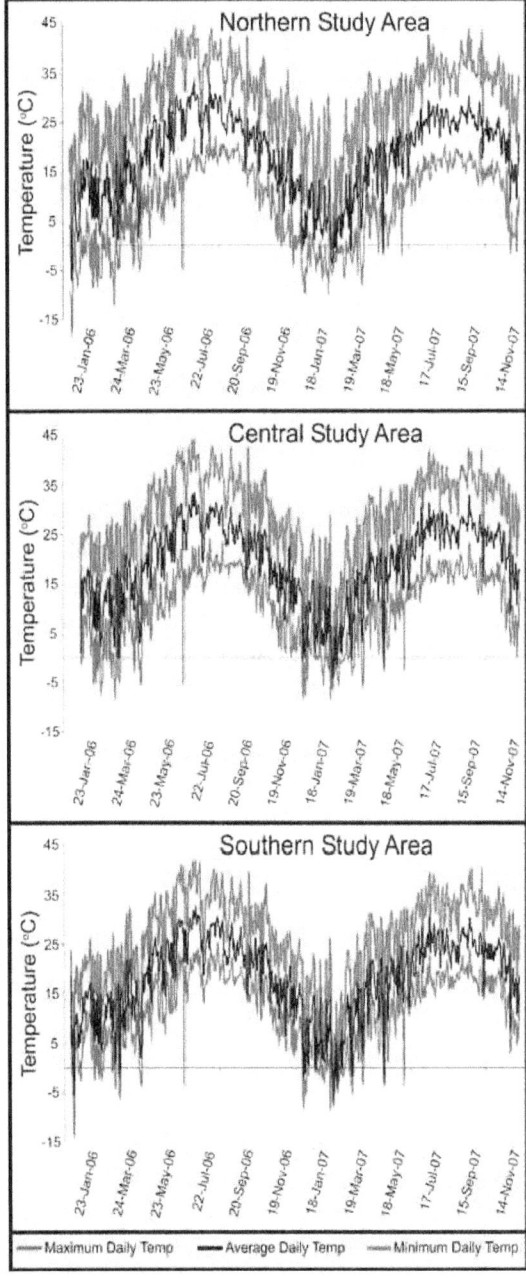

Figure 6. Temperature variation within the study area showing minor latitudinal variations.

rates are somewhat higher in the summer and are more evenly distributed throughout the study area with values ranging from 5-50 cm. The difference in spatial distribution can be correlated with increasing rates of precipitation and humidity, and warmer temperatures throughout the study area.

**Monsoon Season**

The Gypsum Plain receives over 50% of its annual precipitation during the Monsoon Season. During the study period, annual average precipitation was 18.6 cm; 27.6cm in the north, 16.4 cm in the central region and 11.9 cm in the south (Table 1). Temperatures during this season ranged from an average low of 8.9°C to an average high of 27.2°C. Wind patterns showed the greatest shift during this season, with a strong N/NE component predominant in the northern section of the study area (Figure 7). Air masses from the southeast, reported in the central and southern section of the study area could have redirected eastern moving air masses to the north of the Guadalupe Mountains, allowing wind patterns to shift in the northern section. Higher denudation rates are more evenly distributed across the study area, and the increase in moisture accelerates denudation along stream channels and arroyos. Seasonal denudation rates are somewhat higher in the summer and are more evenly distributed throughout the study area with values ranging from 24-80 cm. The difference in spatial distribution can be correlated with increasing rates of precipitation and humidity, and warmer temperatures throughout the study area.

## CONCLUSION

Karst systems associated with sulfate rocks, both surficial and below surface, host some of the most exotic and interesting speleological features. Gypsum karst offers a medium for geochemical, engineering and petrochemical studies, many of which are of interest both within the United States and internationally. The high solubility of gypsum enables the development of extensive karst features much more rapidly, creating many opportunities for current and future studies of landscape evolution.

Within the Castile Formation, surface denudation rates vary both seasonally and spatially, and are directly correlated with available atmospheric and surficial moisture. Dissolution during the Winter Season is lowest; warm, dry air masses flowing from the Chihuahuan Desert reduce available atmospheric moisture. During the Summer and Monsoon

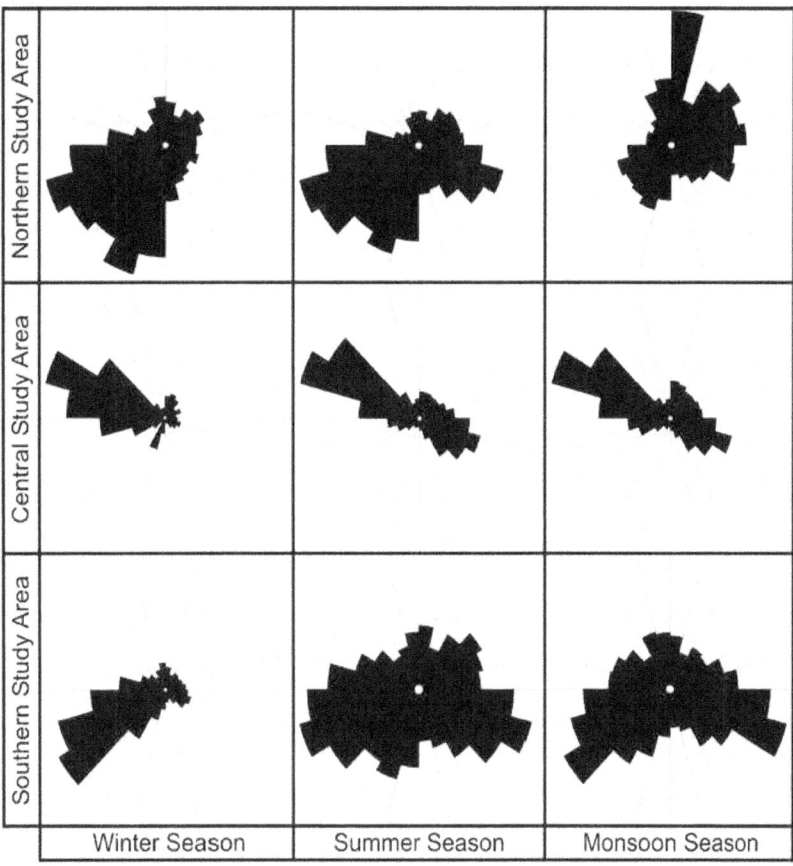

Figure 7. Rose diagrams showing seasonal wind direction variations across the study area (n~80,000 per sampling period). Note: Rose diagrams are displayed in standard azimuth coordinates with north to the top of the page.

Seasons, southeastern winds from the Gulf of Mexico bring warm, moist air increasing atmospheric moisture, precipitation rates and relative humidity, thus inducing greater dissolution. Spatially, higher denudation rates are found in proximity to major stream systems; increased atmospheric moisture proximal to streams results in daily condensation from the diurnal temperature variance.

Precipitation rates can vary seasonally and spatially, during the Winter Season, cool, dry air masses from the northwest and southwest bring in little moisture, and the lowest precipitation rates are reported during this season. Summer and Monsoon Seasons report greater precipitation rates, with the influence of southeastern winds bringing moist, humid air from the Gulf of Mexico. Spatially, higher denudation rates are generally reported in the north and can be directly correlated with an increase in moisture and precipitation. The orographic affect of the Delaware and

Guadalupe Mountains can influence moist air masses and induce precipitation.

True denudation within the Gypsum Plain varies from models developed from standard tablet studies due to variations in surficial deposits; however, standard tablet studies provide a quantitative measurement of the rate at which the gypsum plain is evolving. The evolution of this landscape and variations of karst features within the Castile Formation are an indication that these features and rates of denudation are more complex than can be explained by a standard tablet study. Rates of incision of stream systems, precipitation, wind patterns and exposure of previously buried features by surface denudation all contribute to the geomorphic evolution of the Gypsum Plain.

## ACKNOWLEDGMENTS

The authors are grateful to the numerous individuals who made this study possible.

Generous land access to private ranches in Texas was provided by Jack Blake, Draper Brantley, Stanley Jobe, Lane Sumner and Clay Taylor. John Jasper and Jim Goodbar provided assistance with karst studies on BLM (Bureau of Land Management) public land in New Mexico. The authors are thankful for the useful and insightful comments provided by Wesley A. Brown and Andy Grubbs which helped to improve this manuscript. This research was partially funded through grants from the New Mexico Geological Society (NMGS), the American Association of Petroleum Geologists (AAPG), the Geological Society of America (GSA), with support from the National Cave and Karst Research Institute (NCKRI).

## REFERENCES

Anderson, R.Y., Dean, W.E., Kirkland, D.W. and Snider, H.I., 1972, Permian Castile varved evaporite sequence, West Texas and New Mexico: Geological Society of America Bulletin, v. 83, p. 59-85.

Anderson, R.Y., Kietzke, K.K., and Rhodes, D.J., 1978, Development of dissolution breccias, northern Delaware Basin and adjacent areas: New Mexico Bureau of Mines and Mineral Resources Bulletin 159, p. 47-52.

Hill, C.A., 1996, Geology of the Delaware Basin, Guadalupe, Apache and Glass Mountains: New Mexico and West Texas: Permian Basin Section Society of Economic Paleontologists and Mineralogists, Midland, Texas, 480 p.

Horak, R.L., 1985, Trans-Pecos tectonism and its affect on the Permian Basin, in Dickerson, P.W., and Muelberger, W.R., eds., Structure and Tectonics of Trans-Pecos Texas: Midland, Texas, West Texas Geological Society, p. 81-87.

Johnson, K.S. and Neal, J.T., 2003, Evaporite Karst and Engineering/Environmental Problems in the United States, Norman, Oklahoma Geological Survey: 353 p.

Kirkland, D.W., Rodger, E.D, Dean, W.E., 2000, Parent Brine of the Castile Evaporites (Upper Permian), Texas and New Mexico, Journal of Sedimentary Research, v. 70, no. 3, p. 749-761.

Klimchouk, A., 2007, Hypogene Speleogenesis: Hydrogeological and Morphometric Perspective: Carlsbad, National Cave and Karst Research Institute, Special Paper No. 1, 106 p.

Klimchouk, A., Cucchi, F., Calaforra, J.M., Aksem, S., Finocchiaro, F., and Forti, P., 1996, Dissolution of gypsum from field observations: International Journal of Speleology, v. 25, no. 3-4, p. 37-48.

Leslie, A.B., Kendall, A.C., Harwood, G.M., Powers, D.W, 1996, Conflicting indicators of paleodepth during deposition of the Upper Permian Castile Formation, Texas and New Mexico, Geological Society, London, Special Publications, v. 116, p. 79-92.

Nance, R., 1993, Application of the standard tablet method to a study of denudation in gypsum karst, Chosa Draw, southeastern New Mexico, MS Thesis, University of Northern Colorado, Greely, Colorado, 82 p.

Sares, S.W., 1984, Hydrologic and geomorphic development of a low relief evaporite karst drainage basin, southeastern New Mexico [M.S. Thesis]: Albuquerque, University of New Mexico, 123 p.

Scholle, P.A, Goldstein, R.H., and Ulmer-Scholle, D.S., 2004, Classic Upper Paleozoic Reefs and Bioherms of West Texas and New Mexico, New Mexico Institute of Mining and Technology, Socorro, NM, 166 p.

Scotese, C.R. and Langford, R., 1995, Pangaea and the Paleogeography of the Permian, The Permian of Northern Pangaea, vol. 1, p. 3-19.

Stafford, K.W., Rosales-Lagarde, L., and Boston, P.J., 2008a, Castile Evaporite Karst Potential Map of the Gypsum Plain, Eddy County, New Mexico and Culberson County, Texas: A GIS Methodological Comparison: Journal of Cave and Karst Studies, v. 70, no. 1, p. 35-46.

Stafford, K.W., Nance, R., Rosales-Lagarde, L., and Boston, P.J., 2008b, Epigene and hypogene gypsum karst manifestations of the Castile Formation: Eddy County, New Mexico and Culberson County, Texas, USA: International Journal of Speleology, v. 37, no. 2, p. 83-98.

Stafford, K.W., Ulmer-Scholle, D.S., Rosales-Lagarde, L., 2008c, Hypogene Calcitization: Evaporite Diagenesis in the western Delaware Basin, Carbonates and Evaporites, v. 23, no. 2, 2008, p. 89-103.

# KARST PROGRAM UPDATES
# National Cave and Karst Research Institute:
# Growing Capabilities and Federal Partnerships

By George Veni

National Cave and Karst Research Institute, 400-1 Cascades Avenue, Carlsbad, NM 88220 USA

## Abstract

The National Cave and Karst Research Institute's (NCKRI) growth has accelerated since NCKRI's report to the last Karst Interest Group (KIG) meeting in 2008. This is the result of three primary factors. First, NCKRI's staffing has increased. While the addition of an Education Director and Advancement Director is numerically small, it is significant for a growing organization and the positions are critical to building key NCKRI programs. Second, NCKRI is buying and using significant research equipment. The most important purchase to date is an electrical resistivity survey system, which is being applied to solve field problems and in a NCKRI research program to refine and enhance the resistivity methodology. Third, NCKRI has moved into its newly constructed headquarters. While some rooms await additional funds for completion, all of the public spaces are complete and the building is functional for NCKRI's core needs. NCKRI Headquarters provides a synergistic effect with growing staff and equipment by opening new opportunities for continued growth. For example, exhibits and classroom/meeting space bring in visitors, new partners, and new sources of revenue. Small conferences hosted by NCKRI are being scheduled and perhaps NCKRI will host a future KIG meeting.

NCKRI's federal partners are also growing and strengthening. In March 2009, President Obama signed a bill that removed a legislative impediment to cooperative programs between NCKRI and federal agencies. In May 2010, NCKRI, Bureau of Land Management, National Park Service, U.S. Fish and Wildlife Service, and U.S. Geological Survey gave a joint briefing to the Department of the Interior (DOI) on the status of cave and karst research and management challenges in the U.S. One outcome of the meeting was an initiative to create a DOI-NCKRI Memorandum of Understanding (MOU). Currently in development, this MOU is expected to serve as an umbrella agreement to facilitate NCKRI working with and assisting DOI agencies. As the nation struggled with a major economic crisis during the last three years, NCKRI has continued to grow and will soon be positioned to launch forward when economic prosperity returns.

## INTRODUCTION

The National Cave and Karst Research Institute (NCKRI) was created by the U.S. Congress in 1998 in partnership with the State of New Mexico and the City of Carlsbad. Initially an institute within the National Park Service, NCKRI is now a non-profit 501(c)(3) corporation that retains its federal, state, and city partnerships. Federal and state funding for NCKRI is administered by the New Mexico Institute of Mining and Technology (aka New Mexico Tech or NMT). Funds not produced by agreements through NMT can be accepted directly by NCKRI.

NCKRI's enabling legislation, the National Cave and Karst Research Institute Act of 1998, 16 U.S.C. §4310, identifies NCKRI's mission as to:

1) further the science of speleology;
2) centralize and standardize speleological information;
3) foster interdisciplinary cooperation in cave and karst research programs;
4) promote public education;
5) promote national and international cooperation in protecting the environment for the benefit of cave and karst landforms; and

6) promote and develop environmentally sound and sustainable resource management practices.

Boston and Veni (2008) outlined some of the challenges that faced NCKRI as a young institute that was newly reorganized as a non-profit corporation, as well as some opportunities. This paper reports on NCKRI's progress over the past three years in its programs and federal partnerships. Additional details can be found in NCKRI (2008, 2009, 2010).

## PROGRAM EXPANSION

NCKRI's staffing is increasing. The addition of an Education Director and Advancement Director is numerically small, but significant for a small organization with only one part-time and six full time employees. Likewise, NCKRI is buying critical research equipment and has moved into its newly constructed headquarters. All of these factors are major steps toward building NCKRI's key programs. Status summaries on these programs are provided below.

### Education Program

Dianne Gillespie was hired as NCKRI's first Education Program Director in June 2009. Her first 6 months were spent on a coast-to-coast research expedition across the U.S. She visited 64 sites including show caves, government agencies, universities, museums, parks, and private organizations focused on caves, karst, and related natural resources science. Through her travels she discovered that only 10 locations had a cave education specialist working with school groups. Additional funding and quality educational materials were common needs. Many people requested that NCKRI develop training for informal educators, focused on explaining cave development and karst. Many also asked NCKRI to serve as a clearinghouse for cave and karst education information.

Ms. Gillespie has since developed an Education Plan for NCKRI aimed to reach people of all ages and backgrounds. The plan is currently being reviewed by the NCKRI Board of Directors and approval is anticipated in May 2011, at which time its details will be released.

NCKRI recognizes that effective education requires effective outreach. Education Program activities thus also include overhauling NCKRI's website (www.nckri.org), and establishing the Institute on Facebook and Twitter. Cooperative relationships are also being developed with cave education programs at the American Cave Conservation Association, Carlsbad Caverns National Park, and Ohio State University.

NCKRI is also connecting with people and organizations locally to nationally. For example, NCKRI sponsors a Boy Scouts of America Venture Crew, leads and teaches various cave programs, and serves on a federal White Nose Syndrome committee.

### Academic Program

*Cave and Karst Studies at New Mexico Tech* is NCKRI's academic program and is taught through NMT's Earth and Environmental Sciences Department. Dr. Penelope Boston is the program's director. She teaches a variety of regular and special topic caves and karst courses, several in collaboration with other NMT faculty, including:
- Cave and Karst Systems
- Cave and Karst Laboratory
- Advanced Topics in Speleohydrology
- Moonmilk Research
- Fundamentals of Geobiology
- Extraterrestrial Dissolutional Landforms

Dr. Boston has supervised numerous Ph.D., Masters, and undergraduate senior theses on cave and karst research. Recent topics include:
- Cave micrometeorological modeling
- Salt Basin Karst Aquifer, Texas and New Mexico
- Sulfuric acid caves and sulfur springs of Tabasco, Mexico
- Origin of Moonmilk
- Hydrology and paleoclimatology of Snowy River, Fort Stanton Cave, New Mexico
- Origin and geomicrobiology of fumarolic ice caves, Mount Erebus, Antarctica

### Advancement Program

NCKRI is currently financed primarily through federal and State of New Mexico appropriations. These funds are considered seed monies to establish NCKRI and support its core operations. For long-term stability, growth, and success, NCKRI must increase and diversify its sources of revenue.

On 25 October 2010, Ann Dowdy became NCKRI's first Advancement Director. She administers NCKRI's Advancement Program to develop additional funds and support. This assistance will be used to build programs and partnerships at all levels to enhance the Institute's activities and goals. Fundraising for NCKRI is especially challenging and requires close coordination with the Education Program. Potential donors must first understand what karst is, overcome common misconceptions about caves, and learn how caves and karst are important to daily life.

Ms. Dowdy has been on the job only a short time but is already making a big impact. She is developing a membership program for NCKRI, has written grant requests, and established NCKRI on fundraising websites with some large fundraising programs and potential donors and partners. She has also established a small revenue stream through NCKRI supporters who load the GoodSearch toolbar on their Internet browsers, http://www.goodsearch.com/toolbar/national-cave-and-karst-research-institute. GoodSearch automatically generates small corporate donations to NCKRI with each Internet search and through much online shopping. Future fundraising methods will include online giving, annual and corporate giving, endowments, bequests, special events, and additional appropriation and grant requests.

## Research Programs

NCKRI has basic and applied research programs but has not yet hired a director for either. Nonetheless, substantial research is underway.

Dr. Penelope Boston conducts most of NCKRI's basic research, such as:
- National Science Foundation study of carbonate pool precipitates;
- Study of iron-manganese deposits in caves and desert varnish;
- Geomicrobiological investigation of Cueva de los Cristales in the Naica Mine, Chihuahua, Mexico;

- Evaluating the role of extraterrestrial lava tubes on the Moon and Mars in future space exploration with partners from the Massachusetts Institute of Technology and TechShot, Inc.; and
- NASA funded development of an infrared tunable acousto-optical laser spectrometer to characterize organic components and microbial life—ultimately for extraterrestrial planetary use but being tested on minerals that include cave microbial materials.

Dr. Lewis Land, NCKRI's Research Hydrogeologist, also conducts basic investigations, such as hydrologic characterization of the Roswell Artesian Basin and hydrogeologic mapping of the Southern Sacramento Mountains watershed in southeast New Mexico. Beginning in 2008, his work has become more applied.

NCKRI is purchasing high quality hydrogeological field research equipment, such as flow and multi-meters, to enhance its investigative abilities. The Institute's SuperSting R8/IP 112-electrode electrical resistivity meter and survey-grade Topcon GR3 global positioning system have seen the most demand for use.

In 2008, two brine well cavities collapsed in the Carlsbad area creating sinkholes over 100 m in diameter and 40 m deep. These events raised concerns that a similar cavity at a brine well within the city limits might collapse, and generated interest in high-precision surveys of brine well and natural karstic cavities. NCKRI is being contracted by the State of New Mexico to conduct a resistivity survey of the potential collapse site in Carlsbad. In addition to such problem-solving geophysical investigations, NCKRI has started experiments that may refine the resistivity methodology and interpretation of its data.

NCKRI is also now conducting contract research. This is an important component of diversifying the Institute's sources of income and expanding its programs. However, NCKRI will not function as a consulting company but in a manner akin to the U.S. Geological Survey. The Institute will work on contracted projects that may produce results with broad application that can be published in NCKRI's and other professional literature. While budgetary needs may initially see the Institute work on some projects that lack broad application, those projects will nonetheless provide important data that NCKRI will include in future studies.

## Headquarters Construction

Twenty-seven months to the day after groundbreaking, NCKRI staff moved into its new headquarters on 24 February 2011. This 1,609 m$^2$ building contains a bookstore, museum/exhibit hall, classrooms, laboratory, library, offices, and meeting space. NCKRI Headquarters is highly energy efficient, with innovative design features. Some of these features will serve nationally as models for green building techniques to prevent or minimize adverse environmental impacts.

But NCKRI Headquarters is more than just a building. It is a true partner in advancing the Institute. This beautiful edifice provides:

- NCKRI legitimacy as a national organization in the eyes of many in ways impossible for our previous assemblage of temporary offices;
- Space to easily access our stores of books and research materials, as well as for volunteers, student interns, and visiting scholars to work with us;
- Museum space to better educate the public on the value of caves and karst;
- Educational space for classes, workshops, and small conferences like the Karst Interest Group (which is invited to have its 2014 meeting at NCKRI); and
- Additional sources of funding through rental of rooms and bookstore sales.

An Opening Day Celebration is scheduled for 14 May 2011 as a community event that recognizes the many local and national partners who made the building and NCKRI a reality. While planned as a grand event, it will not be a "grand opening." All public spaces in the building are finished, but the library, laboratory, and associated storage rooms wait for further funding. The Grand Opening will occur once the building is finished and all of the state-of-the-art museum exhibits are installed.

## EXPANDING FEDERAL PARTNERSHIPS

The idea of NCKRI was born from the national cave and karst coordinators of the Bureau of Land Management, National Park Service, and U.S. Forest Service, who were all stationed in Carlsbad in the 1980s. They first envisioned a federal institute dedicated to studying and solving cave and karst problems for federal agencies. By the time Congress authorized NCKRI in 1998, its scope expanded to the current mandates.

In 2006, NCKRI reorganized as a nonprofit to maximize its flexibility to enter into partnerships with other entities, raise funds, and respond quickly to new opportunities. This change was a good thing, giving NCKRI the best of several worlds: national legitimacy though its federal partners and Congressional mandates; excellent administrative, legal, and technical support through its state partners at New Mexico Tech; a wonderful custom built headquarters through the City of Carlsbad; and the business agility achieved through its non-profit status. Yet NCKRI also felt diminished by the loss of its full federal status. That feeling wasn't warranted and didn't last long.

NCKRI's federal connections have grown and strengthened over the past three years. The Institute's Board of Directors has included representatives of the Bureau of Land Management, National Park Service, U.S. Fish and Wildlife Service, U.S. Forest Service, and U.S. Geological Survey, each working to support NCKRI in many ways.

The National Park Service (NPS) remains NCKRI's primary federal partner. Federal funds are channeled to NCKRI, via New Mexico Tech, through the NPS' Geologic Resources Division. The NPS provided architects whose attentive oversight dramatically improved the design of NCKRI Headquarters and assured its quality. The Institute's website is hosted by the NPS and an exhibit for NCKRI's museum was donated by Carlsbad Caverns National Park. In 2010, the Geologic Resources Division provided NCKRI with end-of-year funds to put the finishing touches on the public areas of the Institute's Headquarters. NCKRI is also periodically consulted by the NPS on cave and karst issues, initiatives, and potential collaborative projects.

NCKRI also works regularly with other federal bureaus. NCKRI's Education Director serves on a multi-agency committee on White Nose Syndrome and will co-teach a joint U.S. Forest Service-NCKRI-National Speleological Society workshop on cave and karst management in June 2011. The Institute is sponsoring a similar workshop which will be offered by the Bureau of Land Management hosted by NCKRI at NCKRI Headquarters in March 2011. I serve on a U.S. Fish and Wildlife Service recovery team for several endangered karst invertebrate species. NCKRI also continues to assist the U.S. Geological Survey whenever possible with the National Karst Map and in sponsoring these Karst Interest Group meetings.

Successful federal, state, and NCKRI teamwork was celebrated in March 2009 when President Obama signed a bill into law, removing a legislative impediment to cooperative programs between NCKRI and federal agencies. Building on that, about a year later NCKRI, the Bureau of Land Management, National Park Service, U.S. Fish and Wildlife Service, and U.S. Geological Survey gave a joint briefing to the Department of the Interior (DOI) on the status of cave and karst research and management challenges in the U.S. The meeting led to a cooperative project with the Bureau of Land Management. More significantly, it also started an initiative to create a DOI-NCKRI Memorandum of Understanding. The memorandum is currently in development, but it is expected to serve as an umbrella to facilitate NCKRI working with and assisting all DOI bureaus.

## LOOKING AHEAD

The U.S. is recovering from a severe economic recession that began three years ago. While many organizations have scaled back, laid off employees, and even gone out of business, NCKRI has continued to grow. NCKRI will continue further developing its administrative foundation for the next couple of years, but will now focus on building its education, advancement, and research programs. These will involve new and synergistic partnerships with federal agencies, and should blossom when national economic prosperity returns.

## REFERENCES

Boston, Penelope J, and Veni, George, 2008, National Cave and Karst Research Institute: Partner for the USGS; *in* Kuniansky, Eve, ed., U.S. Geological Survey Karst Interest Group Proceedings, p. 12-12.

National Cave and Karst Research Institute, 2008, National Cave and Karst Research Institute Annual Report 2007-2008; National Cave and Karst Research Institute, Carlsbad, New Mexico, 28 p.

National Cave and Karst Research Institute, 2009, National Cave and Karst Research Institute 2008-2009 Annual Report; National Cave and Karst Research Institute, Carlsbad, New Mexico, 28 p.

National Cave and Karst Research Institute, 2010, National Cave and Karst Research Institute 2009-2010 Annual Report; National Cave and Karst Research Institute, Carlsbad, New Mexico, 32 p.

# The National Karst Map: an Update on its Progress

By David J. Weary and Daniel H. Doctor
U.S. Geological Survey, MS 926A, 12201 Sunrise valley Drive, Reston, VA 20192

## Abstract

Over the last several years, the U.S. Geological Survey (USGS) has worked towards compiling a new map of karst in the United States in cooperation with the National Cave and Karst Research Institute (NCKRI), the National Speleological Society (NSS), and with contributions by various state geological surveys. Our goal is to produce a national-scale map that will have utility across the spectrum of karst research in the United States. The new National Karst Map will replace previous paper maps with a digital, GIS-based product. The geographic extent of the compilation now includes the conterminous United States, Alaska, Hawaii, and Puerto Rico. Resolution of the polygon data range from 1:24,000 to 1:1 million depending on the source data used. This map reflects the presence, at or near the earth's surface, of relatively soluble bedrock (carbonates and evaporites) and late-Tertiary to recent volcanic flows representing areas of potential for the occurrence of karst or volcanic pseudokarst. The compilation is being refined by input from karst researchers, and experts from state geological surveys. In addition to publishing the GIS database, we plan to publish a large paper map, most likely at a scale of 1:5 million. In addition, a USGS fact sheet will be published, and provided online, showing the extent of American karst lands and briefly explaining the importance and diversity of karst in the United States. This will include a downloadable small map graphic suitable for illustration of the distribution of America's karst for presentation purposes. Based on our current data, karst in soluble rocks in the United States comprises about 13.5 percent of the land area. Of that, about 12.4 percent of America's area is underlain by carbonate rocks, and 2.1 percent by evaporite rocks. These numbers understate the extent of shallow karst aquifers, which serve as an important source for drinking water and are often obscured on geologic maps by overlying, non-karstic strata or sediments. Similarly, areas with potential for collapse hazard due to soluble rocks at depth are also understated. An additional 1.3 percent of the nation is underlain by rocks likely to contain volcanic pseudokarst. Future work will entail delineating areas prone to development of other types of pseudokarst, such as piping caves in unconsolidated sedimentary units. Estimating the extent of these areas is extremely difficult as data on the locations of such features as well as the factors contributing to their formation and localization are scarce.

Integration of additional map layers and characteristics into the map database is ongoing. These include the potential extent of buried soluble rocks and paleokarst zones, density of karst features (where databases exist), the extent of mineralized areas that may be linked to hypogenic karst processes, and framework properties of regional carbonate aquifers. Karst areas will be categorized in terms of their geologic and eco-geographic contexts. The map database contains links back to the original geologic stratigraphic units selected from the source data (chiefly digital state geologic maps) so that detailed lithologic information may be extracted for this purpose. Spatial boundaries of many karst areas align reasonably well with Level III ecoregions (using the Omernick Ecoregion System hierarchy) and the database will allow for selection and sorting by these ecoregions. We believe that this organization will help establish not only geographic references to karst regions, but also supply a framework allowing definition of individual regional-scale karst systems characterized by shared genesis and a coherent set of characteristics. In addition, this categorization helps organize karst areas into regions with similar geology, climate, hydrology, vegetation, and wildlife. It is hoped that these regions will provide boundaries useful for multivariate analyses of various questions about the process of karstification and development of karst ecosystems.

# Karst Hydrogeology of the Southern Ozarks

By Van Brahana

University of Arkansas, Department of Geosciences, Fayetteville, AR  72701

## Abstract

The regional karst aquifers of the Salem Plateau of the southern Ozarks are unique in many ways amongst midcontinent U.S. karst aquifer systems. The Ozark region of Missouri and Arkansas contains many more large springs (> 1 $m^3$/sec) than contiguous Paleozoic aquifers elsewhere, including 3 of the 10 largest springs in the county. Eight first-magnitude springs which occur in southern Missouri reflect a much larger median basin size than spring basins further south in the Springfield Plateau of Arkansas. Studies within the past 20 years that have evaluated regional aquifer systems (e.g. Regional Aquifer Systems Analysis [RASA] and National Water Quality Assessment [NAWQA]) have tended to lump the thick Ordovician and older stratigraphic interval of predominantly dolomite and sandstone into the Ozark aquifer.  Stratigraphically, this includes the interval from the base of the Chattanooga or equivalent units (Ozark confining unit) to the top of the Davis Formation (St. Francois confining unit), approximately 800 to 900 m thick.

Generalization is appropriate to a degree for some regional assessment of karst hydrogeology, but for many applied problems dealing with scales ranging from site-specific (hundreds to thousands of $m^2$) to basin ($km^2$ to hundreds of $km^2$) scale, this coarse regionalization is not appropriate. Emerging data indicate that within this thick interval there are strong, localized karst-flow systems that behave independently. Large-scale studies that do not differentiate the entire Ozark aquifer interval into distinct functional karst flow zones tend to oversimplify the hydrogeology, with a concomitant loss of information and understanding. This paper applies a more detailed scale methodology and evaluates the karst attributes of the full stratigraphic range and areal extent of flow distributions in the Ozark aquifer interval, and seeks to build a coherent, refined-scale conceptual model of the evolution of the karst hydrogeology of the Potosi Formation.

Data from numerous municipal wells indicate that the widespread, vuggy porosity, high permeability and the predominant lithology of the Potosi Formation resulted from dolomitization and geochemical alteration early during burial. Well yield and hydraulic testing of individual units likewise establish the Potosi as the most permeable and productive interval within the Ozark aquifer. The concentration of large springs along linear trends that coincide with geophysical anomalies suggests that tectonic reactivation of basement faults fractured overlying brittle carbonate cover into systematic orthogonal joint sets that facilitated vertical flow. Surface faults that appear to truncate the entire hydrostratigraphic section are consistent with this interpretation. The broad distribution of endangered organisms such as *Amblyopsis rosae* and *Cambarus aculabrum* and *Cambarus setosis* also supports this interpretation. Dye tracing over distances of several tens of kilometers, water-level mapping, interbasin stream piracy, borehole flow-meter testing, geochemistry provide additional constraints that explain most of the anomalous hydrogeologic behavior that cannot be accounted for with a single-layer Ozark aquifer model.

# Overview of The Nature Conservancy Ozark Karst Program

By Michael E. Slay, Ethan Inlander, and Cory Gallipeau
Ozark Highlands Office, The Nature Conservancy, 38 W Trenton Blvd., Suite 201, Fayetteville, AR 72701

**Abstract**

The Ozarks contain an underground wilderness of caves, springs and aquifers that over the millennia have formed in the carbonate bedrock of the region. Stretching from northern Arkansas, southern Missouri, and into eastern Oklahoma, subterranean habitats in this landscape harbor bats, salamanders, fish, crustaceans and other invertebrates; including at least 60 species found nowhere else on Earth. The porous and fractured nature of karst terrain makes it susceptible to pollution caused by incompatible land use, and often the animals that live in the caves are negatively impacted by human visitation and vandalism. Because of these threats, The Nature Conservancy created the Ozark Karst Program to work with agencies, partners, and landowners to protect these species and habitats. Since 1978, the Conservancy has acquired 20 caves, installed or repaired two dozen cave gates, and removed 230 tons of garbage near caves or sinkholes. Within the last 10 years, we began a thorough investigation of Arkansas' cave fauna discovering over 15 new species and additional populations of known species. Information about these species and habitats was used to reassess and update karst conservation priorities outlined for Arkansas in the 2003 Ozark Ecoregional Conservation Assessment Plan. We developed a Geographic Information System (GIS)-based index model that assesses site specific threat associated with 57 karst species using threat indicators derived from 25 geospatially available datasets. Threat was assessed for 297 karst habitats (caves, springs, or seeps). The results of the threat modeling can be used to evaluate conservation priorities at site, species, or community levels. We are currently exploring how reserve selection theory may allow us to combine these threat indices with biological information (for example; abundance, richness and endemism) to generate new karst conservation focal areas within Arkansas.

# Microbial Effects on Ozark Karst-Water Chemistry— Medicinal Implications

By John E. Svendsen
Springs of Arkansas, 16101 LaGrande Drive, Suite 101, Little Rock, Arkansas 72223

## Abstract

Many of the springs emerging from the karst regions of the Ozark Plateau have reputed medicinal properties. Promoters, proprietors and "restored visitors" claim these springs are endowed with unique healing properties and that each spring has particular health-giving attributes, yielding different cures and often requiring specific forms of usage. During the late 1800s and early 1900s a large number of resorts and health spas were developed at these healing springs to provide lodging and services to the many people who visited the springs with expectations of being cured through their use. Thousands upon thousands of testimonials suggest that many health-seekers did indeed find relief.

This presentation will examine the role microbial processes may have had in imparting distinct medicinal properties to the groundwater giving rise to these healing springs. It has long been recognized that microbial activity underlies many geochemical processes that influence groundwater chemistry. Microbes play a critical role in the subterranean environments by processing organic matter and facilitating the dissolution of rock and deposition of minerals. Without such biogenic influences, water quality in distinct environments would be significant different, many of the known minerals on Earth would never have formed, and the planet would lack the vast array of chemical diversity needed to sustain complex life forms. The presence of specific microbes and their biogeochemical processing within aquifers likely gives rise to unique water-quality attributes that contribute to the reputed medicinal virtues of these healing springs.

# KARST AQUIFER SYSTEMS
## An Integrated Approach to Recharge Area Delineation of Four Caves in Northern Arkansas and Northeastern Oklahoma

By Jonathan A. Gillip[1], Rheannon M. Hart[1], and Joel M. Galloway[2]
[1]U.S. Geological Survey Arkansas WSC, 401 Hardin Rd., Little Rock, AR 72211
[2]U.S. Geological Survey North Dakota WSC, 821 E. Interstate Ave., Bismark, ND, 58503

## Abstract

A study was conducted from 2004 to 2007 by the U.S. Geological Survey in cooperation with the U.S. Fish and Wildlife Service to assess the characteristics of the local recharge areas for four caves in northern Arkansas and northeastern Oklahoma that provide habitat for a number of unique organisms. An integrated approach was used to determine the hydrogeologic characteristics and the extent of the local recharge areas for Civil War Cave, Wasson's Mud Cave, Nesbitt Spring Cave, and January-Stansbury Cave. This approach incorporated methods of hydrology, structural geology, geomorphology, and geochemistry. Continuous water-level and water-temperature data were collected at each cave for various periods to determine recharge characteristics. Field investigations were conducted to determine surficial controls affecting the groundwater flow and connections of the groundwater system to land-surface processes in each study area. Qualitative groundwater tracing also was conducted at each cave to help define the local recharge areas. These independent methods of investigation provided multiple lines of evidence for effectively describing the flow patterns in complex hydrologic systems.

Civil War Cave is located near the city of Bentonville in Benton County, Arkansas. Civil War Cave is developed in and recharged through the Springfield Plateau aquifer. The daily mean discharge for the period of study was 0.59 cubic feet per second and ranged from 0.19 to 2.79 cubic feet per second. The local recharge area calculated for Civil War Cave was approximately 0.13 to 2.5 square miles using a water balance equation and approximately 1.8 to 3.8 square miles using a normalized base-flow method. Tracer tests indicate water entered Civil War Cave from across a major topographic divide located to the southwest.

Wasson's Mud Cave is located near the city of Springtown in Benton County, Arkansas. Wasson's Mud Cave is developed in the Springfield Plateau aquifer. No flow was observed although the water level did fluctuate. Because there was no detectable flow and tracer tests proved ineffective, the recharge area for Wasson's Mud Cave could not be calculated. The immediate topographic recharge area consists of less than 0.01 square miles around the opening of the cave. The location of the cave indicates that the cave is associated with the shallow groundwater-system (Springfield Plateau aquifer) within the Flint Creek Basin. Therefore, the entire basin potentially contributes recharge to the cave.

Nesbitt Spring Cave is located near the city of Mountain View in Stone County, Arkansas. Nesbitt Spring Cave is developed in the Ozark aquifer and is recharged through the Springfield Plateau aquifer. The daily mean discharge for the period of study was 4.5 cubic feet per second and ranged from 0.39 to 70.7 cubic feet per second. The calculated recharge area for Nesbitt Spring Cave using a water-balance equation ranged from 0.49 square mile to 4.0 square miles. Tracer tests generally showed a portion of water discharging from Nesbitt Spring Cave originates from outside the topographic drainage area.

January-Stansbury Cave is located near the town of Colcord in Delaware County, Oklahoma. January-Stansbury Cave is developed in and recharged through the Springfield Plateau aquifer. The daily mean discharge for the period of study was 1.0 cubic foot per second and ranged from 0.35 to 8.7 cubic feet per second. The calculated recharge area for January-Stansbury Cave using a water-balance equation ranged from approximately 0.04 to 0.83 square mile. Tracer tests generally showed water discharging from January-Stansbury Cave originates from within the topographic drainage area and from an area outside the topographic drainage area to the southwest.

# Groundwater Piracy in Semi-Arid Karst Terrains

By Ronald T. Green[1], F. Paul Bertetti[1], and Mariano Hernandez[2]

[1]Southwest Research Institute®, Geosciences and Engineering Division 6220 Culebra, San Antonio, TX 78238
[2]School of Natural Resources and the Environment, University of Arizona, Tucson, AZ 85721

## Abstract

Adequate characterization of aquifer extent and properties is needed to make informed decisions regarding water resource management. Discharge balancing performed as part of a water budget analysis provides an opportunity to determine the size and extent of groundwater basins. A comprehensive assessment of the hydrogeology of the Edwards-Trinity aquifer was undertaken to determine the extent of groundwater basins in the western portion of the Edwards-Trinity aquifer, a karst limestone aquifer located in central Texas. Recharge was calculated using baseflow estimates calculated from river discharge measurements. Effective recharge rates were calculated by averaging discharge-determined recharge over regions that are believed to have uniform infiltration based on similar precipitation, geology, soil type, and vegetation. The water-budget analysis corroborated the assessment made using a potentiometric-surface map that groundwater basin boundaries extend beyond their overlying surface watershed footprint, thereby support the premise that groundwater piracy in the southern portion of the Edwards-Trinity aquifer where the saturated aquifer is thickest.

## INTRODUCTION

Adequate characterization of aquifer extent and properties is needed to make informed decisions regarding water resource management. Groundwater divides must be identified to accurately designate separate pools or basins within regional-scale aquifers as part of this characterization. Identifying groundwater divides can be problematic, especially when sufficient field data are lacking. Groundwater divides are typically defined using potentiometric maps; however, even with a reasonably representative potentiometric map, the boundaries of groundwater basins, and groundwater divides in particular, are not easily determined. For occurrences where potentiometric maps are either not available or not sufficiently accurate, groundwater basin boundaries are commonly approximated by vertically extrapolating the overlying surface watershed boundaries. This approximation is not always valid. For those instances where groundwater basin boundaries do not coincide with surface watershed boundaries, water resource assessments can be misleading, if not wrong, unless both surface watershed and groundwater basin boundaries are accurately characterized. In the case of karst aquifers, groundwater basins extending into adjoining surface watershed basins is referred to as groundwater piracy (White and White, 2001; White, 2006).

Common methods used to identify aquifer boundaries and delineate karst aquifer basins include hydrogeological mapping, discharge balancing (i.e., water-budget analysis) and dye tracing (Schindel and others, 1997). This study uses discharge balancing to demonstrate groundwater piracy in an eight-county portion of the Edwards-Trinity aquifer in central Texas. The use of discharge balancing demonstrates how to identify groundwater divides within a regional aquifer for which existing hydrogeological data are lacking or relatively sparse. Discharge balancing is determined using measured discharge for gaining rivers, particularly at locations that are proximal to river headwaters. This information sheds insight on effective recharge, the size of the groundwater catchment area that discharges to the river, and ultimately, the area and approximate location of hydraulic boundaries which are interpreted to be groundwater divides.

## BACKGROUND

The Edwards-Trinity aquifer is a regional-scale karst limestone aquifer. The study area is centered over the headwaters for several river watersheds. Springs and seeps in the study area occur mostly where rivers and streams have incised into the aquifer allowing for discharge at locations where permeable media overlying a confining layer are exposed at the surface. There are occasional artesian springs associated with local faulting, but most artesian springs are located to the south of the study area in the Balcones Fault Zone (Brune, 1975). Springs offer an important opportunity to act as gauge points for the upstream basin to measure key components of the water budget (White and White, 2001; White, 2006). In the absence of spring discharge measurements, river gauges downstream from the springs can be used to estimate spring discharge. This approximation is valid if the distance from the spring to the river gauging station is not excessive and if the river does not lose significant water in the reach from the spring to the gauging station. The use of river gauging as a surrogate for spring discharge measurement is necessary in the western Edwards Plateau because discharge at local springs has not been measured on a regular basis (Green and Bertetti, 2010).

## SURFACE WATERSHED BASINS

The western Edwards Plateau is bisected by a surface-water divide between the surface watershed basins of the Colorado River and the Rio Grande (Figure 1). In general, the northeastern half of the western Edwards Plateau lies in the Colorado River surface watershed basin and the southwestern half lies in the Rio Grande surface watershed basin. The western Edwards Plateau is further divided into several small contributing surface watershed basins including the Rio Grande-Amistad, Lower Pecos, Devils, Middle Colorado-Concho, Middle Colorado-Llano, Nueces, and Rio Grande-Falcon (Figure 1).

The study area serves as the headwaters for numerous rivers, streams, and their tributaries. The most notable of these are the Devils, Nueces, Llano, San Saba, and Frio rivers. The upper reaches of these rivers typically flow intermittently. Perennial flow occurs downstream, with the exception of rivers and streams that lose to the subsurface where they flow across recharge zones for aquifers such as the Edwards. River gain/loss surveys compiled by Slade and others (2002) provide quantifiable evidence of where groundwater is discharged into rivers (gain) and where rivers discharge to the subsurface (loss). These studies indicate that virtually all rivers gain in the study area except for the lower San Saba River and various reaches of the Pecos River.

## GROUNDWATER BASINS

Groundwater basins in karst aquifers can only be unambiguously delineated using techniques such as dye tracing, cave surveying, water chemistry sampling and assessment, development of potentiometric maps, and the full use of local geology (White, 2006). In the absence of dye tracer, water chemistry, and cave survey information, an existing potentiometric map of the Edwards-Trinity aquifer is used to characterize groundwater basins in the study area. In this case, the groundwater basins for the study area are approximated in the same way that a surface water basin is determined, except that the potentiometric surface is used to locate groundwater catchment divides.

The potentiometric surface map prepared by Kuniansky (1990) and reproduced by Barker and Ardis (1996) is used to delineate the groundwater basin in the western Edwards Plateau (Figure 2). The surface-water divide between the Colorado River and the Rio Grande watersheds is overlain on the potentiometric surface to compare the extent of the groundwater basins to the surface-water basins. Although the resolution of the map is limited, areas in Upton, Reagan, Sutton, Schleicher, and Edwards counties exhibit groundwater flow toward the Rio Grande from locations within the Colorado River surface watershed basin.

## RIVER DISCHARGE ANALYSIS

River discharge was analyzed as a means to corroborate groundwater basin delineation based on the 1990 potentiometric surface map. Water budgets for individual surface watersheds within the study area were analyzed as a means to balance discharge. The water budget for each of these sub-areas was assessed using available hydrogoelogical information. In the absence of spring discharge measurements, river gauging

124

measurements were used to estimate discharge from the groundwater catchment areas. Fortunately, rivers in the study area have long records of discharge measurements.

River discharge has two principal flow components, baseflow and surface runoff. Baseflow is considered to be the groundwater contribution to stream flow and is interpreted to equal recharge (Arnold and others, 1995; White and White, 2001; White, 2006). Baseflow recession is the rate at which the stream flow diminishes in the absence of recharge. The discharged volume is equated to the amount of recharge to the aquifer that discharges to the river.

Decomposition of stream hydrographs provides estimates of the baseflow and storm flow components to stream discharge. The fraction of river discharge attributed to baseflow was calculated for each river gauging station analyzed in this study. An automated baseflow separation and recession analysis tool, BASEFLOW, was used to estimate the amount of stream flow attributed to baseflow for each sub-area (Arnold and others, 1995; Arnold and Allen, 1999). The automated procedure predicts baseflow recession from the point on the hydrograph where it is assumed that all surface flow has ceased. For those instances when there were gaps in the gauging station time series, recession was analyzed for each individual time series segment. A single value for baseflow fraction for each river watershed was calculated by averaging the baseflow fraction for each time segment weighted by the number of days in the time series subset (Green and Bertetti, 2010). Baseflow fractions of river discharge for the watersheds in the study area are summarized in Table 1.

River gauging station baseflow analyses were used to estimate recharge to their respective watersheds. For example, an average discharge of 125.5 cfs (90,860 acre-feet/year) was measured at the Frio River at Concan gauging station from 1923 to 2010. The baseflow fraction for the Frio River was calculated at 0.75, thus the baseflow component to flow is 94 cfs. The surface drainage area upstream of Concan is measured by the U.S. Geological Survey to be 383 mi$^2$ (245,120 acres). This equates to a recharge rate of 3.36

inch/year if uniformly averaged over the 383 mi$^2$ drainage area. Similar analyses were performed for all watersheds in the western Edwards Plateau that had river gauging stations with sufficient discharge measurements. Recharge calculations for these sub-areas are summarized in Table 1 and plotted at their drainage area locations in Figure 3.

There is a notable contrast in calculated recharge rates between the headwaters of the Nueces and Frio river watersheds and the Llano River watershed. The recharge rates for the Llano River and its north-flowing tributaries are significantly less than those for the south-flowing Nueces and Frio river watersheds and the east-flowing Guadalupe and Medina river watersheds. Inspection of precipitation measured for the western Edwards Plateau (Figure 4) suggests that this significant difference in calculated recharge rates between the north and south cannot be attributed to variations in precipitation observed across the eastern portion of the study area because isohyets trend in a north-south direction.

Significant differences in calculated recharge rates between northern and southern portions of the study area are interpreted to indicate that groundwater catchment areas for the Nueces, Frio, Medina, and Guadalupe rivers extend farther north than the boundaries of the overlying surface watersheds. This feature is indicative of groundwater piracy. The interpretation is consistent with the combined surface watershed and groundwater flow map that indicates that groundwater catchment boundaries do not coincide with surface watershed boundaries (Figure 2). Groundwater piracy from watersheds north of the Rio Grande/Colorado River watershed divide results in lower calculated recharge rates for the watersheds of the Llano River and its tributaries, and higher calculated recharge rates for the watersheds to the south and east (i.e., Nueces, Frio, Medina, and Guadalupe river watersheds). Actual recharge rates are thus greater than the low values to the north and east of the Rio Grande/Colorado River watershed divide and less than the high values to the south and west.

Groundwater piracy is more likely to occur where the affected aquifer is thick and deep. The Edwards-Trinity aquifer is significantly thicker in Sutton, Edwards, Real, Val Verde, and southern Kimble and Schleicher counties relative to the Edwards-Trinity aquifer to the north (Figure 5). Examples of areas from which groundwater appears to be pirated are basins associated with the North Llano River at Junction, the Llano River at Junction, and the West Nueces River at Brackettville. Areas which gain from pirated groundwater will have greater discharge into rivers and streams and higher rates of calculated recharge. Examples include basins associated with the Frio River at Concan, the Nueces River at Laguna, Johnson Creek at Ingram, and the Medina River at Bandera.

The extent to which groundwater is pirated from one surface watershed to another is difficult to measure, but can be estimated using discharge analyses. Factors that influence recharge (i.e., soil and vegetation type, topography) are relatively uniform over the areas where groundwater piracy is thought to occur. The only variable is precipitation. Using this reasoning, recharge rates were averaged along lines of equal precipitation (isohyetals) to estimate a uniform rate of recharge for areas of uniform precipitation. Because the isohyetals are essentially north trending in study area, uniform recharge rates should also be north trending.

Recharge was averaged between the gauging stations on the Frio River at Concan and on the South Llano River at Junction and between the gauging stations on the Nueces River at Laguna and on the North Llano River at Junction (Figure 3). The averages were weight averaged relative to area to conserve mass. In this manner, an average recharge rate of 1.39 inch/year was assigned to the combined watersheds of the upper Frio River and the South Llano River. The combined basins have a total area of 1,468,000 acres and total recharge of 170,500 acre-ft. In this calculation, it is assumed that the area of the combined groundwater basins is the same area as the area of the combined surface watershed. The only differences are the locations of the mutual boundaries of the two systems.

Similarly, an average recharge rate of 1.21 inch/year was assigned to the combined watersheds of the upper Nueces River and the North Llano River. These combined basins have a total area of 1,057,000 acres and total recharge of 107,000 acre-ft. It is also assumed that the

external boundary of the combined watersheds for the upper Nueces River and the North Llano River is the same as the boundary of the combined groundwater basins.

Final recharge rates were calculated for the eight counties in the study by taking an areal average of all recharge rates calculated within that county. Recharge rates for each of the eight counties in the study area are plotted in Figure 6. The recharge rates are relatively uniform within areas of uniform precipitation, with the exception of Menard County. The reason for this discrepancy is that the San Saba River, which runs through the middle of Menard County, is a losing river. Therefore, the basic premise for the river discharge computation method used elsewhere in the study area is violated. Instead of a recharge rate of 0.50 inches/year as indicated in Figure 6, the actual recharge rate should be closer to 1.30 inch/year if the assumption of uniform recharge over areas of uniform precipitation is valid.

The extent to which groundwater basins associated with the Frio and Nueces rivers extend north beyond the surface-water divide between the Rio Grande and the Colorado River is interpreted to be influenced by the saturated thickness of the Edwards-Trinity aquifer. The surface-water divide is overlain on a contour map of the saturated thickness of the Edwards-Trinity aquifer in Figure 5. As illustrated, the greatest potential for groundwater piracy in the study area is northern Edwards, eastern Sutton, most of Kimble, eastern Schleicher, and Kerr counties because the Edwards-Trinity aquifer is locally thicker than in northerly areas. The limit to which groundwater piracy extends to the north can only be definitively determined with field verification such as dye tracer results and water chemistry analysis. In the absence of field verification, the area of potential groundwater piracy is estimated by the 200-ft saturated thickness contour in the Edwards-Trinity aquifer in Figure 5.

## SUMMARY

A hydrogeological assessment was undertaken to approximate the boundaries of groundwater basins in the western portion of the Edwards-Trinity aquifer, a karst limestone aquifer located in central Texas. Eight counties were included in the study; Crockett, Edwards,

Kimble, Menard, Real, Schleicher, Sutton, and Val Verde. The assessment relied on water-budget analyses of hydrologically distinct sub-areas in the study area. Recharge was calculated using baseflow estimates calculated from river discharge measurements. Effective recharge rates were calculated by averaging discharge-determined recharge over regions likely to have uniform recharge owing to similar geology, soil type, vegetation, and precipitation. This average was weighted by area to conserve mass. The final step in the process was to average effective recharge rates over each of the eight counties in the study area. These calculations provide corroborating evidence that groundwater basins extend beyond the extent of their overlying surface watersheds, thereby supporting the premise of groundwater piracy in the southern portion of the Edwards-Trinity aquifer where the saturated aquifer is thickest.

## ACKNOWLEDGMENTS

The authors thank Sutton County Groundwater Conservation District for permission to use the information in this paper. Reviews by Gary Walter, Stuart Stothoff, and Van Brahana improved the quality of the paper.

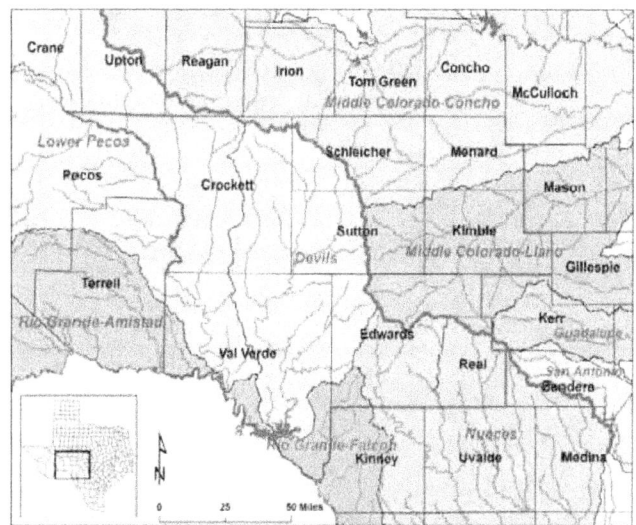

Figure 1. Map of the counties (black labels) and nine watershed basins (blue labels) of the major rivers of the study area (outlined in red). The blue line denotes the watershed divide between the Rio Grande on the southwest and the Colorado River on the northeast.

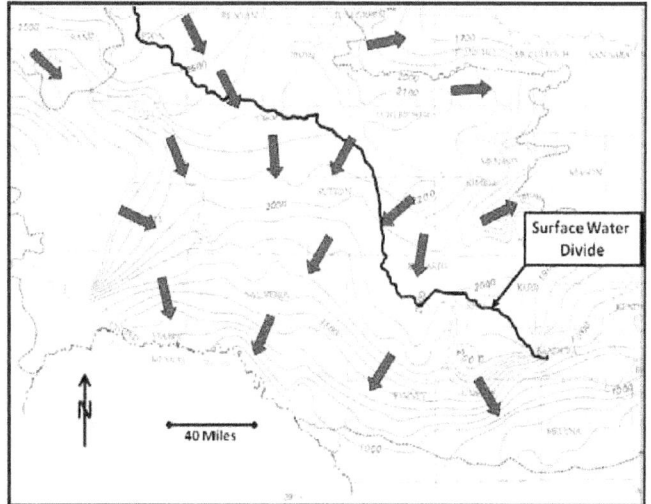

Figure 2. Map of the Rio Grande-Colorado River surface-water divide overlying the groundwater potentiometric surface. Blue arrows are added to denote the direction of groundwater flow assuming isotropic flow. Base map is from Barker and Ardis (1996).

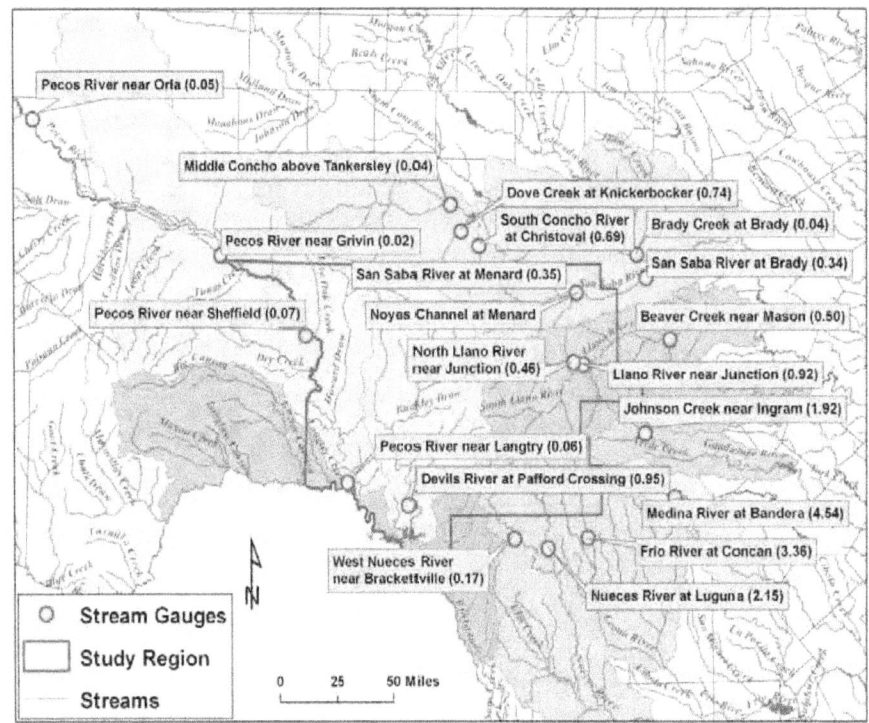

Figure 3. Locations of the U.S. Geological Survey gauging stations in and near the study area. Numbers in parentheses denote recharge rates in inches/year calculated using river discharge rates and corrected for base flow.

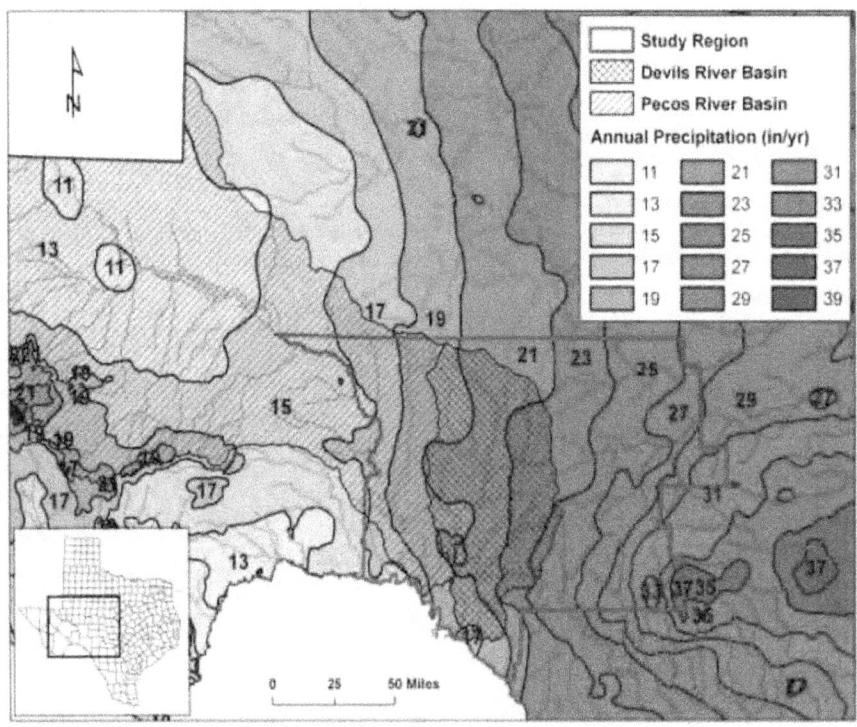

Figure 4. Map showing average annual precipitation (inch/year) for the study area (outlined in blue).

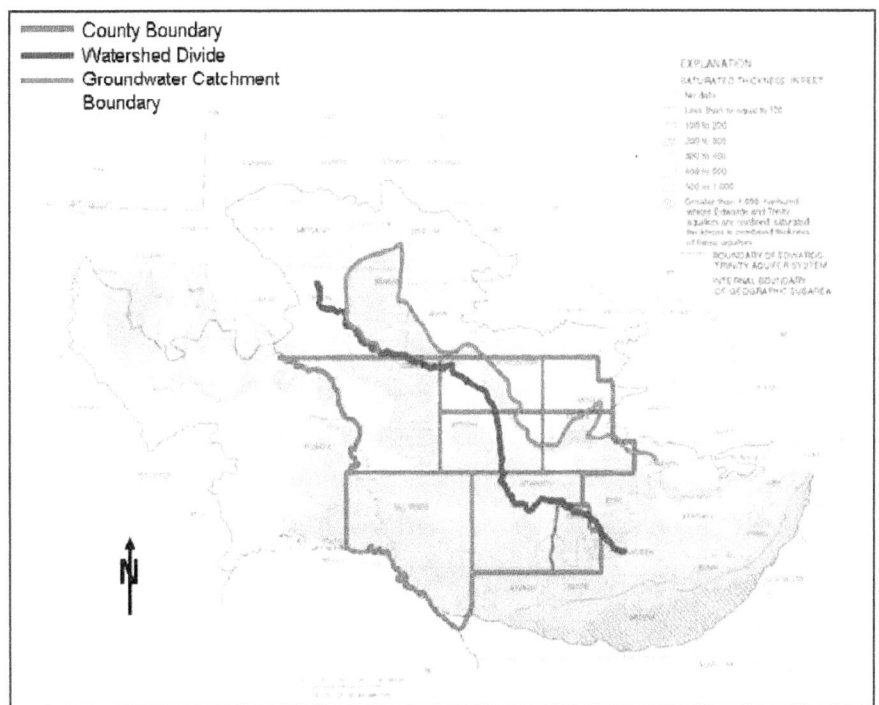

Figure 5. Contour map of the saturated thickness of the Edwards-Trinity aquifer (Kuniansky and Holligan, 1994) with the surface-water divide separating the Rio Grande watershed from the Colorado River watershed (blue line) and the extent of groundwater piracy estimated using the 200-ft saturated thickness contour of the Edwards-Trinity aquifer (green line).

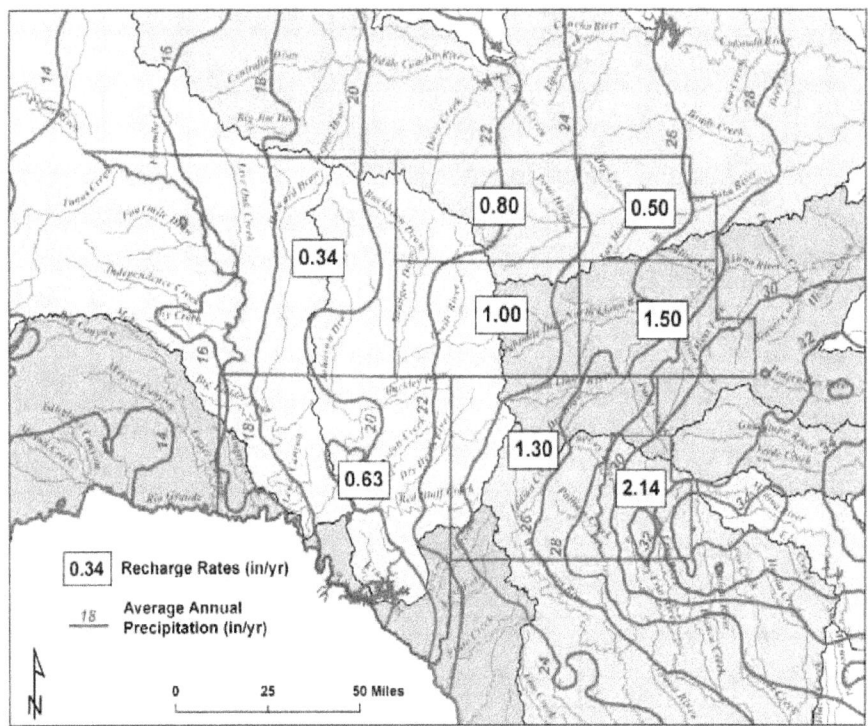

Figure 6. Recharge rates (inch/year) by county calculated for river watershed basins. Thick blue lines denote contours for average annual precipitation (inch/year).

129

Table 1. List of U.S. Geological Survey river gauging stations used in the water budget analysis. Adjusted discharge rates have been corrected for baseflow. All stations are in Texas.

| Station Number | River | Drainage Area (acre) | Gauging Start Date (month/yr) | Baseflow Fraction | Gross Discharge (inch/yr) | Adjusted Discharge (inch/yr) |
|---|---|---|---|---|---|---|
| 08190000 | Nueces River at Laguna | 471,680 | 10/1923 | 0.71 | 3.03 | 2.15 |
| 08190500 | West Nueces River near Brackettville | 444,160 | 9/1939 | 0.25 | 0.67 | 0.17 |
| 08195000 | Frio River at Concan | 284,960 | 10/1923 | 0.75 | 4.45 | 3.36 |
| 08178880 | Medina River at Bandera | 209,920 | 10/1982 | 0.68 | 6.77 | 4.54 |
| 08166000 | Johnson Creek at Ingram | 72,960 | 10/1941 | 0.62 | 3.10 | 1.92 |
| 08145000 | Brady Creek at Brady | 376,320 | 6/1939 | 0.29 | 0.15 | 0.04 |
| 08144600 | San Saba River near Brady | 1,045,120 | 7/1979 | 0.58 | 0.59 | 0.34 |
| 08144500 | San Saba River at Menard | 721,920 | 10/1915 | 0.48 | 0.72 | 0.35 |
| 08150800 | Beaver Creek near Mason | 137,600 | 8/1963 | 0.42 | 1.20 | 0.50 |
| 08148500 | North Llano River at Junction | 584,960 | 10/1915 | 0.46 | 1.00 | 0.46 |
| 08150000 | Llano River at Junction | 1,183,360 | 10/1915 | 0.64 | 1.45 | 0.92 |
| 08128000 | South Concho River at Christoval | 226,560 | 3/1930 | 0.60 | 1.15 | 0.69 |
| 08130500 | Dove Creek at Knickerbocker | 139,520 | 10/1960 | 0.76 | 0.98 | 0.74 |
| 08128400 | Middle Concho River at Tankersley | 714,240 | 4/1961 | 0.21 | 0.17 | 0.04 |
| 08449400 | Devils River at Pafford Crossing | 2,535,040 | 1/1960 | 0.76 | 1.25 | 0.95 |
| 08412500 | Pecos River near Orla | 13,586,560 | 10/1937 | 0.66 | 0.08 | 0.05 |
| 08446500 | Pecos River near Girvin | 18,918,400 | 9/1939 | 0.77 | 0.03 | 0.02 |
| 08447000 | Pecos River near Sheffield | 20,384,000 | 10/1975 | 0.79 | 0.09 | 0.07 |
| 08447410 | Pecos River near Langtry | 28,352,000 | 1/1/1967 | 0.74 | 0.08 | 0.06 |

Table 2. Recharge rates calculated by county for the Edwards-Trinity aquifer

| | Crockett | Edwards | Kimble | Menard | Real | Schleicher | Sutton | Val Verde |
|---|---|---|---|---|---|---|---|---|
| **Area (acres)** | 1,796,480 | 1,356,800 | 800,640 | 576,640 | 448,000 | 839,040 | 929,920 | 2,068,480 |
| **Recharge rate (in/yr)** | 0.25 | 1.30 | 1.50 | 0.50 | 2.14 | 0.80 | 1.00 | 0.63 |
| **Recharge (acre-ft/yr)** | 37,427 | 146,987 | 100,080 | 24,027 | 79,893 | 55,936 | 77,493 | 108,595 |

## REFERENCES

Arnold, J.G., P.M. Allen, R. Muttiah, and G. Bernhardt. 1995. Automated base flow separation and recession analysis techniques. Ground Water 33(6): 1010-1018.

Arnold, J.G. and P.M. Allen. 1999. Automated methods for estimating baseflow and ground water recharge from streamflow records. Journal of the American Water Resources Association 35(2): 411-424.

Barker, R.A., and A.F. Ardis. 1996. Hydrogeologic framework of the Edwards-Trinity aquifer system, West-Central Texas: U.S. Geological Survey Professional Paper 1421-B. 61 p. with plates.

Brune, G. 1975. Major and Historical Springs of Texas. Texas. Texas Water Development Board. Report 189. 94 p.

Green, R.T., and F. P. Bertetti. 2010. Investigating the Water Resources of the Western Edwards-Trinity Aquifer. Contract Report Prepared for the Sutton County Groundwater Conservation District. 79 p.

Kuniansky, E.L. 1990. Potentiometric surface of the Edwards-Trinity aquifer system and contiguous hydraulically connected units, west-central Texas, winter, 1974-75, Water-Resources Investigations Report; 89-4208.

Kuniansky, E.L. and K.Q. Holligan. 1994. Simulations of flow in the Edwards-Trinity Aquifer system and contiguous hydraulically connected units, west-central Texas. U.S. Geological Survey. Water-Resources Investigations Report 93-4039. 40 p.

Schindel, G. M., Quinlan, J. F., Davies, G. and Ray, J. A., 1997, Guidelines for wellhead and springhead protection area delineation in carbonate rocks, U.S. Environmental Protection Agency Tech. Rep. EPA/904/B-97/003.

Slade, R.M., J.T. Bentley, and D. Michaud. 2002. Results of streamflow gain-loss studies in Texas, with emphasis on gains from and losses to major and minor aquifers. U.S. Geological Survey Open-File Report 02–068. 136 p.

White, W.B. 2006. Fifty years of karst hydrology and hydrogeology: 1953–2003. In: Harmon R.S., Wicks C. (eds) Perspectives on karst geomorphology, hydrology, and geochemistry: a tribute volume to Derek C. Ford and William B. White. Geol Soc Am Spec Pap 404:139–152.

White, W.B. and E.L. White. 2001. Conduit fragmentation, cave patterns, and localization of karst ground water basins: The Appalachians as a test case: Theoretical and Applied Karstology, v. 13-14, p. 9-23.

# Ten Relevant Karst Hydrogeologic Insights Gained from 15 Years of *In Situ* Field Studies at the Savoy Experimental Watershed

By Van Brahana

University of Arkansas, Department of Geosciences, Fayetteville, AR 72701

## Abstract

The Savoy Experimental Watershed (SEW) was established nearly 15 years ago, and during that time, has been the focus of applied research, education, and outreach dealing with the practical problems of humans living on a mantled-karst landscape. This report is a synthesis of ten important insights that have been gained during the life of this facility. Although the selection of ten is arbitrary, and much additional information has been learned, these encompass illustrate the wide range of understanding that can be gained from a long-term research site. Each is documented with specific examples and references that will hopefully provide understanding applicable far beyond SEW. These insights are: 1) the purity of the lithology controls the nature of the karst that develops; 2) structural control of brittle-carbonate units is a major determinant of sub-basin boundaries; 3) divergent flowpaths in karst tracing experiments are common; 4) groundwater flow directions commonly are a function of groundwater levels, changing as water levels in the aquifer change; 5) groundwater quality in the shallow aquifers is dominated by mixing; 6) evapotransporative losses from the epikarst are by far the most dominant of the hydrologic budget outflows; 7) flow zones near to one another are frequently hydrologically isolated from each other; 8) the environments and ecosystems of discrete groundwater flow zones exhibit surprising diversity and distribution of organisms; 9) transport of suspended particles and pathogens is a major water-quality concern driven by storm pulses; and 10) the actual complexity of seemingly simple karst systems makes prediction of hydrogeologic response outside the range of conditions for which conceptual models are developed a risky task.

## INTRODUCTION

The Savoy Experimental Watershed (SEW) was established nearly 15 years ago, and during that time, has been the focus of applied research, education, and outreach dealing with the practical problems of humans living on a mantled-karst landscape. Twenty theses and dissertations (table 1) and a comparable number of technical papers (selected references) have been completed describing the karst hydrogeology, groundwater transport phenomena, geochemical processing of nutrients, hydrologic budgets, animal production impacts on karst waters, and development of tools and techniques appropriate to karst studies in general. Numerous classes from the Hydrogeology Program at the University of Arkansas have supplemented the research infrastructure at SEW, and the site currently has more than 100 permanent installations including weirs, flumes, wells, piezometers, lysimeters, rain gages, geophysical arrays, and a weather station.

This report is a synthesis of the major insights that have been gained during the life of this facility, and although the selection of ten is arbitrary, each is documented with specific examples and references providing understanding that will be applicable far beyond SEW. The objective of this paper is to use insights gained in these years of targeted research to characterize the nature of mantled karst. The scope of this report is limited to recent studies at the site and to the mantled karst of the southern Ozarks.

The karst of the SEW is not immediately apparent to most people. The setting is a dissected plateau with steep, dry valleys and few sinkholes. In most of the area, regolith covers the bedrock, leaving a thin, rocky soil that masks the carbonate bedrock beneath. SEW is typical of the Springfield Plateau province of the Ozark highlands. Most of the site (>70%) is covered in second- and third-growth forests, and the remaining land is in pasture. The karst groundwater system at SEW is underdrained by

Table 1. Unpublished theses and dissertations focusing on the Savoy Experimental Watershed.

[Full reference annotation is provided in the list of selected references; M.S., Masters of Science; Ph.D., Doctor of Philosophy; GEOL, Geology; CVEG, Civil Engineering; SOIL, Soils; GEOG, Geography; CENG, Chemical Engineering; ENDY, Environmental Dynamics; BIOL, Biology]

| Author | Date | Degree | Program | Focus of Study |
|---|---|---|---|---|
| Al Qinna | 2004 | Ph.D. | SOILS | Soil water and solute transport in karst |
| Al Rashidy | 1999 | M.S. | GEOL | Hydrogeologic characterization of basin 1; karst |
| Curtis | 2000 | Ph.D. | CVEG | Advective transport model encorporating lineations |
| Dixon | 2001 | Ph.D. | SOILS | Neuro-fuzzy modeling to predict groundwater vulnerability |
| Hamilton | 2001 | M.S. | GEOL | Survival of *E.coli* in stream and spring sediments |
| Hobza | 2005 | M.S. | GEOL | Swine lagoon impact on karst water quality |
| Laincz | 2011 | Ph.D. | ENDY | Nitrogen transport and cycling in karst interflow zone |
| Leh | 2008 | M.S. | BAEG | Rainfall-runoff mechanisms in pasture-dominated watershed |
| Little | 2000 | M.S. | GEOL | Dominant processes; flow and quality; basin 2 |
| Parse | 1995 | M.S. | GEOG | Geomorphic assessment of role of regolith in karst |
| Pennington | 2010 | M.S. | GEOL | Drainage basin analysis using hydrograph decomposition |
| Phelan | 1999 | M.S. | GEOG | GIS and 3-D visualization; static modeling; mantled karst |
| Stanton | 1993 | M.S. | GEOL | Processes and controls affecting Boone-St. Joe aquifer |
| Ting | 2002 | M.S. | CENG | Europium-tagged *E. coli* as karst tracer |
| Ting | 2005 | Ph.D. | CENG | Bacterial transport and storage using lanthanide tracers |
| Unger | 2004 | M.S. | GEOL | Structural controls; boundaries; flow modeling |
| Wagner | 2007 | M.S. | GEOL | Assessment of waste-storage effectiveness using isotopes |
| Whitsett | 2002 | M.S. | GEOL | Sediment and bacterial tracing in mantled karst |
| Winston | 2006 | M.S. | BIOL | Nitrogen sources and processing in aquatic ecosystems |
| Woodstrom | 1999 | M.S. | GEOG | Land use effects on spring water quality |

chert-rich carbonate-rock aquifers that have been selectively dissolved to form an open network of caves, enlarged fractures, bedding planes, conduits, sinking streams, and springs. Flow in these aquifers is typically rapid, flow directions are difficult to predict, interaction between surface and groundwater is typically extensive, and processes of contaminant attenuation that characterize many other groundwater settings are typically absent (White, 1988; Ford and Williams, 2007; Palmer, 2007).

## DISCUSSION

### Insight 1—Lithologic Purity of Carbonates Controls Developing Karst

Field reconnaissance by numerous researchers (Stanton, 1993; Fanning, 1994; Brahana, 1995; Al Rashidy, 1999; Little, 2000) at SEW and throughout the Springfield Plateau indicates the best cave development occurs in rocks that are fairly pure carbonates. The St. Joe, Pitkin, and upper Boone host some of the most karstified rocks in the Ozarks of northwest Arkansas, a fact reflected in not only in cavern development, but in bedding-plane anastomoses, enlarged

joints, in driller's reports, and in wireline geophysical logs. The bulk of the middle and lower Boone Formation, on the other hand, typically has as much as 70% chert and insolubles. Whereas caves in the Boone Formation are common, their size and form typically are small and limited, "pocket caves" as designated by Fanning (1994). The chert and clay plug and occlude nascent porosity and permeability in Boone carbonate. After floods and storms recharge the shallow karst aquifers, flow recedes, velocities decrease, and the suspended sediment load is deposited. Zones previously most permeable become flow restrictions and dams. The Illinois River graben that forms the western boundary of basin 1 at SEW (figure 1) provides an example. This fault is identified by displacement of the St. Joe-Chattanooga contact, an escarpment and the appearance of springs along its base at high flow, and a zone of low permeability. Observations indicate that the fault serves as a groundwater dam localizing the occurrence of Langle and Copperhead Springs. The impurities in the Boone source the weathering products of the regolith. This regolith is responsible for slow release of recharge that sustains epikarst springs and the low flow of base-level springs, and is the reason that typical karst landforms such as dolines, karst windows, ponors, uvalas, and estavelles are not visible here.

Continuous chert layers in the lower Boone Formation serve as confining layers, perching recharge from above (figure 2), and confining phreatic flow moving upward in joints within the underlying St. Joe Formation. The major role played by the chert layers is well-displayed at the two major drains from basin 1. Both Langle and Copperhead Springs resurge from small caves at the base of the lowermost continuous chert layer in the Boone Formation (Al Rashidy, 1999). Downcutting of the Illinois River has allowed vadose entrenchment of these two base-level spring systems.

## Insight 2—Structural Deformation of Brittle Carbonate Rocks Defines Flow Boundaries and Flow Routes

Structural geology and tectonic setting are well-known controls on groundwater in many karst areas (Brahana et al., 1988; Palmer, 2007) Nearly flat-lying formations of the Springfield Plateau in the southern Ozarks are thought to have been significantly impacted by reactivated basement faulting associated with the Ouachita orogeny. Although SEW lies more than 150 km north of the Ouachita Mountains, the evidence that basement faults were active tectonic forces in early Mississippian time is reflected in exposures of olistoliths within 5 km of SEW (Chandler, 2001). Utilizing borehole geophysical logs, field reconnaissance, and surveying, Unger (2004) delineated the top of the Chattanooga Shale.

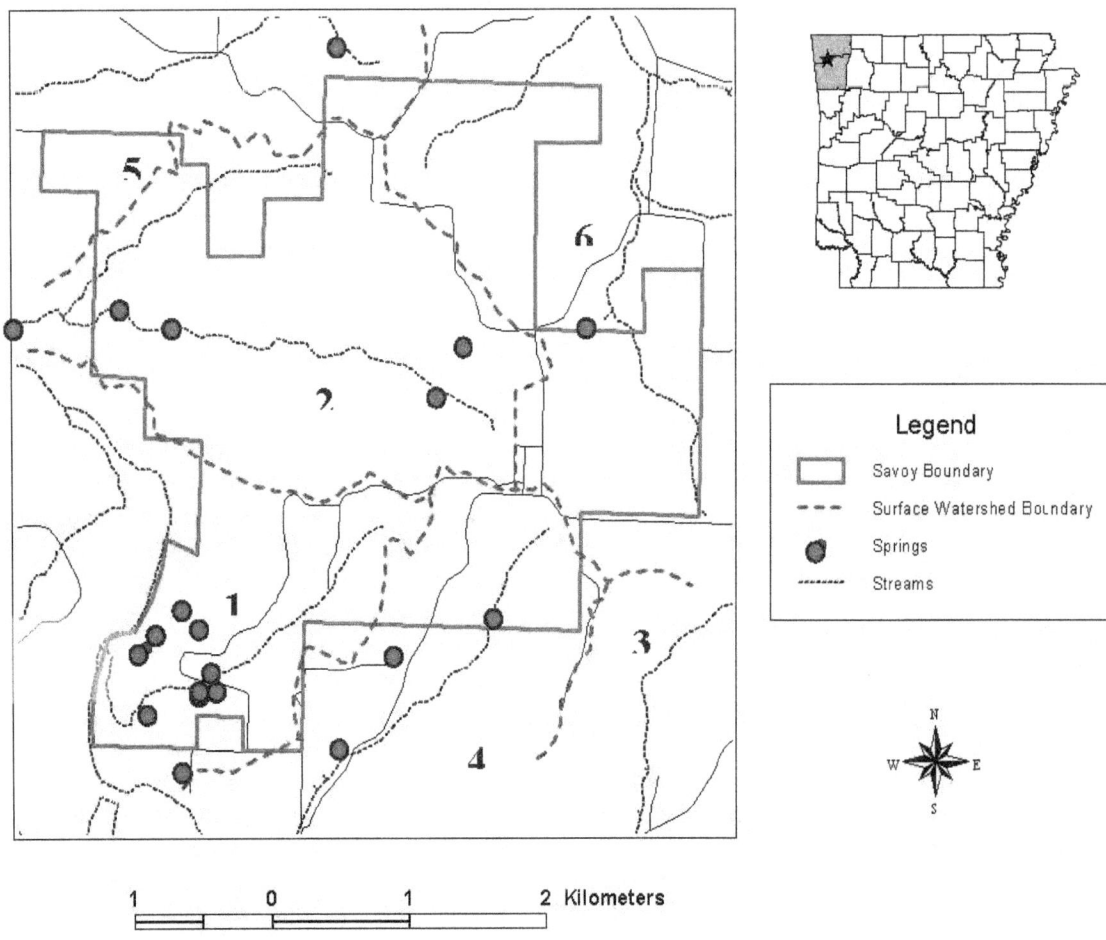

Figure 1. Boundary of the Savoy Experimental Watershed, showing delineation of surface watersheds numbered 1-6, major springs within or draining the site, and major stream valleys. These alluvial valleys only have flowing water after intense precipitation; they are normally dry.

Figure 2. Continuous chert layers perch flow and facilitate dissolution of the soluble limestone directly above the chert at SEW.

This provided an estimate of dip of the lower confining unit, which, like a tilted, impermeable block, controls flow directions toward springs (Bolyard, 2007). Tracing experiments coupled with wireline geophysical logs documented the tilted-block model which has allowed more consistent delineation of spring basins (Brahana, 1997; Ting, 2002; Ting, 2005). Stanton (1993) likewise showed the strong structural control of the SEW and the region with a transect of wireline geophysical logs across Clear Creek Valley, about 5 km east of SEW. These logs identified a previously unmapped graben with a downward displacement of about 30 meters in the valley.

Cavers in the region recognize that vertical shafts and pits are much more dominant in the area of the Buffalo National River (Tennyson et al., 2008), where the St. Joe Formation occurs about 70 meters higher in that area than it does at SEW. This is consistent with uplift along reactivated basement faults, delineated from geologic mapping and from gravity studies. The movement on the basement faults is thought to have created strongly orthogonal joint sets in the overlying, brittle carbonate rocks such as the St. Joe Formation (figure 3). Cave passages in this part of the Ozarks are strongly joint-controlled. Sinking streams such as Spanker and Spavinaw Creeks in northwest Arkansas are typically dry downstream from the intersection of basement fault with the

streambed. The implications of this strong structural control include the need for careful design of water-quality sampling tasks in order to gather representative data, which in most cases means the springs.

## Insight 3—Divergent Pathways Are Typical in Karst Tracing Experiments

Surface streams in the Ozarks typically are dendritic, yet dye tracing in karst aquifers commonly indicates dispersive flow—with one input point, and multiple recovery points (Aley, 2002). Of the more than 25 tracing experiments conducted at SEW, it has been uncommon to have single recovery points.

Figure 3. Orthogonal joints in brittle carbonate rocks serve as pathways for shallow groundwater flow. This site in the overflow channel for Crystal Lake, about 25 km north of SEW, developed as a subsoil karst that was subsequently removed by surface-water erosion diverted into the overflow channel.

The apparent explanation for this is based on multiple lines of evidence. This interconnected, multi-transmissive suite of flow routes (fast-flow along joints, bedding planes, and conduits) typically operates under confinement in the phreatic zone. The dynamic nature of point-source recharge inputs, the turbulent mixing along fast-flow pathways, and the remarkably large transmissivities of the interconnected voids establishes aquifer flow that can mimic a system of multiple and integrated estavelles, pipes, and parallel plates. High permeability allows flow to follow minor head variations, with the result that if voids are connected, the tracer will follow the path of least resistance and disperse. In groundwater systems, unlike unconfined flow in a stream channel at land surface, three-dimensional confinement along the flowpath creates the necessary head variations. Depending on the outlet architecture of the recovery points and the hydrologic conditions at the time of the test, dispersive flow is the norm rather than the exception.

Figure 4. Tracer studies at SEW include fluorescent dyes (such as Rhodamine WT shown here), conservative halogen solutes, lanthanum-series tagged clays, europium-tagged *E. coli*, and thermal pulses.

## Insight 4—Karst Groundwater Flow Paths Are Commonly Related to Water Levels—They Change with Stage and Their Resurgences May Be Covered

Dye tracing and continuous flow monitoring at weirs under varying hydrologic conditions has shown that spring-basin size is related to groundwater levels in basin 1. Langle and Copperhead Springs are known to be outlets from the same base level spring system; Langle is the underflow spring, with a slightly lower outlet elevation (~ 2 cm lower based on total-station surveying). Copperhead is the overflow spring, with greater permeability at higher elevations. For both springs, fast-flow transmissivity is maximum along a set of systematic orthogonal joints in the St. Joe Formation. Langle Spring has higher base flow owing to its lower resurgence point, but Copperhead Spring has greater flood flow because of wider openings at higher elevations (figure 5). Also of relevance is the large number of dye traces undertaken that have not produced positive results with input locations close to these major resurgences. We now believe that buried spring outlets resurge into the Illinois River alluvium, and that our input points were isolated from the main flow paths by continuous chert layers. Ongoing tracer testing is evaluating this hypothesis.

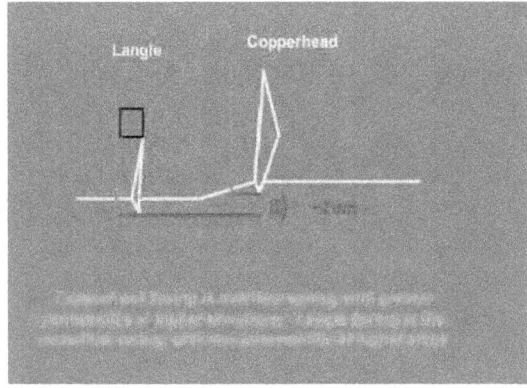

Figure 5. Conceptual model of the Langle-Copperhead Spring system reflecting differences in flow between springs under different water levels.

## Insight 5—Groundwater Quality in Shallow Karst Aquifers Is Dominated by Mixing

Geochemical processes documented in the shallow karst flow systems of SEW include dissolution, mixing, redox reactions, ion exchange, denitrification, volatilization, and precipitation. Of these, the advective transport and fast flow zones of the aquifer favor mixing as the most dominant of the processes observed, although in specific slow-flow zones any of the others may be important as well (Laincz, 2011; Winston, 2006).

Nutrient (nitrate) processing was characterized using stable isotopes, dissolved organic carbon concentration, and bioavailability, and concentration of reactive (nitrate) versus conservative (chloride) species (Laincz, 2011). Additionally, concentration and isotopic composition of $NO_3^-$ was used to determine the extent of denitrification and immobilization of nitrate in the interflow zone. Study results show that as much as 30% of nitrate moving through the interflow zone can be microbially processed within this zone, and the level of processing is highly dependent upon flow-path and hydrologic conditions. Bioavailability of DOC in the interflow is increased relative to the focused-flow zone under high-flow conditions. Nitrogen and oxygen stable isotope data for nitrate suggest denitrification is occurring in the interflow zone. This appears to be a potentially important zone for nitrate attenuation in karst settings, and is an important finding for mantled karst areas where animal production is a major land use.

## Insight 6—Of the Hydrologic-Budget Parameters of Epikarst Springs, Evapotranspiration Losses Are Largest

The measurement of interflow through three springs in basin 1 (Tree, Woodpecker, and Red Dog) provides insight into a component of the karst hydrologic budget that is seldom quantified (Brahana et al., 2005; Leh, 2008; Laincz, 2011). From 15 through 20 July 2005, representing the active plantgrowth period, and from 12 through 17 December 2005, representing the plant dormant period, very precise measurements of

discharge were made in early morning and mid-afternoon. Ten repetitions were averaged for each measurement, and data for diurnal and seasonal intervals were averaged (table 2).

Results indicate that 1) diurnal fluctuation of these epikarst springs is notable during the growing season and minimal during the dormant season when deciduous trees have lost their leaves and transpiration is minimal. This loss of water from the shallow ground-water system during the growing season is interpreted to be evapotranspiration from the groundwater system, and ranges from 5 to 17 mL/s over the course of a diurnal cycle for each spring; 2) during the dormant season, flow losses between morning and afternoon are 1 mL/sec or less; 3) extreme low-flow conditions are manifest differently in different spring basins, and these provide an understanding of the flow mechanisms that may be active in karst settings; 4) the wide range of hydrogeologic response in epikarst springs to identical stresses indicates that our models of these systems are likely grossly oversimplified.

## Insight 7—Closely Spaced Flow Zones in Karst May Not Be Hydraulically Interconnected—Be Careful with Well Data

One of the most surprising observations we have gained studying the karst of SEW and the Ozarks is the degree of anisotropy and heterogeneity present in the bedrock aquifers. The observation is based on numerous surface-cased open-hole wells, some as near to one another as 10 m, that show little or no hydraulic interconnection. Pumping and aquifer testing of these wells, many of which are more than 50 m deep, reveal hydraulic conductivites that are less than 1 cm/d. Approximately half of the open boreholes at SEW show this phenomenon (Pennington, 2010; Al Rashidy, 1999). This indicates that secondary karst development along bedding planes or other near-horizontal fractures is localized and not pervasive throughout the region. Drilling and aquifer testing the St. Joe in the Clear Creek graben area identified by Stanton (1993) indicated that, contrary to expectations, no karstification within the borehole had occurred. Other boreholes are closely interconnected, but this insight is especially valuable when interpreting data collected from wells in these settings. Because the well borehole intersects only a point, and they may or may not be connected to other components of the integrated flow system, data and interpretations therefrom should be carefully considered before flow considerations are made.

## Insight 8—Most Karst Environments and Ecosystems Are Surprisingly Well-Populated

In addition to microbial fauna which have been studied at SEW (Hamilton, 2001; Whitsett, 2002; Ting, 2003 & 2005; Winston, 2006; Laincz, 2011; fig. 6), benthic invertebrates have been characterized by Mike Slay (written commun. The Nature Conservancy, 2010), and macro-invertebrates captured in passive traps set in springs and wells have been described by Mitchell (this volume). Combined assemblages total more than 30 genera, including cave amphipods (*Stygobromus ozarkensis*), troglybitic isopods (*Caecidotea sp.*) in springs, and copopods in wells isolated from surface water. This diversity and numbers of sensitive organisms observed indicate the ecosystems in this karst environment support a wide variety of stygofauna.

## Insight 9—Understanding Transport of Suspended Sediments and Pathogens in Karst Requires a Different Tool Set

In the rapid-flow part of the saturated aquifer, suspended sediments and bacteria are transported more slowly than dissolved tracers, averaging about 2 m/hr during conditions of low flow, as compared to as great as 800 m/hr for conservative solutes. Average velocity may be misleading, however, because it appears the suspended particles are deposited along the subsurface flow path, and do not move until increased recharge from storms increases velocity enough to scour, resuspend, and re-transport the particles. Below a threshold ground-water velocity, turbulent flow does not occur and suspended particles are not in transit.

This is consistent with observations and measurements at the spring resurgences; increases in stage, and *E.coli*, tagged clay and total suspended solids concentrations are rapid and occur around the same time, then decrease quickly. Ting (2002 & 2005) was able to utilize field-scale testing and verify that unique tagging of clays and microbes was a valuable tracing tool in this environment.

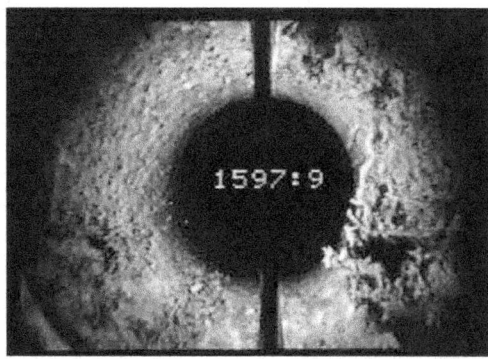

Figure 6. This image is from a borehole televiewer in an uncased well below the regional confining layer (Chattanooga), showing thriving colonies of unidentified facultative bacteria.

## Insight 10—Even the Simplest of Karst Systems Are Complex, and Predictions Should Be Offered with Great Care

The complexities of flow in karst terranes are well-documented throughout the world (Palmer, 2007), and SEW is no different. Continuous monitoring of two springs at the distal end of the groundwater basin 1 provides an illustrative example. Langle Spring, the underflow part of the focused flow system, is about 3 centimeters lower than Copperhead, the overflow spring. During a 33-day period starting in December 1997, five major storms perturbed the stability of the temperature of these springs (fig. 7). These produced an general overall cooling in the discharge waters of Langle, although the trend was by no means linear or gradual. Water from Copperhead, on the other hand, showed both warming and cooling trends, numerous abrupt reversals in heating and cooling, as well as many more fluctuations of about 1° C in the interval from December 13 through December 20. These diurnal variations were ultimately traced to a

[units of discharge in mL/sec; am, 7:00 am measurement;
pm, 3:00 pm measurement; Δ, difference between am and
pm; growing, July; dormant, December]

| \ Spring  Season/Time\ | Tree | WoodPecker | Red Dog |
|---|---|---|---|
| growing/am | 216 | 105 | 99 |
| growing/pm | 199 | 93 | 94 |
| growing/Δ | 17 | 8 | 5 |
| dormant/am | 103 | 92.7 | 78.9 |
| dormant/pm | 102 | 93.0 | 78.0 |
| dormant/Δ | 1 | <1 | <1 |

leaky, ephemeral surface pond that commonly stored water after periods of intense rainfall. The pond was exposed to solar heating during clear weather following storms, and the heated water leaked into the karst aquifer and as it mixed with water from other sources, it imparted the diurnal thermal signature to Copperhead Spring. The signature was not obvious at Langle, owing to its much longer distance from the leaky pond. The wider ranges of temperature in Copperhead Spring suggest that this part of the system is more open to surface water, and less thermally isolated than Langle. Cold water from the storm of 25 December obviously had an impact on the thermal regime of both springs, but the slope of the decrease at Langle supports the hypothesis that much of the water from the northern part of the spring system is more insulated from surface effects. Near identical temperatures from the two springs on 8 December 1997 is interpreted to be caused by point-source input to each spring from fractures nearby the orifices. This effect is repeated on 25 December, and is thought to be a strong indication of the temperature of the precipitation at the time of coincident temperatures.

Documentation by continuous long-term monitoring gives us confidence that system responses cannot be easily simplified in what we initially conceptualized to be a non-complex karst system. Predictive numerical modeling does not appear to be appropriate, although hypothesis testing using models is definitely valuable and meaningful.

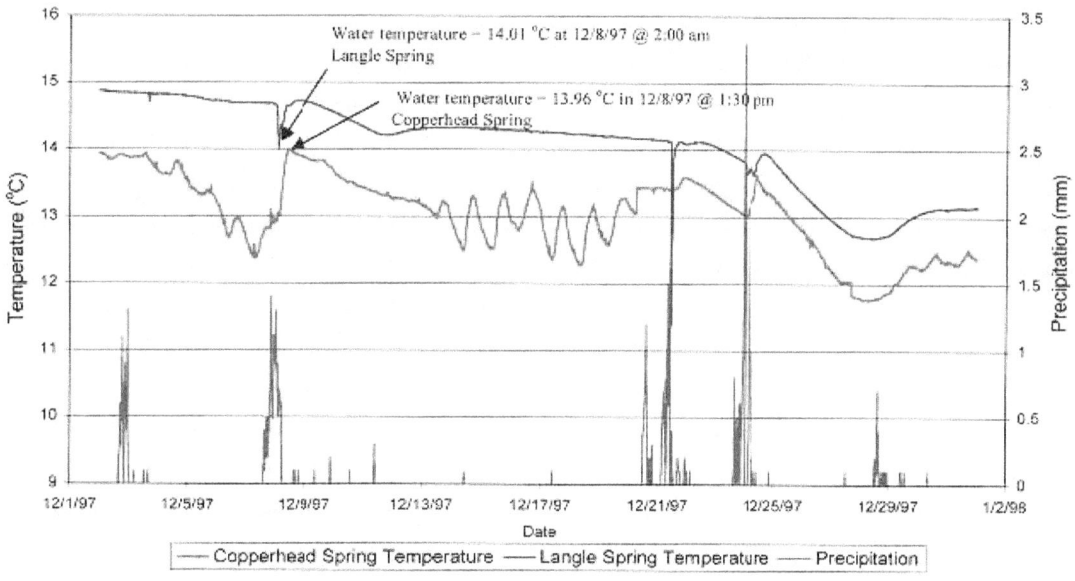

Figure 7. Temperature of two springs, Langle and Copperhead, that resurge from the St. Joe Limestone at the distal end of the ground-water flow path in Basin 1. Temperature variations are natural tracers that show the complexity of input sources and thermal equilibration of this overflow/underflow spring.

## SUMMARY

The ten topics that are presented in this paper represent an arbitrary selection of studies conducted in mantled karst at SEW. The main concepts and insight are not new, but we feel the level of understanding gained contributes significantly to the overall knowledge base of karst science. The establishment of long-term research sites is neither trivial nor inexpensive, yet the opportunities to quantify watershed system response *in situ* to a wide range of natural stresses is essential to ultimately reaching true understanding of the processes that dominate this kind of setting. Time demands and economics will always influence applied research in karst, but the existence of sites such as SEW provides scientists the opportunity to develop new tools and to more fully explore quantifiable aspects of a truly complex system.

## ACKNOWLEDGMENTS

Insight described herein is based primarily on the work of many graduate students, field hydrogeology classes, colleagues, and visiting karst researchers, and I have tried to cite the importance of their many contributions. Unfortunately, limited space has constrained the full description of all who have contributed, and I encourage the interested to seek out published technical reports at the SEW website, which can be accessed at:

http://www.uark.edu/depts/savoyres/index.html

Reviews by Jozef Laincz, Justin Mitchell, and Phil Hays significantly improved this manuscript, and are gratefully acknowledged.

## SELECTED REFERENCES

Aley, Thomas,2002, The Ozark Underground Laboratory's Groundwater Tracing Handbook: Ozark Underground Laboratory, Protem, MO.

Al-Qinna, Mohammed, 2004, Measuring and modeling soil water and solute transport with emphasis on physical mechanisms in karst, Savoy Experimental Watershed, Arkansas: unpublished Ph.D. dissertation, University of Arkansas, 216 p.

Al-Rashidy, Said, 1999, Hydrogeologic controls of groundwater in the shallow mantled karst aquifer, Copperhead Spring, Savoy Experimental Watershed, northwest Arkansas: unpublished M.S. thesis, University of Arkansas, 124 p.

Bolyard, Susan, 2007, Hydrogeology and geochemical processes and water-quality evolution related to the Parsons Landfill near the Beaver Reservoir Area, Arkansas: : unpublished M.S. thesis, University of Arkansas, 69 p.

Brahana, J.V., 1995, Controlling influences on ground-water flow and transport in the shallow karst aquifer of northeastern Oklahoma and northwestern Arkansas: Proceedings Volume, Hydrologic Problems Along the Arkansas-Oklahoma Border, Arkansas Water Resources Center Publication No. MSC-168, p. 25-30.

Brahana, J.V., 1997, Rationale and methodology for approximating spring-basin boundaries in the mantled karst terrane of the Springfield Plateau, northwestern Arkansas: in Beck, B.F. and Stephenson, J. Brad, eds., Sixth Multidisciplinary Conference on Engineering Geology and Hydrogeology of Karst Terranes, A.A. Balkema, Rotterdam, p. 77-82.

Brahana, J.V., Hays, P.D., Kresse, T.M., Sauer, T.J., and Stanton, G.P., 1999, The Savoy Experimental Watershed—Early lessons for hydrogeologic modeling from a well-characterized karst research site: in Palmer, A.N., Palmer, M.V., and Sasowsky, I.D., editors, Karst Modeling: Special Publication 5, Karst Waters Institute, Charles Town, WV, p. 247-254.

Curtis, Darrin L., 2000, An integrated rapid hydrogeologic approach to delineate areas affected by advective transport in mantled karst, with an application to Clear Creek Basin, Washington County, Arkansas: unpublished Ph.D. dissertation, University of Arkansas, 121 p.

Dixon, Barnali, 2001, Application of neuro-fuzzy techniques to predict ground-water vulnerability in northwest Arkansas: unpublished Ph.D. dissertation, University of Arkansas, 262 p.

Fanning, B.J., 1994, Geospeleologic analysis of karst and cave development within the Springfield Plateau of northwest Arkansas: unpublished M.S. thesis, University of Arkansas, 93 p.

Hamilton, Sheri, 2001, Survival of *E. coli* in stream and spring sediments: unpublished M.S. thesis, University of Arkanasas, 48 p.

Hobza, Christopher, 2005, Ground-water quality near a swine waste lagoon in a mantled karst terrane in northwestern Arkansas: unpublished M.S. thesis, University of Arkansas, 76 p.

Laincz, Jozef, 2011 (in progress), Nitrogen transport and cycling in the interflow zone of a mantled-karst terrane, Savoy Experimental Watershed : unpublished Ph.D. dissertation, University of Arkansas, 143 p.

Leh, Mansour, 2006, Quantification of rainfall-runoff mechanisms in a pasture-dominated watershed: unpublished M.S. thesis, University of Arkansas, 98 p.

Leh, M.D., Chaubey, I., Murdoch, J.F., Brahana, J.V., and Haggard, B.E., 2008, Delineating Runoff Processes and Critical Runoff Source Areas in a Pasture Hillslope of the Ozark Highlands: Hydrological Processes, Wiley Interscience, DOI: 10.1002/hyp7021, 15 p.

Little, P.R., 2007, Dominant processes affecting groundwater quality and flow in Basin 2, Savoy Experimental Watershed (SEW): Unpublished M.S. thesis, University of Arkansas, Fayetteville, 93 p.

Palmer, A.N., 2007, Cave geology: Cave Books, Dayton, Ohio, 454 p.

Parse, Matthew, 1995, Geomorphic analysis of the role of regolith in karst landscape development Benton County, Arkansas: unpublished M.A. thesis, University of Arkansas, 177 p.

Pennington, Darrell, 2010, Karst drainage-basin analysis using hydrograph decomposition techniques at the Savoy Experimental Watershed, Savoy, Arkansas: unpublished M.S. thesis, University of Arkansas, 121 p.

Peterson, E.W., Davis, R.K., Brahana, J.V., and Orndorff, H.O., 2002, Movement of nitrate through regolith covered karst terrane, northwest Arkansas: Journal of Hydrology, v. 256, p. 35-47.

Phelan, Terri L., 1999, GIS and 3-D visualization for geologic subsurface static modeling in a mantled karst environment near Savoy, Arkansas: unpublished M.A. thesis, University of Arkansas, 158 p.

Stanton, Gregory P., 1993, Processes and controls affecting anisotropic flow in the Boone-St. Joe aquifer in northwestern Arkansas: unpublished M.S. thesis, University of Arkansas, 212 p.

Tennyson, Rodney, Terry, Jim, Brahana, Van, Hays, Phil, and Pollock, Erik, 2008, Tectonic control of hypogene speleogenesis in the southern Ozarks—Implications and beyond: *in* Kuniansky, E.L., ed., U.S. Geological survey Karst Interest Group Proceedings, Bowling Green, Kentucky, May 27-29, 2008: U.S. Geological Survey Scientific Investigations Report 2008-5023, p. 37-46.

Ting, Tiong Ee, 2002, Development of a bacterial tracer for water quality studies in mantled karst basin using indigenous *Escherichia coli* labeled with europium: unpublished M.S. thesis, University of Arkansas, 106 p.

Ting, Tiong Ee, 2005, Assessing bacterial transport, storage and viability in mantled karst of northwest Arkansas using clay and *Escherichia coli* labeled with lanthanide-series metals: unpublished Ph.D. dissertation, University of Arkansas, 279 p.

Unger, Timothy, 2004, Structural controls influencing ground-water flow within the mantled karst of the Savoy Experimental Watershed, northwest Arkansas: unpublished M.S. thesis, University of Arkansas, 128 p.

Wagner, Dan, 2007, *In-situ* assessment of waste storage effectiveness in karst using stable isotope biogeochemistry: Unpublished M.S. thesis, University of Arkansas, Fayetteville, 58 p.

Whitsett, Kelly S., 2002, Sediment and bacterial tracing in mantled karst at the Savoy Experimental Watershed, northwest Arkansas: unpublished M.S. thesis, University of Arkansas, 66 p.

Winston, Byron, 2006, Land use trends in areas underlain by karst and consequences for N source and processing in aquatic ecosystems: unpublished M.S. thesis, University of Arkansas, 88 p.

Woodstrom, Freya A., 1999, The effects of landuse on the spatial and temporal variations in the water quality of selected springs in Washington County, Arkansas: unpublished M.A. thesis, University of Arkansas, 120 p.

# Traps Designed to Document the Occurrence of Groundwater Stygofauna at the Savoy Experimental Watershed, Washington County, Arkansas

By Justin R. Mitchell
Department of Geosciences, 113 Ozark Hall, University of Arkansas, Fayetteville, AR 72701

## Abstract

The capture and documentation of stygofauna (obligate aquatic cave dwellers) in the Springfield Plateau of the Arkansas Ozarks has been typically limited to the physical access of caves, springs, and sinkholes. This project has expanded the potential search area for stygofauna to include wells through the use of specially designed traps. It was hoped that this study could find an unknown population of cave crayfish; however no *Cambarus* species were observed. Baited traps were set in three springs and seven wells at the Savoy Experimental Watershed, a University of Arkansas Experimental Station located in Washington, County Arkansas. Traps were checked three times over a three week period in November 2010. These traps were proven effective in actively capturing various zooplankton species, lotic leaches, and the epigean crayfish *Orconectes neglectus*. Cyclopoid copepods (small shrimp-like crustaceans) were repeatedly captured in two deep wells, whereas five deep wells produced no faunal specimens. Cyclopoid copepods as well as five other epigean (surface dwelling) aquatic taxa were captured in the three springs sampled. The use of these combination traps has demonstrated an effective active technique for documenting aquatic organisms occurring in the groundwater of wells and surface springs at the Savoy Experimental Watershed.

## INTRODUCTION

Six distinct species of stygobitic crayfish of the genus *Cambarus* (Decapoda: Cambaridae) are known to exist in the Ozark Plateau region of Arkansas, Missouri, and Oklahoma (Koppelman and Figg, 1995). These unpigmented sightless crayfish with long slender claws are the largest invertebrate members of Ozark cave ecosystems (Koppelman and Figg, 1995). The Benton County cave crayfish, *Cambarus aculabrum*, is a unique aquatic resident of Northwest Arkansas' cave systems and a federally listed endangered species (Graening and others, 2006) (Fig. 1).

The documented range of this troglobitic crustacean is limited to four locations within the Springfield Plateau of the Arkansas Ozarks, all in Benton and Washington counties (Graening and others, 2006). However, due to the geology of the extensive karst environment in which these crayfish are found, they are likely to occur elsewhere in the ground water of the region. Except for caves and springs at the distal end of karst flow systems, researchers are physically too large to access much of the hypothetical habitat of these animals. Therefore, specimens are almost always observed in caves and springs, explaining the common name of cave crayfish.

Figure 1. *Cambarus aculabrum*, The Benton County Cave Crayfish, Photo by Michael E. Slay, The Nature Conservancy.

One proven historical method of documenting *Cambarus* in locations inaccessible for direct human exploration was through the use of specially designed traps lowered into wells (Purvis and Opsahl, 2005). This method was implemented in Georgia, U.S.A. and successfully extended the range of a karst dwelling *Cambarus* species.

Similar traps were manufactured for this study and modified from the original design to potentially capture a variety of stygofauna in addition to crayfish. Numerous other organisms exhibiting exclusively subterranean aquatic life styles are known to occur in the Ozarks. These karst aquifer residents include cavefish, salamanders, snails, isopods, amphipods, and other invertebrates (Elliott, 2007). This report describes the effectiveness of these modified traps and investigates the existence of stygofauna at a rural location in Washington County, Arkansas (Figure 2).

Figure 2. Deep well 72 and surrounding landscape, Savoy Experimental Watershed, October 2010.

## Study Area

The site for this study was the Savoy Experimental Watershed (SEW) located 24 kilometers west of the University of Arkansas campus in Fayetteville, Arkansas. Situated within the Springfield Plateau of the Ozarks, the SEW is an approximately 1250-hectare property

that serves as a scientific research area for the University of Arkansas' Departments of Animal Science, Agronomy, Geosciences, and Poultry Science, along with various state and federal entities (Brahana and others, 2001). The landscape is hilly with well dissected plateaus cut by dry valleys. Vegetation is composed of mixed deciduous forests and grassy pastures. Limited cattle grazing, confined swine production, and poultry operations occur within discrete parts of the property, associated with Animal Science research. Most of the SEW, however, is undeveloped second-growth forest.

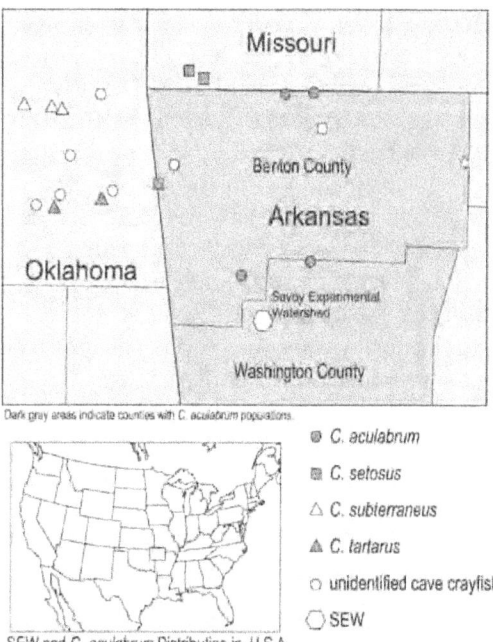

Figure 3. Location of the Savoy Experimental Watershed and the distribution of *Cambarus aculabrum* with other cave crayfish populations in adjacent states-. Cave crayfish distributions modified from Graening and others, (2006).

The SEW is located in northern Washington county and is relatively close to documented *Cambarus* populations (Fig. 3) Although no *Cambarus* species are known from SEW, a cave isopod *Caecidotea* sp., and a cave amphipod *Stygobromus ozarkensis* occur in Copperhead Spring, indicating suitable habitat for stygobitic organisms (Graening and others 2005; Slay, 2000). A *Caecidotea* species was also observed in Dead Cow Spring. A total of twenty-nine unique taxa have been collected in four SEW springs using a D-ring kick net following substrate disturbance (Slay, 2000). The taxa richness of the four springs studied was found to range from fourteen to twenty-two distinct species per spring (Slay, 2000). The biodiversity discovered previously in the SEW springs justifies continued investigations of groundwater habitats and their inhabitants occurring within the property.

The SEW is characterized as a mantled karst terrain and drains into the adjacent Illinois River. The cave-forming carbonate rock underlying the area is generally covered in a layer of regolith of variable thickness, consisting of insoluble chert and clays. Beneath the lower Mississippian Boone and St. Joe limestone formations, exists a layer of upper Devonian shale, the Chattanooga formation, which acts as a regional confining layer, preventing the vertical movement of groundwater (Brahana, 1997). Dissolution of the limestone occurs at this contact zone, along joints, and within overlying bedding planes influenced by the presence of insoluble chert (Brahana, 1997). This chemical weathering process enlarges paths for the movement of water, forms caves, and provides potential habitat for stygofauna.

The existence of numerous deep test wells and springs on one contiguous tract of land with easy access was convenient and ideal for this study. The wells are in close proximity to natural springs and caves and can be visited in a relatively short period of time. The deep wells at SEW were drilled 30.5 meters to 80.5 meters deep. Geophysical logs of these wells have provided valuable information about the subsurface geology and hydrology at SEW (Stanton and others, 1998) (Fig.4). Most deep wells penetrate through the regolith, Boone, St.Joe, and Chattanooga formations and terminate in the Ordovician dolomite beneath, accessing two karst aquifers separated by the Chattanooga confining layer. Fluid resistivity and fluid conductance logs indicate bedding plane openings in the Boone formation and voids in the contact zone between the St. Joe limestone and Chattanooga shale (J. Van Brahana, 2010, Personal Communication, Department of Geosciences, University of Arkansas, Fayetteville, Arkansas; Keys, 1990). These cavities facilitate the storage and

movement of ground water and are possible habitats for Ozark stygofauna.

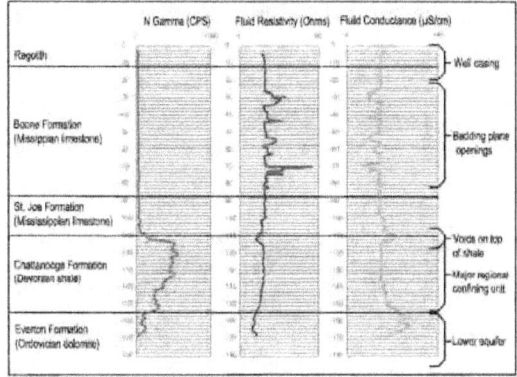

Figure 4. Geophysical log of Deep Well 72 at the SEW-. Well log modified from Stanton and others, (1998).

## Materials and Methods

The working principle of the traps used is simple and non lethal: the bait attracts organisms to the trap, allows organisms to enter, and restricts an easy exit. The traps used for this study were made from polycarbonate sports drink containers similar to the design used by Purvis and Opsahl. However, for this study, improvements were made in the construction and modifications were made to facilitate the capture of smaller organisms including zooplankton (Fig.5). Only one specific type of bottle would work for this new design owing to the ribbed shape of the container. The top 6 cm of the bottle was cut off and slowly heated until it contracted and became malleable. The hot upper portion of the bottle was then pressed into the lower portion of the bottle and lodged between the ribs. This technique created a compression fitting between the upper and lower portions. No glue was required for attachment which allowed for easy opening and closing of the trap in the field. Small 4 mm holes to facilitate air and water flow were melted in the bottom and upper portion of the trap with a soldering iron. Holes were also melted in the sides for the attachment of three 15 cm nylon leader lines, used to lower and retrieve the trap.

Leader lines were attached using plastic cable ties. Four 15 g glass weights were glued to the sides using 100% silicone adhesive. Instead of using lead sinkers, glass weights were used to avoid introducing lead into the ground water. A large plastic cable tie was threaded through holes in the bottom and left protruding out of the bottom. This forced the trap to fall or lean on its side instead of resting flat on its bottom. A custom designed conical polyester plankton net was then attached to the base with a cable tie. The plankton net was easily removed and refitted in the field. The traps were baited with 5 g of raw chicken liver placed in small polyester bags and fastened with a small cable tie to the base of the trap.

Traps were placed in seven deep wells and allowed to sink to the bottom. Traps were also set in three springs. Two springs sampled, Copperhead Spring and Langle Spring, emerge at the contact zone of the Boone and the St. Joe limestone formations (Al Rashidy, 1999). The other spring sampled, Red Dog Spring, is an epikarst spring that emerges from the contact zone between the regolith and the Boone formation. Traps were set at all 10 locations October 27, 2010 and checked November 6, 10, and 22, 2010. There was not any significant rain during the sampling period and the springs flowed at typical low flow discharges (Al Rashidy, 1999).

Each time a trap was checked, large organisms were removed from the bottle and a sample was collected from the plankton net. The net was detached from the trap, inverted over a collection container, and the contents rinsed with 70% ethanol. The specimens present were killed and preserved in this process. These samples were later examined in a lab using a dissection microscope. Organisms were identified to a basic taxonomic level using Pennak (1989), sorted, and preserved in 70% ethanol for later species identification.

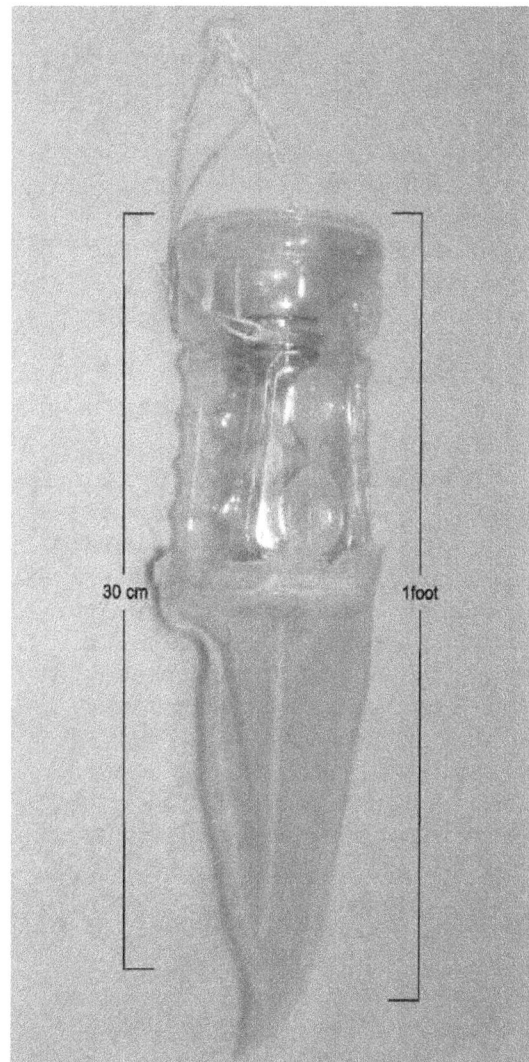

Figure 5. Combination Crayfish/ Zooplankton Trap

## RESULTS

Aquatic invertebrates were captured in all three springs and two of the seven deep wells. Five wells did not yield any faunal organisms. The target species, *Cambarus aculabrum*, was not found at any location. The stygobitic *Caecidotea* sp. and *Stygobromus ozarkensis* were not captured. However, the trap was able to capture zooplankton in both wells and springs. Cyclopoid copepods were captured in two wells: well 70 (2 individuals) and well 72 (27 individuals). Cyclopoid copepods were also captured in the three springs: 9 from Copperhead Spring, 223 from Langle Spring, and 248 from Red Dog Spring. Ostracods were found in two springs: 174 in Red Dog Spring and 3 in Langle Spring. Samples from Red Dog Spring also yielded 95 *Daphnia* and 1 *Hydra*. In addition to zooplankton, larger invertebrates were also captured in the springs. 80 gastropods (snails) and 3 *Hirudinea* (leeches) were collected in Red Dog Spring. *Orconectes neglectus*, the epigean crayfish commonly known as the ringed crayfish (Pflieger, 1996), was captured once in Copperhead Spring and once in Langle Spring.

## DISCUSSION

The well sampler originally designed by Purvis and Opsahl (2005) appears to be a useful tool for sampling non-typical groundwater habitats (i.e. wells), and the trap modifications presented here allow the trap to capture aquatic organisms of a variety of sizes. *Cambarus aculabrum* is a federally listed endangered species and any discoveries of its existence or the existence of other stygofauna outside of known locations would be useful for conservationists and beneficial for the species. This project serves as a model study that could be reproduced, improved, and implemented in other wells and springs in the region or elsewhere in karst environments around the world. Due to the negative anthropogenic impacts on cave habitats and ground water quality throughout the region, it is important to locate, document, and protect stygobitic species for their continued survival (Graening and others, 2006). Cave animals are sensitive to the water quality, chemistry, geology, and hydrology of their unique habitat. The presence (or absence) of some sensitive species can indicate changes in water quality over time (Palmer, 2007). The Purvis and Opsahl trap, along with the modifications, can potentially help ecologists with population and distribution studies, cave fauna inventories, and aid in new discoveries. Little characterization of organisms found in karst aquifers has been undertaken and more extensive work needs to be done to identify, understand, and protect the stygofauna of the Ozarks and beyond. Such studies translate into good stewardship of our precious groundwater resources.

The copepods captured in deep well 72 and deep well 70 are of some significance and possibly cave forms due to their apparent isolation from the surface (Michael E. Slay, 2010, Personal Communication, The Nature Conservancy, Fayetteville, Arkansas). Well 72 is 172 feet deep and at the time of this study the water level was 91 feet beneath the surface. Well 70 is 103 feet deep and the water level was 7 feet from the surface. The wells are always capped except for brief intervals of water-quality sampling and trap insertion and removal. The capture of these copepods on two separate occasions in these wells during a period of base groundwater flow indicates the existence of groundwater invertebrates living deep below the surface, relatively far from any associated surface orifice at SEW. Plans call for specimens collected in this study to be sent to a copepod taxonomist for species identification.

Even though two crayfish were captured, larger crayfish, most likely *Orconectes neglectus*, appeared to be able to feed on bait and escape from this trap. One insecurely fastened bait bag was completely removed from a trap in Copperhead Spring (Fig.6) Another bait bag from a trap in Langle Spring was found torn with the bait removed. This problem has been considered and a new escape prevention modification consisting of a collapsible mesh tube attached to the inside of the trap's entrance has been designed. This modification is currently being tested for functionality in the field.

Figure 6. Copperhead Spring, SEW, October 2010

## ACKNOWLEDGMENTS

Thanks to Mike Slay of The Nature Conservancy (TNC) for suggesting the plankton net addition to create this combination trap. Slay provided support, materials, and assistance with identification of the collected specimens for this project and he facilitated permission from The Arkansas Game and Fish Commission to proceed with this project. Thanks to Jerry Anne Mitchell for sewing the plankton nets and bait bags, Jimmy Mitchell Sr. for suggesting glass for non-toxic weights, Erin Timmons-Mitchell for her continuing support of my education and scientific endeavors, and Professor J. Van Brahana for his continuing encouragement, assistance, and support for this project.

## REFERENCES

Al Rashidy, S. 1999. Hydrogeologic controls of groundwater in the shallow mantled karst aquifer, Copperhead Spring, Savoy Experimental Watershed, Northwest Arkansas. Unpublished M.S. Thesis, Department of Geosciences, University of Arkansas, Fayetteville.

Brahana, J. V. 1997. Rationale and methodology for approximating spring-basin boundaries in the mantled karst terrane of the Springfield Plateau, Northwestern Arkansas. The Engineering Geology and Hydrogeology of Karst Terranes. Beck & Stephenson eds. Balkema, Rotterdam. p.77-82.

Brahana, J. V., T. L. Phelan, P. D. Hays, T. J. Sauer, R. K. Davis, and J. H. Cole. 2001. Savoy Experimental Watershed and field research facility- a long-term karst research site. University of Arkansas, Fayetteville. poster.

Elliott, W. R. 2007. Zoogeography and biodiversity of Missouri caves and karst. Journal of Cave and Karst Studies. v. 69, no. 1, p. 135-162.

Graening, G. O., M. E. Slay, and J. R. Holsinger. 2005. Annotated checklist of the amphipoda of Arkansas with an emphasis upon groundwater habitats. Journal of the Arkansas Academy of Science. v. 59, p. 80-87.

Graening, G. O., M. E. Slay, A. V. Brown, and J. B. Koppleman. 2006. Status and distribution of the endangered Benton cave crayfish, *Cambarus aculabrum* (Decapoda: Camaridae). The Southwestern Naturalist. v. 51, no. 3, p. 376-439

Harrod, J. 2006. Saving the last great places under earth. The Nature Conservancy.

http://www.nature.org/wherewework/northameric a/states/Arkansas/preserves/art24994.html. accessed October 2010.

Keys, W. S. 1990. Techniques of water-resources investigations of the United States Geological Survey, Chapter E2 Borehole Geophysics Applied to Ground-Water Investigations. U.S. Department of the Interior, U.S. Geological Survey. Denver, Colorado. p. 9-22.

Koppelmann, J. B. and D. E. Figg. 1995. Genetic estimates of variability and relatedness for conservation of an Ozark cave crayfish species complex. Conservation Biology. v. 9, no. 5, p. 1288-1294.

Palmer, A. N. 2007. Cave Geology. Cave Books. Dayton, Ohio.

Pennak, R. W. 1989. Fresh-Water Invertebrates of the United States Protozoa to Mollusca 3$^{rd}$ Ed. John Wiley & Sons, Inc. Wiley – Interscience Publication. New York.

Pflieger, W. L. 1996. The Crayfishes of Missouri. Missouri Department of Conservation. Jefferson City, Missouri.

Purvis, K. M. and S. P. Opsahl. 2005. A novel technique for invertebrate trapping in groundwater wells identifies new populations of the troglobitic crayfish, *Cambarus cryptodytes*, in Southwest Georgia, USA. Journal of Freshwater Ecology. v. 20, no. 2, p. 361-365.

Slay, Michael. 2000. Aquatic faunal inventories of four springs on the Savoy Experimental Watershed. Unpublished Karst Hydrogeology term paper, Department of Biological Sciences, University of Arkansas, Fayetteville.

Stanton, G. and Students. 1998. Geophysical log data collected by the Summer Field Hydrogeology class, Department of Geosciences, University of Arkansas and the U.S. Geological Survey Southeast Region Logger Service. *in* Al Rashidy, S. 1999. Hydrogeologic controls of groundwater in the shallow mantled karst aquifer, Copperhead Spring, Savoy Experimental Watershed, Northwest Arkansas. Unpublished M.S. Thesis, Department of Geosciences, University of Arkansas, Fayetteville.

# Assessment of Sinkhole Formation in a Well Field in the Dougherty Plain, near Albany, Georgia

By Debbie Warner Gordon

U.S. Geological Survey, 3039 Amwiler Rd, Suite 130, Atlanta, GA 30360

## Abstract

To assess the causes of sinkhole formation and determine if groundwater withdrawals are affecting the frequency of sinkhole development, the U.S. Geological Survey, in cooperation with the Albany Water, Gas, and Light Commission, conducted hydrologic studies at an Upper Floridan aquifer well field southwest of Albany, Georgia, during 2009–2010. The Upper Floridan aquifer is the main water-supply source in southwestern Georgia and consists of limestone that is overlain by 30 to 60 feet (ft) of sand and clay (undifferentiated overburden). Borehole data, including geophysical, lithologic, and video, indicate two zones of high secondary permeability (dissolution features) within the Upper Floridan aquifer at the well field. These zones are cavernous and may be vulnerable to collapse. Loading from ponds excavated during well-field construction and water-level fluctuations due to well-field pumping increase the potential for sinkhole development.

Hydrologic data, including groundwater levels, production-well pumping records, and rainfall, were compared to subsurface data to assess the formation of sinkholes at the well field. From October 2003 (initiation of well-field pumping) through December 2009, 23 sinkholes developed at the well field. Most of the sinkholes formed in or adjacent to storage ponds located along the western part of the well field (fig. 1). These manmade ponds increase the load on the underlying sediments, which may increase the potential for sinkhole formation (Tihansky, 1999). A cluster of eight sinkholes southwest of the ponds formed while a nearby production well was pumping, so the well was removed from production. Other sinkholes formed farther away from the ponds and may be linked hydrologically to high-permeable zones in the aquifer.

Sinkholes may form in response to changes in water levels relative to the top of the limestone aquifer or relative to the top of a cavity in the limestone. A comparison of water levels in well 12L382 during October 2003–December 2009 to the depths to the top of the Upper Floridan aquifer and top of the upper permeable zone shows the relation of sinkhole formation to groundwater levels in the Upper Floridan aquifer (fig. 1). The relation between groundwater level fluctuations and sinkhole formation at the well field is not completely understood. In 2003 and 2005, water levels were almost 20 ft above the top of the aquifer for most of the year, yet two sinkholes formed during each of those years. During 2007, water levels oscillated above and below the top of the Upper Floridan aquifer and 12 sink holes formed; however, no sinkholes formed during 2008, even though water levels and pumping patterns were similar to those during 2007. Ongoing monitoring of groundwater levels, with the possible addition of pond stage, may provide additional insight into factors controlling sinkhole formation at the well field.

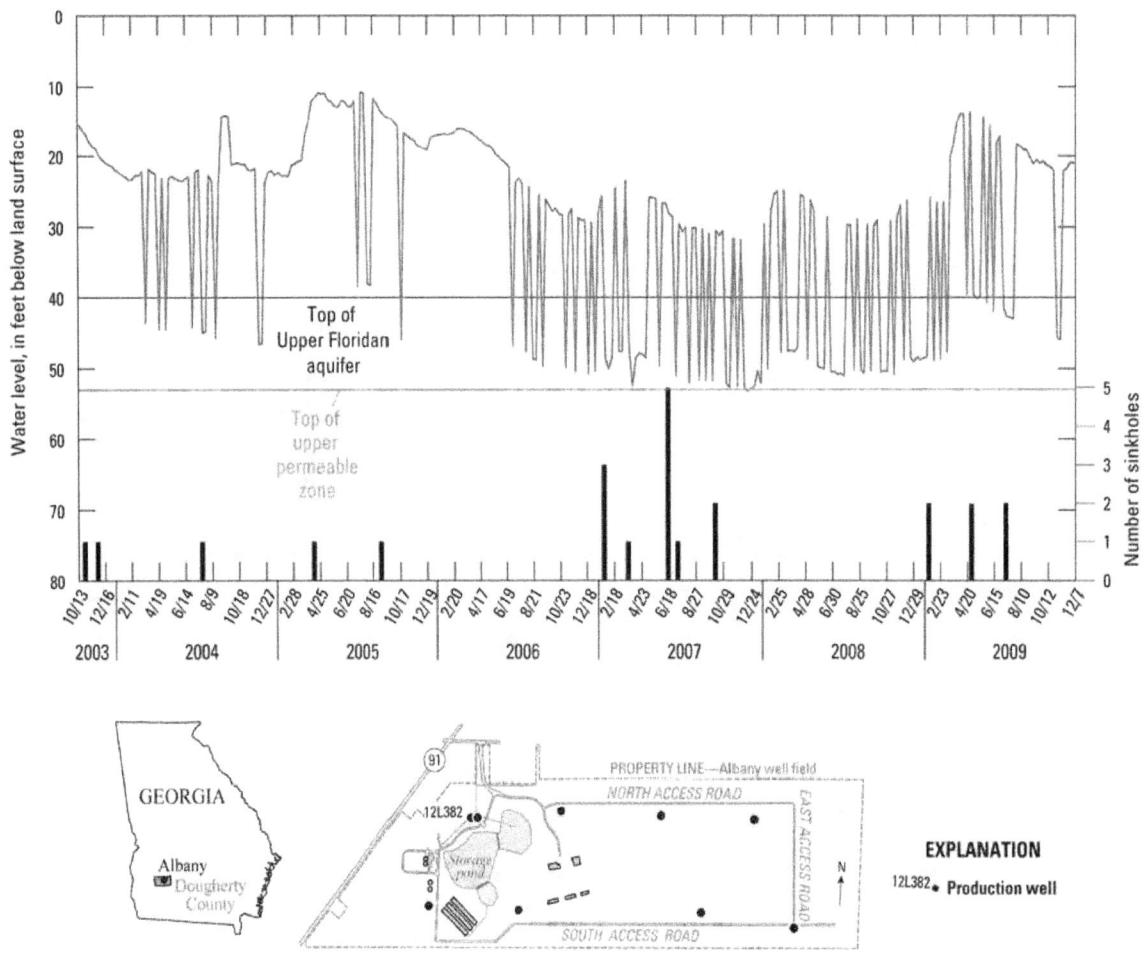

Figure 1. The number of sinkholes that formed in the Albany well field, Dougherty County, Georgia, daily depth to water in well 12L382, and the depth to the top of the Upper Floridan aquifer and upper permeable zone at well 12L382, October 2003–December 2009.

## REFERENCE CITED

Tihansky, A.B., 1999, Sinkholes, west-central Florida, *in* Galloway, Devin, Jones, D.R., Ingebritsen, S.E., eds., Land subsidence in the United States: U.S. Geological Survey Circular 1182, p. 121–140.

# Detecting Karst Conduits through their Effects on Nearby Monitoring Wells

By Fred Paillet and Terryl Daniels
Geosciences Department, University of Arkansas, 113 Ozark Hall, Fayetteville AR 72701

## Abstract

Karst aquifers transmit flow through high-permeability conduits that comprise a very small fraction of the porosity of the total rock mass. These conduits are very difficult to locate and characterize by the random drilling of boreholes. The characterization of conduits at the University of Arkansas Savoy wellfield is especially difficult because the conduits are developed along vertical joints in a formation overlain by competent chert beds that mask the surface expression of the subsurface solution openings. In this study we examine the conditions under which water level fluctuations and transient vertical flow in observation wells can indicate the presence of conduits in the surrounding rock mass. Slug tests in Savoy wells 69 and 70 indicate total well transmissivity in the range of 0.1 to 0.3 ft$^2$/day. We adopt a model where one or more bedding plane openings with T = 0.1 ft$^2$/day transmit flow to boreholes from hydraulic head increases of about a meter in a nearby karst conduit, with a linear increase over 24 hours followed by an exponential decline thereafter. The model (Paillet, 1998) uses an infinite bedding plane of specified T and S embedded in low-permeability rock and connected to a reservoir subject to fluctuating water levels. The model uses the slug test response integrated over time to compute water level and flow in the observation well as the water level in the rock surrounding the bedding plane is driven by hydraulic head changes in the karst conduit. The observation borehole water level reflects the full conduit head change modulated by the significant flow losses incurred in moving water into or out of the well bore. Model runs show that the low transmissivity of the bedding plane causes the response in the observation well to lag behind the head changes in the nearby conduit whenever bedding plane T is less than 100 ft$^2$/day. When bedding plane T = 0.1 ft$^2$/day and we use S = 10$^{-4}$, a conduit head change of 1.5 m over 24 hours produces water level fluctuations that are about 20% of the conduit increase (readily measurable), and produces vertical flow in and out of casing storage of about 0.003 gpm (below the resolution of the HP flowmeter but measurable with dilution methods). If there are two such bedding planes and only one connected to the conduit, the model predicts very little change in the response, with outflow into the unconnected bedding plane having almost no effect on the response. If there are two bedding planes connected to two different conduits, the observation well response will give the T-weighted average of the heads in the two conduits. If, however, an unconnected plane has significant transmissivity (for example, T = 10 ft$^2$/day; probably not the case at Savoy), virtually all of the inflow driven by head changes in the karst conduit goes into that other more permeable "thief" zone. In that case there would be no measurable change in water level in the observation well, while the flow between bedding planes would be too small to measure with the HP flowmeter and could only be detected by dilution methods. On the basis of these results we propose a program at Savoy where we use dilution methods to identify the depth and T of inflow zones in observation wells, and then water level fluctuations after rainfall events to infer water level fluctuations in karst conduits to which those planes are connected. If we suspect that two karst conduits and two or more bedding planes are involved, we can use dilution methods to estimate flow in the observation well during response to rainfall events to estimate the head changes in the two separate conduits. We believe that this model approach will allow us to use monitoring well responses to infer hydraulic head changes in the karst conduits in the vicinity of observation wells even when the well does not actually penetrate the conduit opening.

## REFERENCE

Paillet, F.L., 1998, Flow modeling and permeability estimation using borehole flow logs in heterogeneous fractured formations, *Water Resources Research, 34,* 997-1010.

# Spring Hydrology of Colorado Bend State Park, Central Texas

By Kevin W. Stafford, Melinda G. Shaw, and Jessica L. DeLeon
Department of Geology, Stephen F. Austin State University, 1901 N. Raguet, Nacogdoches, TX 75962

## Abstract

Karst development in Ellenburger carbonates near Colorado Bend State Park in central Texas exhibits complex polygenetic origins, with porosity development dominated by an early hypogene phase that has subsequently been overprinted to varying degrees by epigene processes. Quarterly physicochemical and continuous thermal monitoring analyses of eight springs in the study area indicate that modern groundwater flow paths are highly variable. Springs exhibit patterns that range from shallow, distributed recharge into diffuse-flow dominated systems, to focused recharge into well-connected conduit systems, to deep-circulation systems that equilibrate with bedrock. All springs, except Sulphur Spring, exhibit physicochemical characteristics indicative of proximal epigenic groundwater flow through Ellenburger carbonates, while Sulphur Spring shows elevated temperature and dissolved-ion concentrations indicative of longer groundwater flow paths through deeper strata. The polygenetic nature of karst development in the Colorado Bend State Park has created an enhanced porosity structure which forms a complex modern groundwater flow network.

## INTRODUCTION

Colorado Bend State Park (CBSP) is located on the northern edge of the Texas Hill Country on the flank of the Llano Uplift (Figure 1, 2). Here Ordovician Ellenburger carbonates crop out along a highly entrenched segment of the Colorado River, immediately upstream from Lake Buchanan. Proximal to the river, numerous springs discharge including subaqueous springs, springs within a few meters of the river and springs that discharge hundreds of meters from the river (Figure 1). Most springs discharge with normal epigenic karst chemistries; however, one spring in the region, Sulphur Spring, discharges with a slightly thermal component, an elevated sulfate content and easily discernable odor of hydrogen sulfide.

CBSP is located approximately 180 kilometers northwest of Austin, Texas in San Saba and Lampasas Counties. The park covers 21.6 square kilometers including a seven kilometer long stretch of the Colorado River. The area is located along the boundary between subtropical steppe climate and subtropical subhumid climate, with average annual temperature of 20°C and minimum and maximums of 8°C

and 30°C respectively (Estaville and Earl, 2008). Annual precipitation averages 30 cm, with most precipitation occurring during Spring (March – May) and Fall (September – November).

More than 400 karst features have been identified within CBSP and surrounding properties, including more than 100 physically mapped caves. Most caves exhibit characteristics of complex, polygenetic origins,

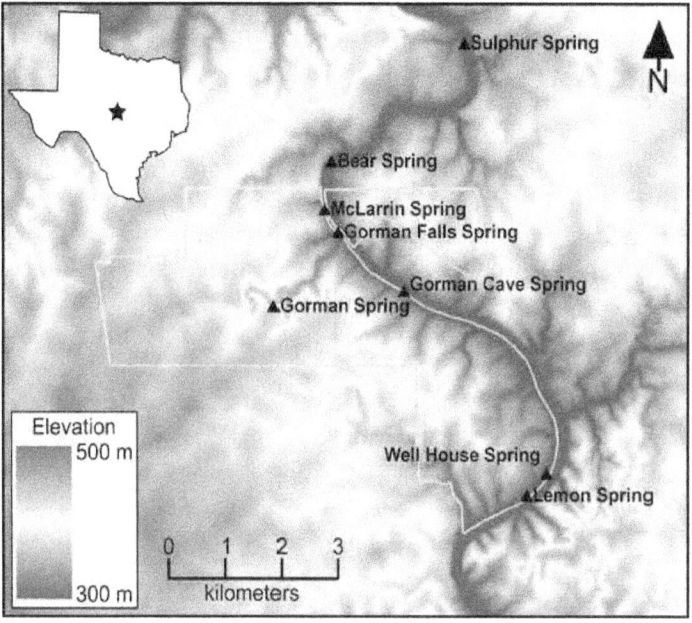

Figure 1. Map of study site showing location of springs, outline boundary of CBSP in yellow and approximate location of study site with reference to the state of Texas.

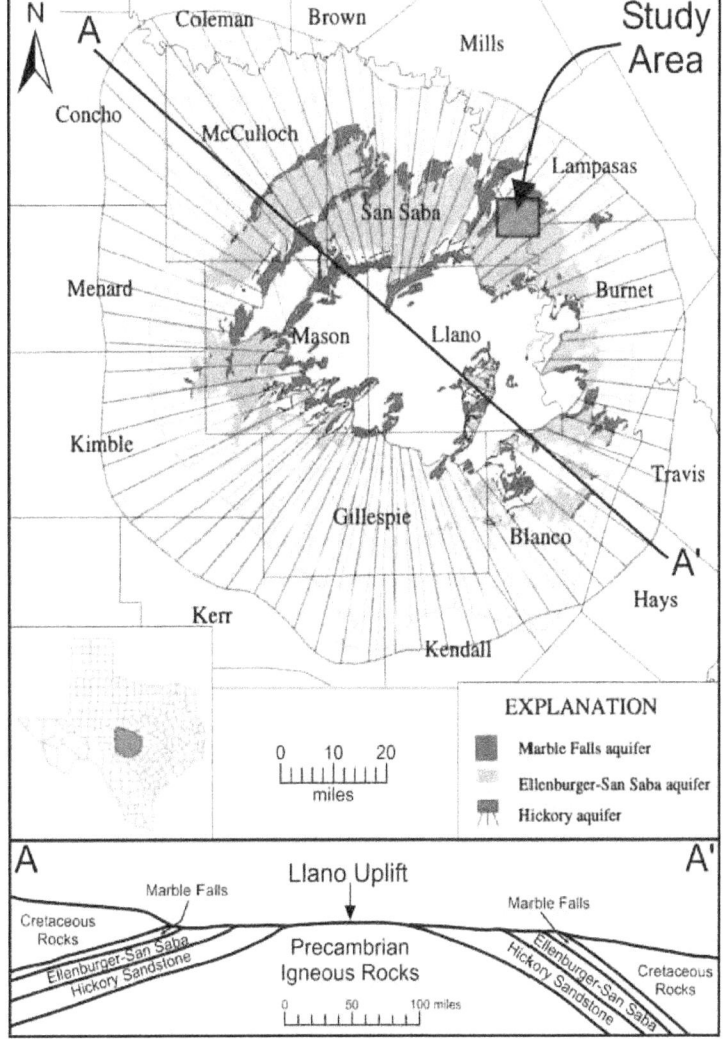

Figure 2. Spatial distribution of Llano Uplift aquifers in relation to project study area (red box) with ~10X vertically exaggerated cross section (modified from Preston and others, 1996).

development with abundant springs that discharge from subaqueous position in the Colorado River to more than sixty five meters above river level suggests a complex hydrogeologic system that has not completely equilibrated with current climatic and geomorphic conditions.

## GEOLOGIC OVERVIEW

The Llano Uplift, and greater Texas Hill Country, is dominated by Precambrian (~1.0 bya - ~1.2 bya) basement rocks which form a large structural dome overlain unconformably by Cambrian and Ordovician clastics and carbonates (Figure 2) (Sellards and others, 1932). Ordovician strata are unconformably overlain by Carboniferous rocks that are subsequently overlain unconformably by Cretaceous strata that compose the northern extension of the Edwards Plateau (Rose, 1972). Precambrian strata were emplaced as part of the Grenville orogenic event, while these and overlying Paleozoic rocks in the study area were modified by tectonism associated with the Ouachita Orogeny, primarily resulting in minor tilting and faulting (Standen and Ruggiero, 2007). Throughout the Cenozoic, all strata within the region have seen additional brittle deformation and minor tilting as a result of uplift of the Edwards Plateau and down-dropping of the gulf coastal plain, with intense faulting along the Balcones Fault Zone, approximately 80 kilometers east of the study area (Collins, 1995).

Karst development within the study area is largely limited to Ordovician Ellenburger carbonates, including cave and spring development in all three Ellenburger units, oldest to youngest — Tanyard, Gorman and Honeycutt Formations. The Tanyard Formation is ~170 meters thick and consists of fine- to coarse-grained, irregularly bedded dolomite deposited as high-energy, restricted, subtidal facies, including common ooidic zones and cryptalgal laminae (Kerans, 1990). The Gorman

including many that exhibit classic hypogene MSRF (Morphologic Suite of Rising Flow) (Klimchouk, 2007) characteristics with varying degrees of epigenic overprinting, while other caves are more purely epigene in origin. DeLeon (2010) showed that at least two thirds of caves developed within the CBSP region exhibit hypogene origins with variable degrees of epigene overprinting, while less than one fourth of all caves showed clear dominance of epigene origins. Eight known perennial springs discharge subaerially within the park, while two other springs discharge upstream within the commercially operated Sulphur Springs fish camp (Figure 1). Stream mapping conducted by Mitchell and others (2011) suggests that at least ten additional springs also discharge subaqueously within the Colorado River in this region. The polygenetic nature of karst

Formation is ~130 meters thick and consists of micro-granular dolomite associated with deposition in a low-energy, restricted-shelf environment, including common macrofossils, distributed zones of intense burrowing and cryptalgal laminae, as well as zones indicative of subaerial exposure that include rip-up clasts and siliciclastic sediments (Kerans, 1990). The Honeycutt Formation is ~40 meters thick and consist of thinly interbedded limestones and dolomites that were deposited in an open, shallow-water shelf environment, including common structures indicative of periods of brief subaerial exposure (e.g. desiccation cracks, rip-up clasts) and structures indicative of variable current energies (e.g. ooids, current ripples, cryptalgal laminae) (Kerans, 1990).

Within the study area, three minor aquifers collectively referred to as the Llano Uplift aquifers are developed in the Paleozoic rocks, including from bottom to top: Hickory aquifer, Ellenburger-San Saba aquifer, and Marble Falls aquifer (Preston and others, 1996). The aquifer system dips gently into the subsurface away from the central Llano Uplift, dipping mainly to the north in the study area. The Hickory aquifer is developed in the Cambrian Hickory Sandstone, which is underlain by Precambrian basement rocks. The Ellenburger-San Saba aquifer includes all three formations of the Ellenburger Group plus the San Saba member of the underlying Wilberns Formation. The Ellenburger-San Saba aquifer is compartmentalized in regions due to local and regional block faulting, with significant

others, 1996). The Marble Falls aquifer is developed in corresponding Carboniferous limestone, which exhibits highly variable permeability due to well-developed secondary porosity. All three of the Llano Uplift aquifers show gradual increases in total dissolved solids in the down-dip direction away from the main Llano Uplift dome, with deep, distal components containing total dissolved solids greater than 10,000 mg/L (Preston and others, 1996).

In the CBSP area, Ellenburger strata are exposed at the surface and most associated springs exhibit normal, epigene karst chemistries, while one spring exhibits anomalous characteristics. The seven springs that exhibit normal epigene karst chemistries, include Bear Spring, Gorman Cave Spring, Gorman Falls Spring, Gorman Spring, Lemon Spring, McLarrin Spring and Well House Spring (Figure 1, Table 1). Sulphur Spring, as the name implies, exhibits anomalous patterns. Bear Spring discharges from a solutionally widened fracture in the Honeycutt Formation that is located four meters above and twenty five meters away from the river. Gorman Cave Spring is associated with Gorman Cave and discharges from the Gorman Formation. Gorman Cave Spring discharges directly into the Colorado River through alluvial sediments, but the spring is accessible for sampling in a stream passage in Gorman Cave, which is more than one hundred meters from the river and six meters above. Gorman Falls Spring is located immediately adjacent to the Colorado River with discharge from a decimeter-scale conduit in the Gorman Formation. Gorman Spring is located at the headwaters of Gorman Creek with discharge occurring as artesian flow from a vertical fracture more than a kilometer from the river and sixty five meters above (England and others, 2010). Lemon Spring discharges horizontally from a thin bedding

Table 1. Average physiochemical characteristics of springs with standard deviation (stdev) based on three month sampling.

|  | TDS | | pH | | Conductivity | | Sulphate | |
| --- | --- | --- | --- | --- | --- | --- | --- | --- |
|  | ppm | stdev | pH | stdev | mV | stdev | ppm | stdev |
| Bear Spring | 345 | 41 | 6.8 | 0.1 | -8 | 9 | 1.3 | 0.4 |
| Lemon Spring | 386 | 139 | 6.8 | 0.2 | -16 | 3 | 1.0 | 0.2 |
| Gorman Cave Spring | 325 | 88 | 6.6 | 0.2 | -5 | 10 | 1.3 | 0.5 |
| Gorman Falls Spring | 344 | 26 | 6.7 | 0.2 | -16 | 12 | 1.2 | 0.6 |
| Gorman Spring | 351 | 22 | 6.7 | 0.3 | -9 | 19 | 0.7 | 0.4 |
| McLarrin Spring | 309 | 46 | 6.9 | 0.2 | -17 | 13 | 0.4 | 0.4 |
| Sulphur Spring | 2067 | 235 | 6.6 | 0.1 | -1 | 14 | 3.0 | 0.3 |
| Well House Spring | 361 | 8 | 6.8 | 0.1 | -13 | 12 | 1.1 | 0.3 |

solutional overprinting and is in some regions locally connected hydrologically with the overlying Marble Falls aquifer (Preston and plane in the Tanyard Formation approximately seventy five meters from the river and six meters above. McLarrin Spring discharges through a conduit in the Gorman Formation within twenty meters of the river and less than two meters above. Well House Spring has been encased and is used as the primary water supply for CBSP. Well House Spring discharges under artesian pressure from the Tanyard Formation approximately one hundred meters from the river and five meters above. Sulphur Spring discharges under artesian pressure from the Honeycutt Formation and is channelized into an artificial impoundment for recreation. Sulphur Spring is located fifty meters from the river and five meters above.

## SPRING MONITORING

Springs in the CBSP area were monitored in order to evaluate spatial and temporal variations in groundwater discharge from the Ellenburger karst system along the Colorado River. Onset HOBO Pendant Temperature Data Loggers were installed at each of the springs to record thermal variations through the course of the study. At three-month intervals, starting in March 2009 and concluding in June 2010, the temperature data loggers were downloaded and redeployed. At each of these intervals, chemistry of spring discharge was measured in the field with Oakton Portable meters (i.e. Oakton pH 300 and CON 400 meters) in order to evaluate spring pH, conductivity and total dissolved solids. Also at each sample period, a water sample was collected for sulfate analysis, which was conducted in the Stephen F. Austin State University Geochemistry Lab with use of an Agilent 8453 UV-Vis Spectrophotometer through precipitation of sulfate ions as barium sulfate induced by the addition of barium chloride (methodology after Shaw, 2006).

Throughout the study, gaps in data exist for various springs. Gaps in thermal data collected with data loggers occur as a result of loss of loggers either from theft or as a result of being dislodged from springs and lost due to natural conditions. Gaps in data collected with hand-held meters are the result of temporary inaccessibility to spring sites either as a result of bat hibernation (e.g. Gorman Cave Spring) or seasonal hunting activity (e.g. Bear Spring).

However, in the case of all springs, except Bear Spring from which deployed data loggers were lost in every sample interval, sufficient data were collected to provide initial characterization of each springs based on temporal and spatial variability.

## RESULTS

The seven springs with typical epigenic karst characteristics (i.e. Bear, Gorman, Gorman Cave, Gorman Falls, Lemon, McLarrin and Well House) exhibited average total dissolved solid values of 346 ppm, average pH values of 6.7, and average conductivities of -11.8 mV (Table 1). Sulfate concentrations in these seven springs were more variable with an average of 0.98 ppm, with slightly elevated concentrations for Gorman Spring, Gorman Falls Spring and Bear Spring and significantly lower concentrations in Gorman Cave Spring and McLarrin Spring (Table 1). In contrast, Sulphur Spring consistently exhibited sulfate levels with an average of 2.98 ppm and correspondingly elevated total dissolved solids averaging 2067 ppm, elevated average temperature of 23.0°C, elevated average conductivity of -1.3 mV and a slightly lower pH of 6.6 (Table 1).

Temporal monitoring of spring temperature (Figure 3) coupled with monitoring of surface temperature and precipitation (Figure 4) showed greater variability amongst springs as expected from quarterly spot sampling with hand-held meters. Springs exhibited variable seasonal fluctuation (Figure 3, Table 2), with maximum fluctuation observed at Gorman Falls Spring and virtually no seasonal fluctuation at McLarrin and Sulphur Springs. Lags in seasonal fluctuation of one to two days occurred with Gorman Cave Spring and Gorman Spring, one to two weeks in seasonal fluctuation was observed in Lemon, Forman Falls and McLarrin Springs; while no data were collected for Bear Springs because of logger loss. Thermal shifts associated with precipitation events show that Lemon Spring, Well House Spring and Sulphur Spring waters thermally equilibrate completely with host rock and do not show thermal response to individual precipitation events. Daily thermal fluctuations are observed in Well House Spring because the data logger was installed in a parkwater

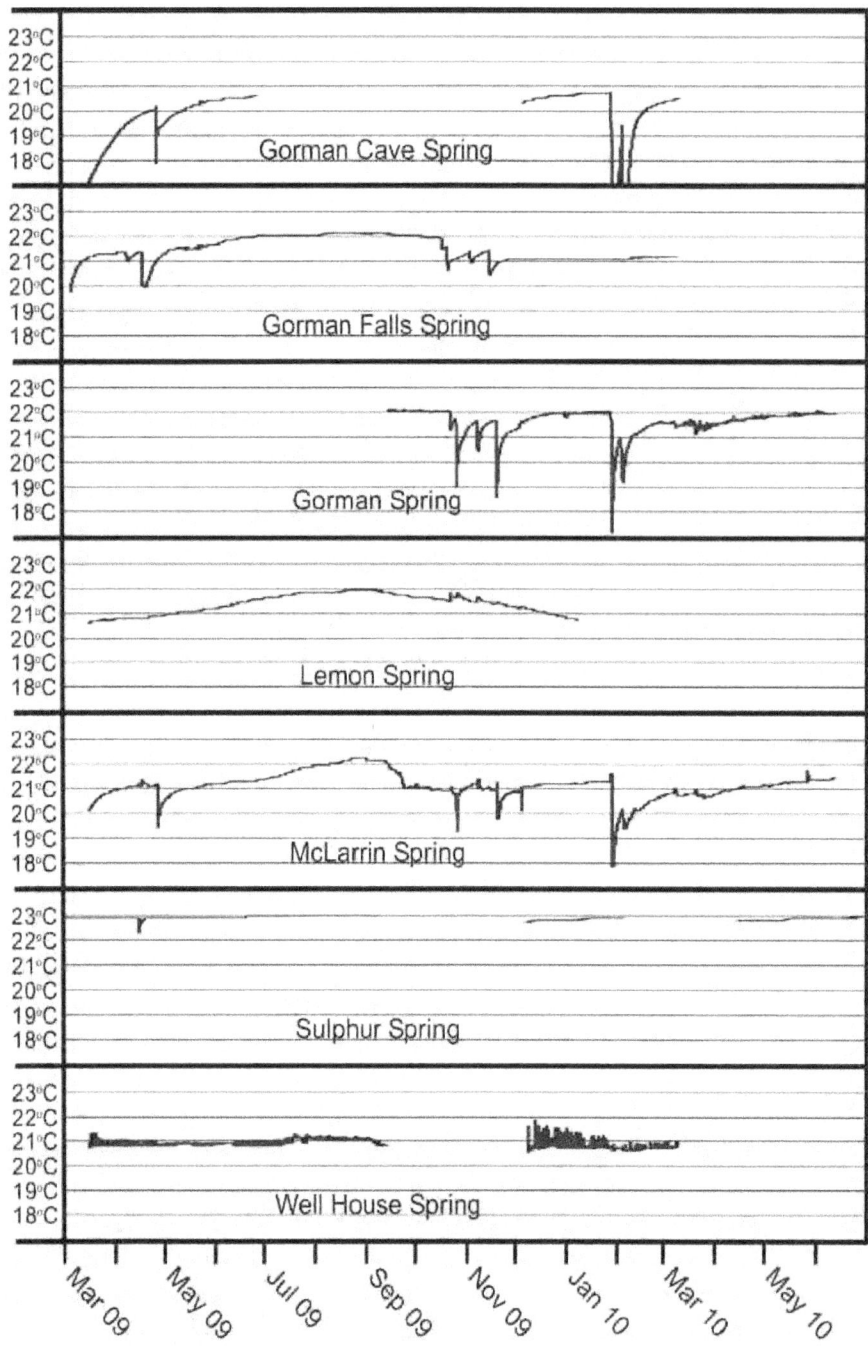

from one degree of equilibration to ten degrees of equilibration, respectively (Table 2).

## DISCUSSION

Significant variability in chemical and thermal characteristics is observed in the springs within the CBSP study area. Physicochemical characteristics measured with portable meters suggest that all of the springs in the study area except Sulphur Spring are associated with relatively shallow flow through Ellenburger Strata, because they exhibit consistent values for temperature and ion activity. In contrast, Sulphur Spring exhibits elevated concentrations of dissolved ions, including significantly greater sulfate content, which combined with the mild thermal component, ~2°C higher than the other springs, suggests a deeper / longer relative flow path

Figure 3. Thermal spring data for springs in study. Note that gaps in data represent lost data loggers, data logger malfunction or inability to retrieve data during hunting seasons.

overflow discharge pipe that is exposed to diurnal heating that overprints spring behavior.

Gorman Cave Spring, Gorman Spring, McLarrin Spring and Gorman Falls Spring, show varying degrees of thermal equilibration between precipitation recharge and discharge, ranging

(Table 1, 2).

Thermal patterns provide insight into the relative flow paths of the groundwater systems associated with heat-transfer between water and conduit walls. Various researchers have made great progress towards characterizing karst

Table 2. Thermal variability of spring discharge, including seasonal response and response to storm events with temperature variations significant enough to show responses in spring temperature. Note that no data is available for Bear Spring because the data logger was lost during all sampling periods.

| | Annual Temperature (°C) | | | Sesonal Variation | | Precipitation Event Response | |
|---|---|---|---|---|---|---|---|
| | avg | min | max | °C Change | lag time (days) | avg °C temp shift | residence time |
| Bear Spring | n/a | n/a | n/a | n/a | n/a | n/a | n/a |
| Gorman Cave Spring | 21.0 | 20.5 | 21.5 | 0.5 | 2.0 | 1.2 | hours |
| Gorman Falls Spring | 21.8 | 20.9 | 22.8 | 0.9 | 14.0 | 10.5 | hours |
| Gorman Spring | 21.8 | 21.5 | 22.1 | 0.3 | 1.0 | 4.8 | hours |
| Lemon Spring | 21.3 | 20.7 | 21.9 | 0.6 | 11.0 | equilabrated | days |
| McLarrin Spring | 21.1 | 20.6 | 21.6 | 0.5 | 10.0 | 7.0 | hours |
| Sulphur Spring | 23.0 | 22.9 | 23.0 | 0.1 | 30.0 | equilabrated | weeks |
| Well House Spring | 20.9 | 20.7 | 21.0 | 0.1 | 21.0 | equilabrated | weeks |

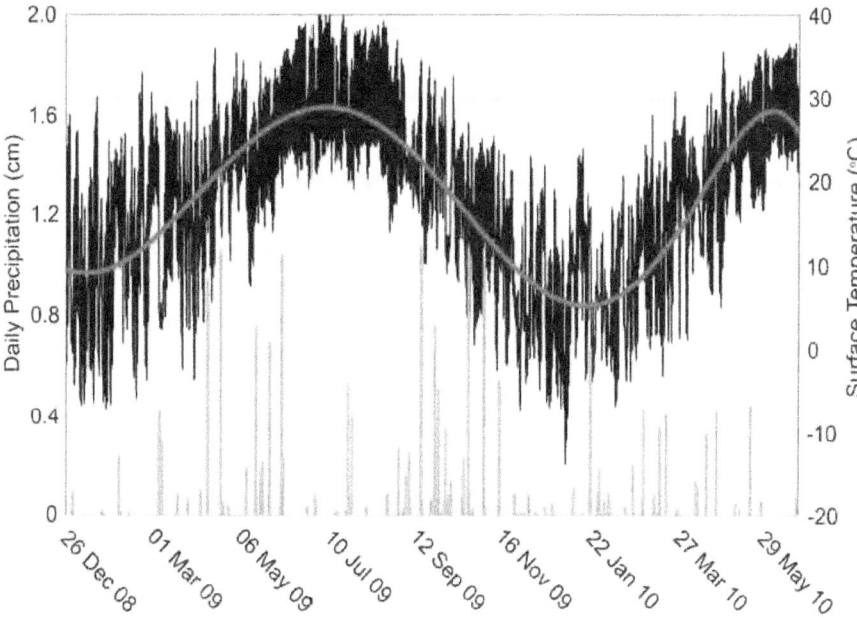

Figure 4. Surface temperature with daily averages (black) and seasonal average (red) compared with precipitation events (blue).

discharge are often not possible because of location and economics.

Thermal patterns analyses indicate that at Lemon Spring, Sulphur Spring and Well House Spring effective heat exchange occurs on the flow paths which is associated with distributed recharge. Well House and Sulphur Spring appear to be largely decoupled from surficial processes and represent deep-circulation flow paths with artesian discharge. Lemon Spring is a shallow karst system influenced by seasonal temperature fluctuations and likely dominated by matrix flow. Gorman Cave Spring, Gorman Falls Spring, Gorman Spring and McLarrin Spring all exhibit ineffective heat exchange along flow suggesting that these systems are controlled by localized recharge through well-integrated fractures and conduits. Gorman Falls Spring and McLarrin Spring show significant thermal response to individual storm events, indicating that flow velocities through these systems are sufficiently high to prevent thermal equilibration with conduit walls. Gorman Cave Spring and Gorman Springs show

systems with temperature being used as a groundwater tracer (e.g. Long and Gilcrease, 2009; Manga, 2001); however, the complexities of karst systems and the relative effectiveness of heat-transfer in different karst systems greatly complicate interpretation. Luhmann and others (*in press*) have developed qualitative methods for comparing karst systems based on thermal patterns observed at spring discharge, in which effectiveness of heat-transfer exchange is evaluated. Storm hydrographs, coupled with thermal monitoring, can provide significant insight into the configuration of karst systems (Birk and others, 2006), but the installation of weirs and level loggers for monitoring of greater thermal equilibrium with individual storm events indicating longer flow paths than Gorman Falls Spring and McLarrin Spring.

## CONCLUSION

Karst development in Ellenburger carbonates proximal to CBSP is of complex polygenetic origin, with current groundwater flow paths exhibiting highly variable behavior. Two springs, Sulphur and Well House Springs, are associated with deep-circulation flow paths associated with distal, distributed recharge. Lemon Spring is a shallow, karst system associated with shallow, distributed recharge and diffuse groundwater flow that is being focused as discharge along a bedding plane. Gorman Falls Spring and McLarrin Spring are associated with well-connected, conduits and localized recharge that rapidly respond to storm events and seasonal fluctuations. Gorman Cave Spring and Gorman Spring also exhibit well-connected conduit systems, but localized recharge appears to be significantly distal to spring discharge based on the degree of equilibration between storm-event water and conduit channels.

These four observed patterns of thermal response suggest that groundwater circulation within the study area likely occurs across multiple horizons of groundwater flow that cross one another at different depths in the subsurface as flow ultimately directed to the potentiometric low imposed by the locally deep incision of the Colorado River. Variability in flow paths may be associated with varying degrees of epigenic overprinting of previously existing hypogene conduits, where unconfined groundwater flow is now occurring through solutional paths established during confined conditions.

In conjuction with thermal patterns, physicochemical characteristics indicate that all spring flow paths, except Sulphur Spring, are effectively limited to flow within the Ellenburger carbonates, even though thermal patterns indicate that the Well House Spring system is associated with deeper / longer circulation flow paths. Elevated temperature and dissolved ions in Sulphur Spring discharge, coupled with degassing of hydrogen sulfide, indicates that this deep-circulation system is

likely associated with regional groundwater flow in contact with mineralized zones near to underlying basement rocks. Reported occurrences of sulfide minerals are associated with lead mineralization proximal to basement rock (Barnes, 1956), which could provide the source for observed fluid chemistry at Sulphur Spring.

Currently, investigations in the study area are attempting to delineate the stratigraphic and structural controls on local and regional karst development. Spring monitoring continues and isotopic studies are planned, including tritium analysis, in order to better delineate the age of water discharging from Sulphur Springs. Heavy metal and trace element analyses are planned to determine the source of sulfur associated with anomalous springs in the Llano Uplift and Texas Hill Country region.

## ACKNOWLEDGMENTS

The authors are grateful to Colorado Bend State Park, and specifically Cory Evans and Kevin Ferguson, for their assistance and access to research sites within the study area. The authors also thank Wesley Brown, Phillip Hays and Andy Grubbs for their constructive reviews of this manuscript which helped to improve it.

## REFERENCES

Barnes, V.E., 1956, Lead Deposits in the Upper Cambrian of Central Texas, Report of Investigations No. 26; University of Texas, Bureau of Economic Geology, Austin, Texas, 68 p.

Birk, S., Liedl, R., and Sauter, M., 2006, Karst spring response examined by process-based modeling; Groundwater, Vol. 44, No. 6, p. 832-836.

Collins, E.W., 1995, Structural framework of the Edwards Aquifer, Balcones fault zone, central Texas; Transactions – Gulf Coast Association of Geological Societies, Vo. 45, p. 135-142.

DeLeon, J.L., 2010, Characterizing the Temporal and Spatial Variability of Spring Discharge Along the Colorado River Near Colorado Bend State Park, Central Texas [Master's Thesis]; Stephen F. Austin State University, Nacogdoches, TX, 156 p.

England, J. Suter, A., and Stafford, K., 2010, Tufa diagenesis in a carbonate karst fluvial environment; Geological Society of America Abstracts with Programs, Vol. 42, No. 2, p. 88.

Estaville, L.E. and Earl, R.A., 2008, Texas Water Atlas; Texas A&M University Press, College Station, TX, 129 p.

Kerans, C., 1990, Depositional Systems and Karst Geology of the Ellenburger Group (Lower Ordovician), Subsurface West Texas, Reports of Investigations; University of Texas, Bureau of Economic Geology, Austin, TX, 63 p.

Klimchouk, A., 2007, Hypogene Speleogenesis: Hydrolgeologic and Morphometric Perspective; National Cave and Karst Research Institute, Carlsbad, NM, 106 p.

Long, A.J. and Gilcrease, P.C., 2009, A one-dimensional heat-transport model for conduit flow in karst aquifers; Journal of Hydrology, Vol. 378, p. 230-239.

Luhmann, A.J., Covington, M.D., Peters, A.J., Alexander, S.C., Anger, C.T., Green, J.A., Runkel, A.C., and Alexander, E.C., *in press*, Classification of thermal patterns at karst springs and cave streams; Groundwater.

Manga, M., 2001, Using springs to study groundwater flow and active geologic processes; Annual Reviews in Earth and Planetary Sciences, Vol. 29, p. 201-228.

Mitchell, K., Dornak, S., and Stafford, K.W., 2011, Geochemical spatial variability of the Colorado River associated with karst springs, Colorado Bend State Park, Central Texas; Geological Society of America Abstracts with Programs, Vol. 43, No. 3, p. 10.

Present, R.D., Pavilcek, D.J., Bluntzer, R.L., and Derton, J., 1996, The Paleozoic and Related Aquifers of Central Texas, Report 346; Texas Water Development Board, Austin, TX, 76 p.

Rose, P.R., 1972, Edwards Group, Surface and Subsurface, Central Texas; University of Texas, Bureau of Economic Geology, Austin, TX, 198 p.

Sellards, E.H., Adkins, W.S., and Plummer, F.B., 1932, The Geology of Texas, v. 1, Stratigraphy; University of Texas, Bureau of Economic Geology, Austin, TX, 1007 p.

Shaw, M.G., 2006, Geologic Controls of Stream Water Composition in Cherokee, Smith and Rusk Counites, Texas [Master's Thesis]; Stephen F. Austin State University, Nacogdoches, TX, 193 p.

Standen, A. and Ruggiero, R., 2007, Llano Uplift Aquifers: Structure and Stratigraphy; Texas Water Development Board, Austin, TX, 78 p.

# Analysis of Long-Term Trends in Flow from a Large Spring Complex in Northern Florida

By Jack W. Grubbs

U.S. Geological Survey, 2639 North Monroe Street, Suite A-200, Tallahassee, FL 32303

## Abstract

Nonparametric regression analysis of historic flow and rainfall data was used to estimate declining flows in a river draining a large spring complex in northern Florida, USA. The analysis indicated that flow declined by an estimated 23 percent from 1900 to 2009. The rate of decline appeared to increase over time, from about 0.8 cubic foot per second per year during the period from 1930-1970, to about 1.1 cubic feet per second per year over the period from 1970-2009. The estimated decline for the period prior to 1980 is consistent with evidence indicating groundwater withdrawals to the east of the study area have diverted groundwater that formerly flowed toward the Ichetucknee River under predevelopment conditions.

## INTRODUCTION

The Ichetucknee River is a tributary to the Suwannee River and drains an area of karst topography in northern peninsular Florida, USA. Flow in the 5.2-mile long river is sustained by 8 named springs and spring complexes that discharge groundwater from the highly transmissive Floridan aquifer system. The purpose of this paper is to describe a long-term trend of declining flow in the Ichetucknee River, and changes in the groundwater flow system in the vicinity of the river that have most likely contributed to this trend.

## REGIONAL GROUNDWATER FLOW SYSTEM

Groundwater in the Floridan aquifer system generally flows toward the Ichetucknee River from the region north and east of the springs. This pattern of flow is evident in potentiometric-surface maps of conditions prior to the initiation of substantial groundwater withdrawals from the Floridan aquifer system in the late 1800s (predevelopement conditions), as well as conditions subsequent to groundwater development, in May 1980 (Johnston and others, 1980; Johnston and others, 1981; figs. 1 and 2, respectively). These two potentiometric-surface maps were chosen because they are the only maps that had been published for the entire Floridan aquifer system, and therefore make it possible to evaluate patterns of the groundwater flow system over large areas that could affect the discharge of groundwater to the Ichetucknee River. The flow lines shown in figures 1 and 2 are drawn so that (1) they are directed from areas of higher water levels toward areas of lower water levels, and (2) points of intersections between the flow lines and equipotential (contour) lines in the potentiometric surface form right angles. In addition to representing directions of groundwater flow, the flow lines can also be used to define the approximate location of important boundaries within the groundwater flow system, as well as the areas that contribute groundwater flow to springs and rivers.

A key feature in the regional groundwater flow system of the Floridan aquifer system in northern Florida is the groundwater flow line that runs roughly northwest to southeast and defines the boundary between groundwater flowing eastward (toward the Atlantic Ocean) and westward (toward the Suwannee, Santa Fe, and Ichetucknee Rivers) (figs.1 and 2). The boundary is formed by two flow lines. The first originates on a high, dome-shaped area of the potentiometric surface of the Floridan aquifer system, near the city of Valdosta, in southern Georgia. The second flow line originates on another high, dome-shaped area of the potentiometric surface of the Floridan aquifer system east of the Ichetucknee River, near Keystone Heights, Florida. These flow paths meet at one or more points where groundwater flow direction diverges and splits (stagnation

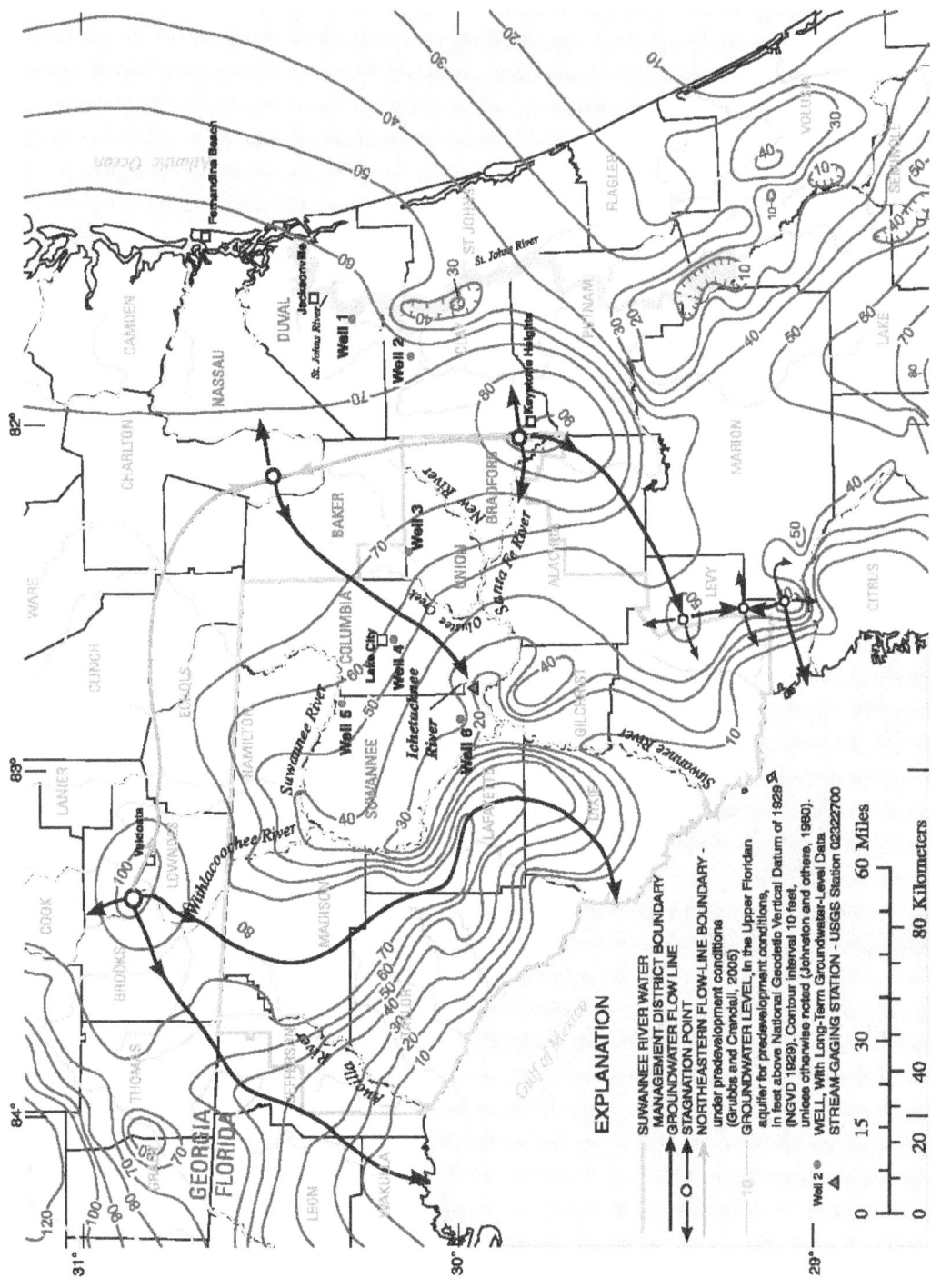

Figure 1. Estimated potentiometric surface of the Upper Floridan aquifer during predevelopment conditions, including key flow line boundaries (contour lines are from Johnston and others, 1980; modified from figure 19 in Grubbs and Crandall, 2005).

EXPLANATION

SUWANNEE RIVER WATER
MANAGEMENT DISTRICT BOUNDARY
GROUNDWATER FLOW LINE
STAGNATION POINT
NORTHEASTERN FLOW-LINE BOUNDARY -
under predevelopment conditions
(Grubbs and Crandall, 2005)
GROUNDWATER LEVEL in the Upper Floridan
aquifer for predevelopment conditions,
in feet above National Geodetic Vertical Datum of 1929
(NGVD 1929). Contour interval 10 feet,
unless otherwise noted (Johnston and others, 1980).
WELL, With Long-Term Groundwater-Level Data
STREAM-GAGING STATION - USGS Station 02322700

Well 2

0    15    30    40    60 Miles

0    20    30    80 Kilometers

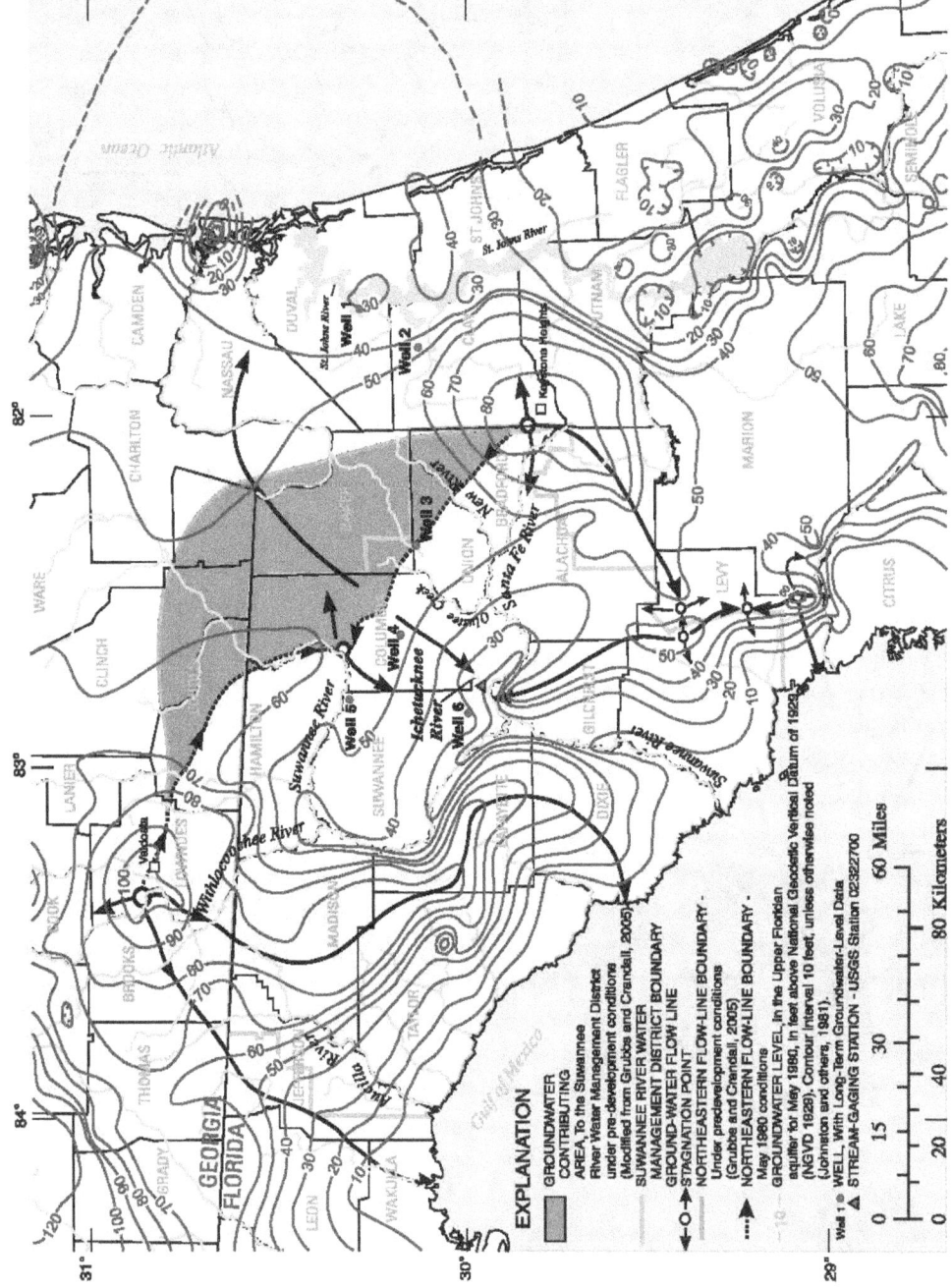

Figure 2. Potentiometric surface of the Upper Floridan aquifer during May 1980 showing: (1) key groundwater flow lines, (2) the northeastern flow-line boundary, and (3) the estimated area where groundwater flow has been diverted away from the Suwannee, Ichetucknee, and Santa Fe Rivers (contour lines are from Johnston and others, 1981; modified from figure 18 in Grubbs and Crandall, 2005).

point) into eastward- and westward-directed flow paths. In this paper, this divide is referred to as the northeastern flow-line boundary.

The location of the northeastern flow-line boundary is evident in the potentiometric-surface map representing predevelopment conditions, as well as in maps made subsequent to groundwater development. The configuration of the potentiometric-surface map corresponding to predevelopment conditions (Johnston and others, 1980) suggests that the northeastern groundwater flow boundary passed through an area near the eastern boundaries of Baker and Bradford Counties in Florida (fig. 1; Grubbs and Crandall, 2007, fig. 19, page 26) prior to groundwater development. This area coincides with the Trail Ridge, a physiographic feature with the highest elevation in the area between the Suwannee and the Atlantic Ocean.

The configuration of the May 1980 surface suggests that the northeastern groundwater flow boundary migrated westward after development of the groundwater flow system (fig. 2; Grubbs and Crandall, 2007). This movement is consistent with the historic patterns of groundwater development (2011, http://fl.water.usgs.gov/infodata/wateruse/counties.html), in which groundwater withdrawals have been substantialy larger to the northeast (in Duval and Nassau Counties) than in the four counties closest to Ichetucknee Springs (Columbia, Union, Baker, and Suwannee) (fig. 3). Comparison of the two potentiometric-surface maps also indicates that groundwater-level declines between predevelopment conditions and May 1980 were negligible near the Ichetucknee River, but became more pronounced toward the northeast. For example, the maps indicate that groundwater levels fell by approximately 10-12 feet in Baker and northern Union Counties, approximately 25 feet in eastern Duval and Nassau Counties, and by as much as 90 feet near Fernandina Beach, where paper mills withdrew approximately 50 cubic feet per second ($ft^3 s^{-1}$) from the Floridan aquifer system in 2000 (Richard Marella, U.S. Geological Survey, written commun., 2010). This geographic pattern of groundwater-level decline and groundwater withdrawals suggest that larger withdrawals, by lowering groundwater levels over a larger area, shift the

northeastern flow-line boundary and create a larger contributing area to capture greater amounts of groundwater recharge. Westward migration of the boundary has also resulted in a contraction of the area of westward-flowing groundwater toward the Suwannee, Ichetucknee and Lower Santa Fe Rivers (fig. 2). It should be noted that this area where the groundwater contributing area has changed (fig. 2) was originally delineated in figure 18 in Grubbs and Crandall (2007). The area shown in figure 2 differs from that originally shown in figure 18 in Grubbs and Crandall (2007) because it reflects corrections that were made to the contour lines that were shown in the original figure.

The declines in groundwater levels of the Floridan aquifer system observed at individual wells are consistent with the differences seen in the predevelopment and May 1980 potentiometric-surface maps. Data from selected wells (fig. 4) indicate that average-annual groundwater levels declined by 2 to 10 feet in wells east and west of the northeastern groundwater flow divide from 1960 to 1980, and by approximately 4 to 12 feet in these same wells from 1960 to 2009. Longer-term data from well 1 near the metropolitan area of Jacksonville indicated a decline of more than 30 feet from 1930 to 2009. Kendall's tau tests (Conover, 1980) for negative correlations between groundwater-levels and time were significant for all of the time series shown in figure 4, with the highest levels of significance (p-value < 0.0001) associated with wells 1-5 (well 6 had a one-sided p-value of 0.048). The computed values of Kendall's tau also became progressively more negative (and increased in absolute value) for wells closer to Duval and Nassau Counties, indicating a closer correspondence between groundwater level and time as one moves from the Ichetucknee River toward these two counties.

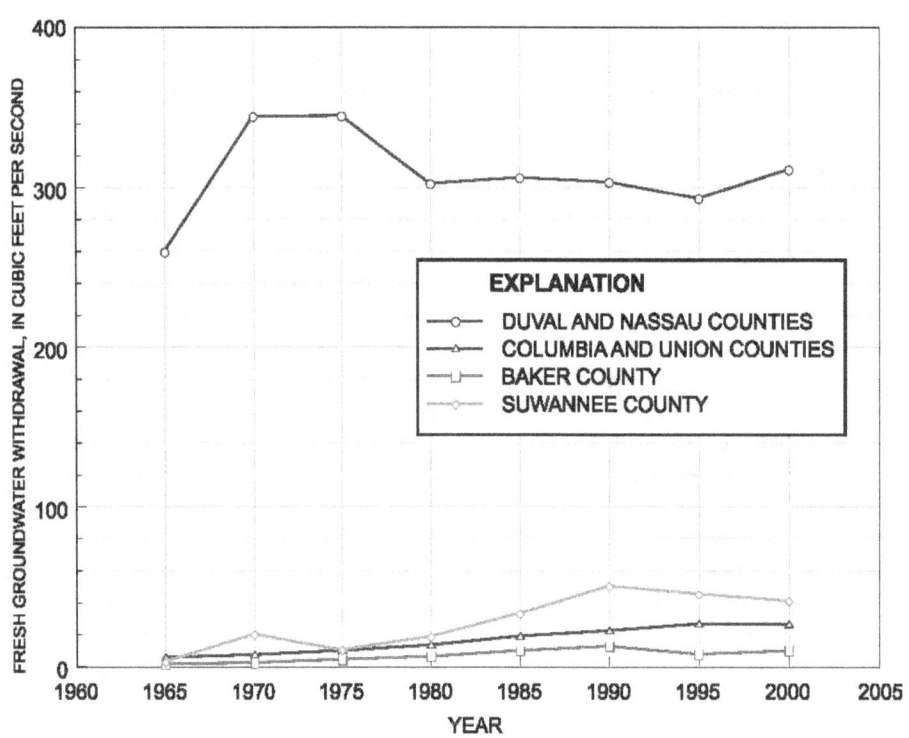

Figure 3. Time-series of historic groundwater withdrawals from selected countines in northern peninsular Florida, 1965-2000.

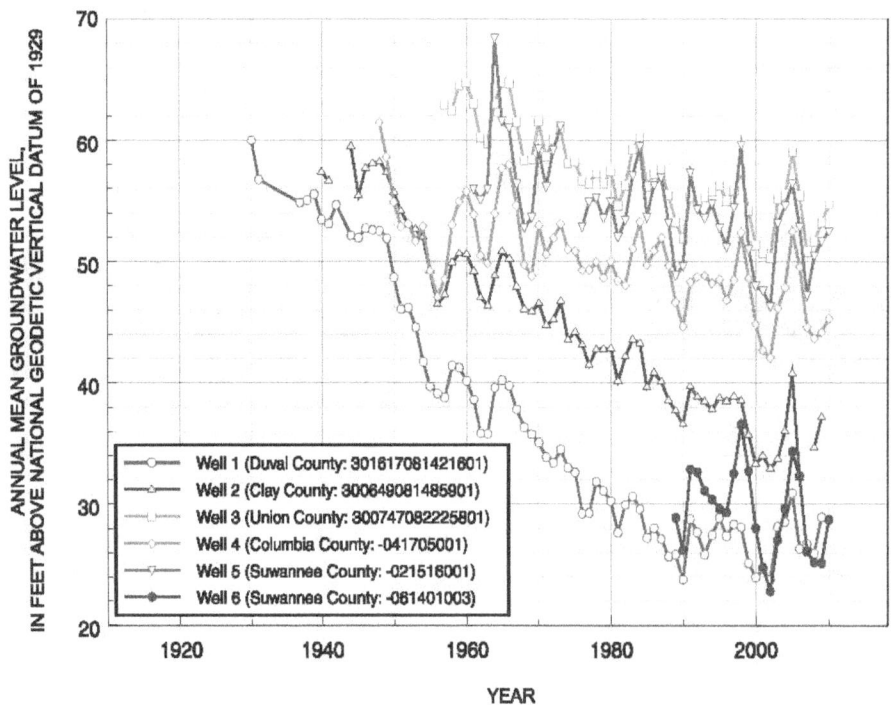

Figure 4. Time-series of groundwater levels from selected wells in northern peninsular Florida.

## TRENDS IN THE FLOW OF THE ICHETUCKNEE RIVER

Between December 1898 and September 2009 the U.S. Geological Survey (USGS) made 538 individual field measurements of flow at a cross section of the Ichetucknee River downstream from its major springs, where it passes under the bridge at U.S. Highway 27 (USGS station 02322700). Comparison of time-series plots of the flows measured at this cross section and rainfall totals accumulated at Lake City (National Weather Service coop station 084731) for the 24 month period prior to (and including) the month during which each flow measurement illustrate the relation between river flow, rainfall, and time (fig. 5). As expected, measured flows generally increased with increasing antecedent rainfall totals, and decreased as these rainfall amounts decreased. For example, flows in the river generally increased as conditions became wetter from late 1963 to late 1965, and subsequently decreased as conditions became drier from late 1965 to late 1968. In addition, periods of extreme low flow coincided with periods of deficient rainfall in the

mid 1950s, 2000-2003, and 2007-2009. Evidence for a possible relation between time and flow is suggested by noting that the average flow from 1930-1965 (346 $ft^3s^{-1}$) was higher than the average flow for the period after 1970 (326 $ft^3s^{-1}$), even though the average 24-month rainfall accumulation for measurements made in the earlier period was lower than that of the later period (100 inches versus 111 inches, respectively). Similarly, the low flows during dry periods in 2000-2003 and 2007-2009 were generally lower than those during the dry period in the mid 1950s, even though the 24-month antecedent rainfall totals in the 2000-2003 and 2007-2009 dry periods were much higher than the rainfall totals during the mid 1950s (fig. 5). These observations suggest that flow in the Ichetucknee River may have declined over time from causes that are unrelated to rainfall.

To assess the significance of the time trends in the flow in the Ichetucknee River while controlling for the effects of rainfall, a multivariate, locally-weighted scatterplot smoothing, (LOWESS; Cleveland and others, 1988) regression model was created to estimate flow in the Ichetucknee River at U.S. Highway

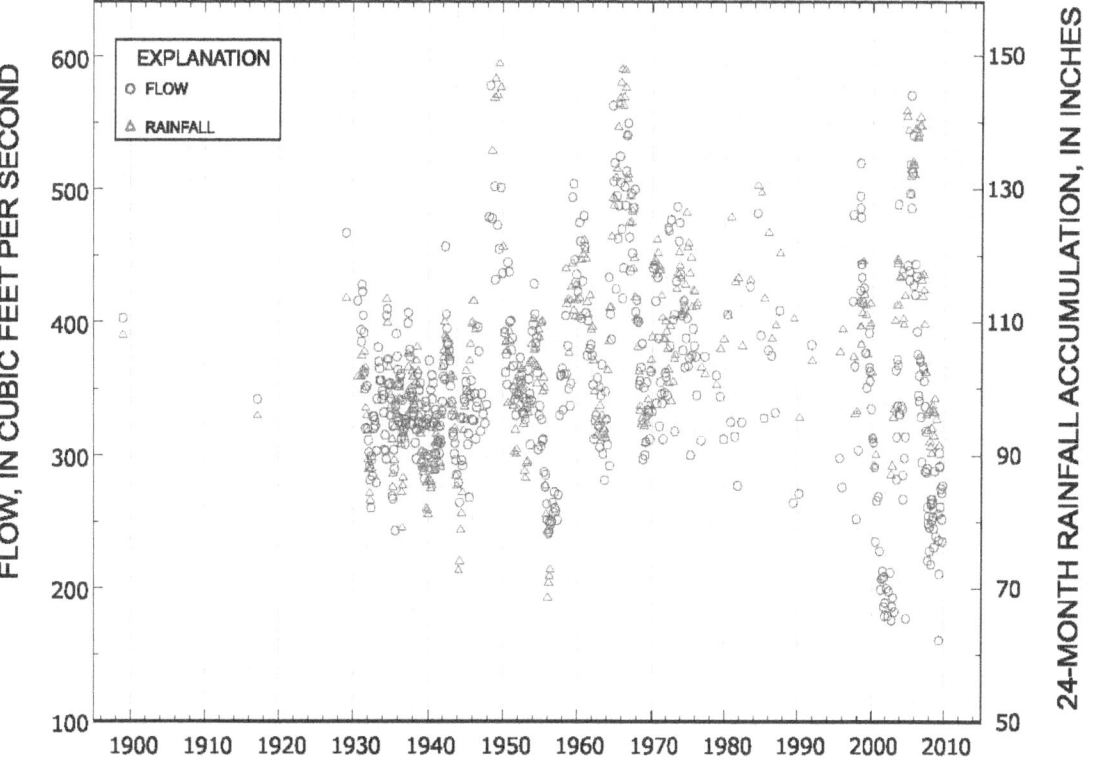

Figure 5. Time series of historic stream-flow measurements in the Ichetucknee River at USGS gaging station 02322700, and associated values of 24-month antecedent rainfall accumulations from the National Weather Service Weather Station, Lake City 2 E (COOP ID 084731).

27 as a function of two explanatory variables: time and 24-month antecedent rainfall. LOWESS regression was chosen because it requires no prior assumptions about the form of the relation between the response and explanatory variables, such as whether the relation is linear or nonlinear. The data used to fit the model were obtained by grouping the individual flow measurements by the year in which each measurement was made, and then computing the average of the flows, 24-month rainfall accumulations, and dates associated with each measurement made in a given year. This was done to obtain more reliable estimates of the model error and confidence limits for the flow values produced by the model. The overall model fit was highly significant (p-value < 0.001). Nested F-tests indicated that, although both explanatory variables were significant at a confidence levels of 0.96 or better, the addition of the explanatory variable associated with the 24-month rainfall accumulation was responsible for a greater degree of model improvement (p-value < 0.007) than that associated with time (p-value < 0.04).

The regression model made it possible to evaluate the magnitude and significance of the time trend over selected time intervals by using it to estimate river flow at the beginning and end of a given time interval (using the same antecedent rainfall amount for computing both estimates), then computing changes in flow over the interval as the difference between the two estimates. Table 1 shows the results from an analysis of changes in the Ichetucknee River flow under conditions of typical rainfall (24-month rainfall accumulation was set equal to the median of 24-month rainfall accumulation values used to fit the regression model). In this analysis, the estimated flow declined by 92 $ft^3s^{-1}$ from 1900 to 2009 (table 1). The analysis also suggests that the stream-flow losses have probably accelerated over time. For example, the estimated slope of the trend for the period between 1930 and 1970 was about -0.8 cubic foot per second per year (downward), but steepened to about -1.1 cubic feet per second per year in the period between 1970 to 2009.

## DISCUSSION

The estimated decline in the flow of the Ichetucknee River from 1900 to 1980 is approximately 60 $ft^3s^{-1}$ (table 1), with a 95 percent confidence interval of 23 to 96 $ft^3s^{-1}$. The configuration of the May 1980 potentiometric surface and a more recent detailed study of the surface in the vicinity of the Ichetucknee River (Sepulveda and others, 2006) suggest that the area contributing groundwater to the river during this period was probably located within an area roughly comprising the southern half of Columbia County and perhaps half of Union County. The estimated total fresh groundwater withdrawals from the Floridan aquifer system in these two counties was 13 $ft^3s^{-1}$ in 1980 (2011, http://fl.water.usgs.gov/infodata/wateruse/counties.html). Thus it seems unlikely that estimated decline in the flow of the Ichetucknee River from predevelopment conditions to 1980 can be explained entirely by groundwater withdrawals within the post-development groundwater contributing area to the river. The most likely alternative explanation is that estimated flow declines were caused by the westward migration of the northeastern groundwater flow boundary (and the concurrent contraction of the groundwater contributing area to the Ichetucknee River).

This explanation is also consistent with estimates of the magnitude of groundwater flow that was diverted from the river in response to the contraction of its contributing area, computed as the product of change in the size of the contributing area and groundwater recharge rate to the Floridan aquifer system. The potentiometric-surface maps suggest a plausible range of approximately 450 to 850 square miles for the reduction in the size of the groundwater contributing area to the Ichetucknee River from predevelopment conditions to May 1980. Recharge rates of approximately 1 to 1.8 inches per year over an area of this magnitude are sufficient to produce a flux of 60 $ft^3s^{-1}$, and are also consistent with plausible rates of recharge (for example, Sepulveda, 2002) in this area where groundwater flow directions have been reversed. Therefore it seems likely that some (if not most) of the estimated reductions in flow in the Ichetucknee River from predevelopment conditions to 1980 resulted from changes in the groundwater flow system arising from withdrawals outside of the area that contributed groundwater to the river in and prior to 1980.

Table 1. Estimated Changes in Flow in the Ichetucknee River at USGS Station 02322700 for Typical Rainfall Conditions and Selected Time Intervals.

[Unit abbreviations: $ft^3s^{-1}$, cubic feet per second; $ft^3s^{-1}yr^{-1}$, cubic feet per second per year]

| Time Interval | Time Span ($\Delta$t), in years | Estimated Decline in Flow ($\Delta$Q), in $ft^3s^{-1}$ | Standard Error of $\Delta$Q, in $ft^3s^{-1}$ | Lower limit of 95% confidence interval for $\Delta$Q, in $ft^3s^{-1}$ | Upper limit of 95% confidence interval for $\Delta$Q, in $ft^3s^{-1}yr^{-1}$ | Estimated Rate of Decline in Flow, in $ft^3s^{-1}yr^{-1}$ |
|---|---|---|---|---|---|---|
| 1900-2009 | 119 | 92 | 20 | 52 | 132 | 0.84 |
| 1930-1970 | 40 | 31 | 11 | 10 | 53 | 0.78 |
| 1970-2009 | 39 | 42 | 11 | 20 | 65 | 1.08 |
| 1900-1980 | 80 | 60 | 18 | 23 | 96 | 0.74 |
| 1980-2009 | 29 | 32 | 11 | 10 | 55 | 1.12 |

## REFERENCES

Cleveland, W.S., and Devlin, S.J., 1988, Locally-weighted Regression: An Approach to Regression Analysis by Local Fitting. J. Am. Statist. Assoc., Vol. 83, pp 596-610.

Conover, W. J., 1980, Practical Nonparametric Statistics. 2nd. ed. New York: Wiley

Grubbs, J.W., and Crandall, C.A., 2007, Exchanges of Water between the Upper Floridan Aquifer and the Lower Suwannee and Lower Santa Fe Rivers, Florida: U.S. Geological Survey Professional Paper 1656-C, 83 p.

Johnston, R.H., Healy, H.G., and Hayes, L.R., 1981, Potentiometric surface of the Tertiary limestone aquifer system, southeastern United States, May 1980: U.S. Geological Survey Open-File Report 81-486, 1 sheet.

Johnston, R.H., Krause, R.E., Meyer, F.W., Ryder, P.D., Tibbals, C.H., and Hunn, J.D., 1980, Estimated potentiometric surface for the Tertiary limestone aquifer system, southeastern United States, prior to development: U.S. Geological Survey Open-File Report 80-406, 1 sheet.

Sepulveda, A.A., Katz, B.G., and Mahon, G.L., 2006, Potentiometric Surface of the Upper Floridan Aquifer in the Ichetucknee Springshed and Vicinity, Northern Florida, September 2003: U.S. Geological Survey Open-File Report 2006-1031, 1 sheet.

Sepulveda, Nicasio, 2002, Simulation of Ground-Water Flow in the Intermediate and Floridan Aquifer Systems in Peninsular Florida: U.S. Geological Survey Water-Resources Investigations Report 02-4009, 130 p.

# KARST MODELING
## Modifications to the Conduit Flow Process Mode2 for MODFLOW-2005

By Thomas Reimann[1], Steffen Birk[2], and Christoph Rehrl[2]
[1]Technische Universitaet Dresden, Institute for Groundwater Management, D-01062 Dresden, Germany
[2]University Graz, Institute for Earth Sciences, Heinrichstraße 26, A-8010 Graz, Austria

## Abstract

Generally, groundwater flow is assumed to be slow and therefore laminar. However, karst aquifers exhibit highly conductive features resulting from rock dissolution processes that cause strong heterogeneity and anisotropy. Groundwater flow can become turbulent within these structures. It is a challenging task to address these karst features in numerical models of groundwater flow. One approach is to couple the continuum model with a pipe network that represents the conduit system, for example the Conduit Flow Process Mode 1 (Shoemaker and others, 2008). These models exhibit an enhanced parameter demand as well as increased numerical efforts. In contrast, the Conduit Flow Process Mode 2 (CFPM2) computes turbulent flow by adapting the hydraulic conductivity within the linear Darcy law used by the MODFLOW continuum model (Shoemaker and others, 2008). Consequently, this approach allows computation of turbulent flow within the continuum model. In doing so, CFPM2 reduces the practical as well as the numerical efforts for simulating turbulent conduit flow.

Contrary to laminar flow, which is usually described by linear equations like the well known Darcy formula, turbulent flow can be described by nonlinear gradient functions. A universal formulation is:

$$\partial h/\partial x = \beta_1 q + \beta_2 q^m \tag{1}$$

where $h$ is the hydraulic head, $x$ is the spatial coordinate, $q$ is the specific discharge, $\beta_1$ as well as $\beta_2$ are parameters related to fluid and rock properties, and $m$ represents the constant exponent of a power law. Laminar flow is covered by the linear term in equation (1) whereas the nonlinear term describes turbulent flow by a power law. For turbulent flow conditions with high specific discharge the linear term in equation 1 becomes small and therefore can be neglected.

The original formulation within CFPM2 was for large pore aquifers where the onset of turbulence occurs at low Reynolds numbers (1 to 100) and not for conduits. Benchmark computations demonstrate that CFPM2 considers turbulent flow according to the universal power law with $m = 1.5$ (Reimann and others, 2011). In addition, the existing CFPM2 needs several time steps for convergence because of iterative adjustment of the hydraulic conductivity without additional control of closure. Modifications to the existing CFPM2 were made by implementing a generalized power function where the exponent $m$ is user-defined (Reimann and others, 2011). This allows for matching turbulence in porous media as well as pipes and eliminates the time steps required for iterative adjustment of hydraulic conductivity. The modified CFPM2 successfully reproduced simple validation test problems.

## REFERENCES

Reimann T., Birk, S,, Rehrl, C., and Shoemaker, W. B., 2011, Modifications to the Conduit Flow Process Mode 2 for MODFLOW-2005: Ground Water, (online early, March, 2011).

Shoemaker, W.B., Kuniansky, E. L., Birk, S., Bauer, S., and Swain, E. D., 2008, Documentation of a Conduit Flow Process (CFP) for MODFLOW-2005: U.S. Geological Survey Techniques and Methods 6-A24, 50 p.

This work was funded by the German Research Foundation (DFG) under grants no. LI 727/11-1 and SA 501/24-1 and the Austrian Science Fund (FWF) under grant no. L576-N21. The authors express their appreciation to Martin Sauter and Tobias Geyer from the University Göttingen as well as to Rudolf Liedl from the TU Dresden for thoughtful discussions. Sadly, we have to report that our colleague and friend Dr. Christoph Rehrl died on 21 March 2010. His dedication and person were an inspiration to us all.

# Comparison of Three Model Approaches for Spring Simulation, Woodville Karst Plain, Florida

By Eve L. Kuniansky[1], Josue J. Gallegos[2], and J. Hal Davis[3]

[1]U.S. Geological Survey, 2850 Holcomb Bridge Road, Norcross, GA 30092
[2]Graduate Student, Department of Geological Sciences, Florida State University, Tallahassee, FL 32306
[3] U.S. Geological Survey, 2639 North Monroe Street, Suite A-200, Tallahassee, FL 32303

## Abstract

Simulation of flow in karst aquifers with conduits is difficult because the presence and exact locations of conduits are frequently unknown. The conduits typically convey as much as 99 percent of the fast flow (Palmer, 2007). The Wakulla-Leon Sinks submerged cave system of the Woodville Karst Plain is an exception as many conduits are mapped. Extensive mapping of the conduit system has been accomplished over the past twenty years by cave divers of the Global Underwater Explorers as part of their Woodville Karst Plain Project. As a result of their pioneering work, this karst drainage system is believed to be moderately well characterized. Their maps served as the basis for three model approaches used to simulate the system. The three model approaches for simulation of the groundwater-conduit system are: (1) single continuum laminar flow only (Davis and others, 2010); (2) single continuum that allows for turbulent and laminar flow (model of Davis and others, 2010 with Conduit Flow Process mode 2; Shoemaker and others, 2008 as modified by Reimann and others, 2011); and (3) hybrid model—that consists of a single continuum coupled to pipe-flow network capable of both turbulent and laminar flow (Gallegos, 2011 with Conduit Flow Process mode 1; Shoemaker and others, 2008 ).

The presentation of field applications of MODFLOW-2005 (Harbaugh, 2005) and the Conduit Flow Process (Shoemaker and others, 2008) examines spring discharge under various time discretization and flow conditions. Model approach 1,the subregional model of Davis et al. 2010 was a transient simulation from January 1, 1966, when sprayfield operations began through 2018, when effects of system upgrades should have occurred. The stress periods were mostly annual. Calibration data for water levels and spring discharge were available for November 1991 and May to early June 2006, plus tracer test datasets conducted in 2006 and 2007 by Hazlett-Kincaid, Inc (Davis and others, 2010). Model approach 2 uses Model 1 and was not calibrated. Model approach 3, the hybrid model, was calibrated to stressed steady-state conditions for the same calibration datasets as model 1 (Gallegos, 2011). Models (1) and (3) simulated observed average spring discharge at Wakulla and Spring Creek Springs within 10 percent and head residuals were within the calibration criteria of plus or minus 5 ft. Thus, for average spring discharge and tracer test time of travel, the models produce similar results and both are considered acceptable calibrations.

In order to observe differences between the model approaches, a transient period with daily springflow observations and a storm event was simulated (rising and falling of spring flow, no recalibration). For the 52 day storm event, which was simulated with daily time steps, none of the models simulated the observed peak discharge well, defined as within 10 percent of observed. Model approach 2 used the identical input for MODFLOW-2000 from Davis and others (2010) and the transient recharge data prepared for the 52-day period, but with the implementation of the turbulence in standard MODFLOW as derived by Reimann and others (2011) using constant values of pipe diameter 20, lower critical Reynolds number of 3,000 and upper critical Reynolds number of 10,000. The greatest differences between the shape of the simulated storm hydrographs for Wakulla Springs and Spring Creek were between the hybrid model (3) and the continuum model (1). The continuum model with turbulence, model 2, had similar hydrographs as the hybrid model. Methods 1 and 3 were in error by approximately 30 percent in simulation of Wakulla Spring peak discharge and method 2 by 10 percent. However, if the total volume of discharge for the 52 day period at Wakulla Springs is compared, model 3 matched observed (within 0.01 percent); model 1by 23 percent; and model 2by 16 percent. Model 1 over-simulated peak observed discharge and models 2 and 3 under-simulated peak observed discharge at Wakulla Spring. For Spring Creek, model 3 was in error by 75 percent at peak discharge; model 1 by 42 percent; and model 2 by 7 percent (considered a good match). Models 1 and 3 over-simulated peak observed for Spring Creek. However, if the total volume of discharge for Spring Creek is compared, model 3 was in error 22 percent; model 2 by 27 percent; and model 1 by 41 percent. The computation time required by the hybrid model (3) was more than 100 times longer than the single continuum model 1 (over 72 hours on a personal computer, which precludes using parameter estimation to improve model fit). This test of hybrid models is perhaps the largest with over 1,000 pipes and nodes. The hybrid approach requires that 2 models are iteratively solved until both converge at each one-day stress period and the pipe network is very slow to converge for this application.

A recent study (Hill and others, 2010), comparing the hybrid model approach with the single continuum model approach for two other springs systems in Florida, concluded that the hybrid model more closely simulated observed transient spring discharge. This study concludes that for average conditions neither model approach 1 or 3 is distinctly better. For the storm hydrograph, only model 2 met the calibration criteria of matching peak spring discharge within 10 percent and better calibration with parameter estimation techniques was not possible for the hybrid model approach as a result of the long simulation time. However, the hybrid model (3) did outperform the continuum models (1 and 2) for simulation of total discharge and matched total discharge at Wakulla Springs and mathematically could mimic nature more accurately if fully calibrated. Thus, it is unclear if the extra effort required to use a hybrid model, both in data preparation and computation time, is justified based on these two apparently conflicting case studies. Two additional considerations of this study need to be examined further; the first- the lack of inclusion of turbulence in the single continuum model by Davis and others (2010) did not result in a greater peak discharge for Spring Creek than model 3 as it did for Wakulla Springs, which was expected since turbulent flow was simulated in the pipe network (Gallegos, 2011). The second was that change in temporal discretization for simulation of a storm event appeared to result in the need for recalibration of the models, indicating that the models are not suitable for simulation of storm hydrographs and should only be used for annual average conditions.

Note the U. S. Geological Survey version of CFP mode 2 that incorporates the new algorithm by Reimann and other (2011) is a beta version and full testing of the code is not complete at this time, but will be incorporated in version 2.

## References

Davis, J.H., Katz, B.G., and Griffin, D.W., 2010, Nitrate-N Movement in Groundwater from the Land Application of Treated Municipal Wastewater and Other Sources in the Wakulla Springs Springshed, Leon and Wakulla Counties, Florida, 1966-2018: U.S. Geological Survey Scientific Investigations Report 2010-5099, 90 p.

Gallegos, J.J., 2011, Modeling Groundwater Flow in Karst Aquifers: an evaluation of MODFLOW-CFP at a laboratory and sub-regional scale. Master's Thesis, Florida State University, Tallahassee, Florida.

Harbaugh, A.W. 2005. MODFLOW-2005, the U.S. Geological Survey modular ground-water model -- the Ground-Water Flow Process: USGS Techniques and Methods 6-A16.

Hill, M. E., Stewart, M. T. and Martin, A., 2010, Evaluation of the MODFLOW-2005 Conduit Flow Process: Ground Water, vol. 48, no. 4, pp. 549–559.

Palmer, A.N., 2007, Cave Geology: Cave Books, Dayton, Ohio, 453 p.

Reimann, Thomas., Birk, Steffen., Rehrl, Christoph., and Shoemaker, W.B., 2011, Modifications to the Conduit Flow Process for MODFLOW-2005 Mode 2: Ground Water (online early, March 2011).

Shoemaker, W.B., Kuniansky, E. L., Birk, Steffen, Bauer, Sebastian, and Swain, E.D., 2008, Documentation of a Conduit Flow Process (CFP) for MODFLOW-2005, U.S. Geological Survey Techniques and Methods Book 6, Chapter A, 50p.

# Synthesis of Multiple Scale Modeling in the Faulted and Folded Karst of the Shenandoah Valley, Virginia and West Virginia

By Kurt J. McCoy[1], Mark D. Kozar[2], Richard M. Yager[3], George E. Harlow[1], and David L. Nelms[1]

[1]U.S. Geological Survey, 1730 East Parham Road, Richmond, VA 23228
[2] U.S. Geological Survey, 934 Broadway, Tacoma, WA 98402
[3] U.S. Geological Survey, 30 Brown Road, Ithaca, NY 14850

## Abstract

Groundwater flow models were developed at the basin scale, 3,300 square miles ($mi^2$); sub-basin scale , 300 $mi^2$; and watershed scale, 10 $mi^2$; in the Shenandoah Valley to evaluate geologic controls on groundwater flow patterns and to describe water budgets at various management scales. At basin scale, ground-water flow in the folded and faulted sedimentary rocks of the Shenandoah Valley is largely controlled by anisotropy along bedding. Refinement of the conceptual model at sub-basin and watershed scales highlights the role of faults, fault orientation, heterogeneity in carbonate unit aquifer properties, and epikarst to control the rates of direction of groundwater flow. A synthesis of these modeling efforts shows how scale-related changes in conceptual models provide hydrogeologic insight into differing geologic features of importance to water resource managers at varying scales.

# A Hydrograph Recession Technique for Karst Springs with Quickflow Components That Do Not Exhibit Simple, Zero-Order Decay

By Darrell W. Pennington[1] and Van Brahana[2]
[1]FTN Associates, Ltd., 124 W. Sunbridge Drive, Suite #3, Fayetteville, AR 72703
[2]Geosciences, University of Arkansas, 113 Ozark Hall, Fayetteville, AR 72701

**Abstract**

Recession analysis of karst-spring hydrographs provides valuable information about the hydrodynamic performance of the system following storm events. Responses of karst aquifers to recharge events depend upon numerous controlling factors which generate a variety of observed recession forms. Hydrograph decomposition techniques using exponential curve fitting can identify up to three, sometimes more, straight-line segments. Observations of springs from karst regions throughout the world have shown that in some instances the intermediate slope (or recession coefficient) is steeper than the initial slope, producing a convex form on the quickflow curve. Inasmuch as discharge is proportional to head, the discharge is expected to decline most rapidly (i.e. steepest slope segment ) immediately following the peak when greater water volume provides driving force. Additionally, in semi-log space, the quickflow component is expected to decay in a linear manner, and the baseflow component is expected to decay in an exponential manner, producing a concave shape on the quickflow curve. Whereas numerous recession methodologies are described in the literature, almost all of these approaches are designed for springs that exhibit simple zero- or first-order decay and are not applicable to recession coefficients that increase with time. We revised the methodology of Mangin to evaluate two karst springs that exhibit concave and convex quickflow components located at the Savoy Experimental Watershed in the southwestern Ozark Plateaus. Three functions are used to simulate the hydrograph recessions: 1) Mangin's linear equation for quickflow components that exhibit concavity, 2) a new, empirically derived equation, proposed herein, for quickflow components that exhibit convexity, and 3) Maillet's exponential for baseflow components. Comparisons of results show that despite a similar hydrogeologic setting, these two springs have very different transmission behaviors that are attributed to differences in flow path properties and hydraulic connection with the epikarstic zone. Based on available monitored parameters, the recession form appears to be a function of antecedent saturation in aquifer components prior to rainfall, available recharge, and hydraulic control and interaction among the reservoir zones.

## INTRODUCTION

The temporal and spatial variation of hydraulic properties in karst aquifers under different flow regimes makes quantification and prediction of groundwater movement through these systems using traditional field methods designed for flow in porous media difficult (Baedke and Krothe, 2001). The distribution and frequency of a high-velocity conduit network (quickflow) embedded in a slow-velocity rock matrix (baseflow) creates dualistic, heterogeneous, and anisotropic hydraulic processes (i.e. variable infiltration, flow, and discharge rates) that affect the overall hydrodynamic performance of the system (Atkinson, 1977; Király, 2003). In these settings, well data gathered from conventional aquifer tests cannot be extrapolated far beyond the borehole owing to unknowns about the geometry, changes in aperture size, and interconnectivity of the conduits (Baedke and Krothe, 2001).

In contrast to wells, karst springs integrate flow over an area and show the response of the entire basin to recharge events (Padilla et al., 1994; Baedke and Krothe, 2001). Analytical spring-hydrograph techniques provide realistic estimates of hydraulic parameters of karstic aquifers and reveal information regarding the void size and spatial configuration of the conduit network (Kovács and Perrochet, 2008). The principal influences of precipitation, storage conditions, basin characteristics (such as size and slope), drainage network density, geological variability, vegetation, and soil affect the lag time between input and output response as well as the output pattern; therefore, spring hydrographs can show a wide range of forms (Ford and Williams, 2007).

## Karst-Spring Recession Analysis

Traditional karst-spring hydrograph decomposition techniques using exponential curve fitting can identify up to three, sometimes more, straight-line segments (Figure 1). The reason for these segments is debated where one group of researchers propose that the segments represent discharge components from three individual (or parallel) reservoirs: conduit network, fracture network, and low-permeability matrix (Atkinson, 1977; Padillia et al., 1994; Shevenell, 1996). These three segments appear due to changes in hydraulic conductivities (K) where water levels in aquifer components with lower hydraulic resistance respond more quickly to recharge. Accordingly, transient flow through each of these reservoirs proceeds at rates controlled by K, causing drainage among the three reservoirs to occur in an out-of-phase sequence that is easily recognizable on a hydrograph (White, 1988).

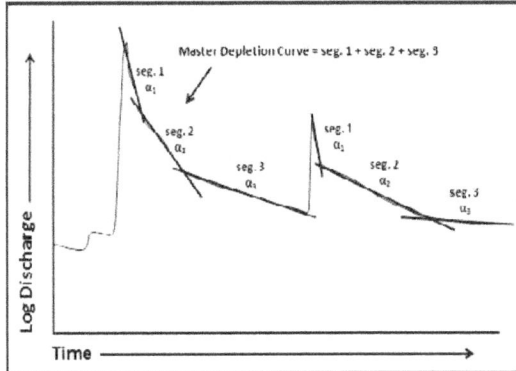

Figure 1. Fitting of exponential segments to karst-spring recession limbs. The combination of each segment represents the master depletion curve (after Shevenell, 1996).

A major limitation to this argument is that it is based on Darcy's Law, which is only applicable to laminar flow and is not valid for high velocity flows in karst systems that produce turbulence (Baedke and Krothe, 2001). In the interpretation of this theory, each exponential segment (or recession coefficient) is thought to represent the depletion of a particular reservoir component. In reality, these three reservoir components are not isolated from each other and drain concomitantly, and formulae designed to

determine the K, transmissivity (T), and storage (S) of each reservoir component separately could result in gross miscalculations (Király, 2003).

Other researchers advocate that the straight-line segments represent water flowing through a series of large conduits characterized by constrictive pathways or debris-filled passages, a concept referred to as reservoir-constricted flow (Bonacci 1987 and 1993). Halihan et al., (1998) numerically modeled the storm-event responses of a karst-spring system, Devil's Ice Box in Missouri, characterized by reservoir-constricted flow. Results suggested that hydrograph forms were controlled by changes from phreatic to epiphreatic conditions in a conduit, in conduit geometry, or from multiple springs draining the same system. Models by Covington et al., (2009) suggest that recession hydrographs only reveal information about conduit geometry when reservoir constrictions are present along the flow path. Owing to these changes in geometry, flow is delayed through the conduits by different amounts of time, and the constrictions and voids of varying sizes produces a shape that may be characteristic only for a particular spring (Baedke and Krothe, 2001).

An unusual situation for recession curves occurs when the intermediate recession coefficient $(\alpha_2)$ is steeper than the initial slope $(\alpha_1)$, producing a convex form on the master depletion curve (see Figure 1). Inasmuch as discharge (Q) is proportional to head (H) [Q $\propto$ H], the discharge is expected to decline the quickest (i.e. the slope to be steepest) immediately following the peak when greater water volume provides driving force (Bonacci, 1993). Additionally, in semi-log space, the quickflow component is expected to decay in a linear manner, and the baseflow component is expected to decay in an exponential manner, producing a concave shape on the quickflow curve (Padilla et al., 1994).

Recession coefficients that increase with time have been observed in southern France (Bailly-Comte et al., 2010), Greece (Soulios, 1991), the Dineric Karst (Mangin, 1975; Bonacci, 1987 and 1993), and, recently, in northwest Arkansas. Three models have been proposed for recession coefficients that increase with time: 1) Spring basins with two lithologies of contrasting permeabilities, which creates a lag-time for the inflow of water to the spring orifice (Soulios, 1991), 2) changes in storage-outflow relations created by temporarily flooded cave passages or ponor retention ponds that are drained by smaller diameter conduits (Bonacci, 1987 and 1993), and 3) water exchange and pressure transfer between the matrix and conduit when antecedent water levels in the aquifer matrix are medium to high compared to those of the conduits prior to rainfall (Bailly-Comte et al., 2010).

Bonacci (1987 and 1993) proposed that when the water levels in a cave chamber or a ponor retention pond rise above the surrounding ground-water table in the aquifer, the capacity of the conduits draining these features is exceeded. At this point, the conduits are under pressure and discharge occurs according to a form of the Darcy-Weisbach equation. While the water levels in the flooded passages remain elevated above the surrounding groundwater table, the discharge of the conduit is independent of the hydraulic head (i.e. the discharge can only occur according to the conduit capacity where continued in-flow will cause the water level in the flooded features to rise without a corresponding increase in discharge). This situation will cause a flattening in the initial recession coefficient ($\alpha_1$). Once the flooded cave chamber, or ponor retention pond, is drained to a level equal to, or below, that of the water table, the discharge of the conduit is no longer under pressure, and spring flow will, once again, be proportional to the head ($Q \propto H$). This transition can cause a break in the recession slope, producing an intermediate recession coefficient ($\alpha_2$) characterized by a steeper slope with respect to the initial recession coefficient ($\alpha_1$), creating a quickflow curve with a convex form (Bonacci, 1987 and 1993).

Bonacci (1987) observed this behavior and hydrograph form in the discharge of springs draining flooded ponors in the Dinaric karst. In one case, the discharge of the spring ceased to increase beyond approximately 22 $m^3$/s, while the ponor retention pond continued to rise another 5 m above land surface. Mangin (1975) was the first to notice this phenomenon and called it a "trop-plein" hydrograph, meaning "too-full".

Bailly-Comte et al. (2010) observed a similar hydrograph form in the Véne Spring near Montpellier, France. Véne Spring is an ephemeral spring fed by the sinking of the Coulazou River into the karstified Aumelas Causse Aquifer. In that study, the spring, a nearby well with no known connection to karst features, and several conduits of a cave (Puits de l'aven) were monitored for stage responses to storm events. All of the monitoring stations had a direct hydraulic connection to the Coulazou River.

Stage fluctuations between the well (representing the matrix flow regime (MFR) and the ground-water table) and the springs and cave conduits (representing the conduit flow regime (CFR)) indicated that various recession forms were produced on the master depletion curve depending upon antecedent water levels in the MFR. When pre-storm water levels in the MFR were medium to high with respect to the water levels in the conduits, a convex form developed on the master depletion curve. The authors observed that following a storm event, focused recharge to the conduits would cause water levels to rapidly increase to a stage above that of the MFR. At some time after the increase in stage, the spring hydrograph would peak and enter into a linear decrease while the stage of the well would continue to rise for a period of time. When the water level in the conduits equaled the water level in the matrix, an inflexion point occurred on the hydrograph, after which time the discharge decreased exponentially. The authors suggested that while the stage in the conduits exceeded the water level in the well, the conduits were recharging both the spring and the MFR. When the water levels in the MFR and CFR became equal, there was an abrupt modification within the karst drainage network where the flow regime transitioned from a CFR to a MFR. This transition manifested as an intermediate recession coefficient that was steeper than the initial recession coefficient owing to water exchange and pressure transfer (Bailly-Comte et al., 2010).

## Study Location

The Savoy Experimental Watershed (SEW) is a University of Arkansas property that encompasses about 1250 ha in the mantled-karst of the Springfield Plateau in the southwestern

Ozark Plateaus (Figure 2). SEW is a long-term, integrated, and multidisciplinary field laboratory where processes, controls, and hydrologic and nutrient–flux budgets for surface water, soil, and shallow ground-water environments may be evaluated in a well-characterized and highly-instrumented setting.

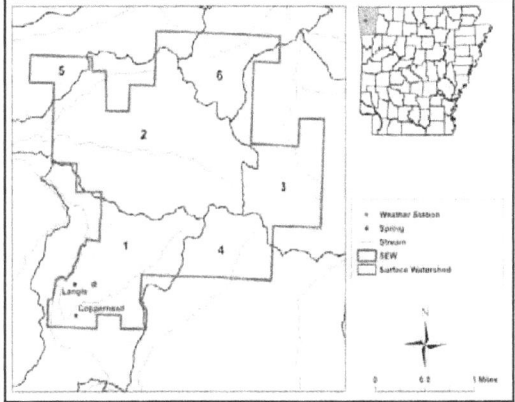

Figure 2. SEW site location map with property (red line) and watershed boundaries (black lines with sequential numbering) delineated. Basin-1 is the focus of this paper.

The karst hydrogeology in this area includes carbonate aquifers covered by a thin, rocky soil, and a variable thickness of regolith. The carbonate rocks have been dissolved to form an open network of caves, enlarged fractures, bedding planes, conduits, sinkholes, swallets, sinking streams, and springs. Flow in these aquifers is typically rapid, flow directions are difficult to predict, interaction between surface and ground water is typically extensive, and surface watersheds and ground-water basins do not always coincide (Brahana et al., 1999)

### SEW Conceptual Model

The previous studies at the SEW established a three-component conceptual model consisting of the interface (soil-regolith) zone, interflow (soil-epikarst) zone, and focused flow (phreatic) zone. The focused flow system is embedded in a block matrix characterized by low porosity (1%), low hydraulic conductivity ($6 \times 10^{-6}$ ft/d) and a low storage capacity, where matrix contribution to flow is considered to be negligible.

Permeability contrasts between these zones and within the aquifer are the primary control

for ground-water flow, with structural features, such as joints and fractures, forming a secondary control. Recharge occurs during precipitation events, where fast-flow conduits govern ground-water movement that is highly variable and exhibits appreciable surface-groundwater inter-action. Surface-water and groundwater basins do not coincide, and interbasin transfer is common. An example of this is spring recharge zones changing with naturally varying water levels. Basin-1 is under-drained by two major springs, Langle and Copperhead (Figure 2), which issue from the phreatic zone and form an underflow/overflow system. Copperhead, the overflow spring, resurges at an elevation of 314.21 m mean sea level (msl) with an estimated spring-basin area of 0.30 km², and is characterized by a flashy discharge (0.85 to 379.16 l/s over the period of record), a low-flow discharge of 0.91 l/s, and temperature and conductivity profiles indicative of shallow flow paths with a high degree of surface-ground water interaction. Langle Spring, the overflow spring, resurges at an elevation of 314.05 m msl with an estimated spring-basin area of 1.40 km², and is characterized by a lower discharge range (1.13 to 89.48 l/s over the period of record), a low-flow discharge of 2.01 l/s, and temperature and conductivity profiles indicative of deeper flow paths with less surface-ground water interaction than Copperhead (Al-Rashidy, 1999; Brahana et al., 1999).

### Approach

From October 1, 2007 through 2008, continuous monitoring of water levels at Copperhead and Langle Springs was reinitiated using HOBO® U20 submersible data loggers equipped with pressure transducers. The pressure transducers record the absolute pressure values, which consist of atmospheric pressure and water head. Barometric effects were com-pensated for by using one additional data logger as a reference station placed near land surface. After compensation, the pressure sensor has a typical error of 2.1 cm and a maximum error 6.5 cm. These stations were programmed to record measurements every 15 minutes. In addition to the data loggers, a weather station located at the SEW (Figure 2) was programmed to record pre-cipitation measurements every 10 minutes. The weather station consists of a tipping bucket rain gauge with a 0.254 cm resolution and a Campbell Scientific® CR10X Measurement and Control System.

## Recession Methodology

Numerous recession methodologies are described in the literature, and Brodie and Hostetler (2005) provide a detailed summary of published techniques. Almost all of the proposed approaches are designed for recessions that exhibit zero- or first-order decay and are not applicable to recession coefficients that increase with time. Review of available literature has revealed only one proposed method for recession limbs that exhibit a convex form. Bonacci (1993) proposed a system of functions for simulating hydrographs with recession coefficients that increased with time. However, these equations required either the collection (or assumption) of hydrological and climatological parameters, such as evapo-transpiration and soil moisture retention as a function of time. These parameters are seldom, if ever, available or cost-effective for most spring-basin studies. A new method is proposed herein requiring only discharge data to simulate the recession. This methodology relies on an adaptation of the recession functions published by Mangin (1975) as described in Padilla et al. (1994) to handle convexity and requires no assumed variables.

Mangin (1975) proposed that, during the recession of a karst spring, the discharge at time $(Q_t)$ can be expressed with the following formula:

$$Q_t = \Phi_t + \Psi_t \qquad \text{eqn. (1)}$$

where $\Psi_t$ is a function that translates the effects of recharge through the unsaturated zone to the spring, modified by transport through the phreatic zone; and $\Phi_t$ is described by Maillet's formula (1905):

$$\Phi_t = Q_t^{bf} = q_0^{bf} e^{-\alpha t} \qquad \text{eqn. (2)}$$

where $Q_t^{bf}$ is the baseflow discharge at time t; $q_0^{bf}$ is max baseflow discharge as extrapolated from time of interest $(t_i)$ – when flow transitions from quickflow to baseflow – at the onset of the recession as shown in Figure 3; t is the time increment for which the equation is being eval-

uated; and $\alpha$ is the baseflow (or recession) coefficient (Ford and Williams (2007):

$$\alpha = \frac{\log q_0^{bf} - \log q_{tf}}{0.4343\,(\Delta t)} \qquad \text{eqn. (3)}$$

where $q_{tf}$ is the discharge at time final $(t_f)$, which equals the end of the recession period; and $\Delta t$ is the difference in time between $q_0^{bf}$ and $q_{tf}$.

Maillet (1905) observed that recession curves from basins consisting of porous media during dry periods followed first-order decay that could be treated linearly in semi-log space. Although equation 2 was developed for homogeneous and isotropic media, numerous studies have successfully applied forms of this equation to karstified and fissured media (Atkinson, 1977; Bonacci, 1987 and 1993; Padilla et al., 1994).

The parameter $\Psi_t$ is an empirical function that simulates quickflow and is expressed by the following formula for a recession limb with a concave form (Padilla et al., 1994):

$$\Psi_t = Q_t^{qf} = q_0^{qf} \left( \frac{1 - nt}{1 + \varepsilon t} \right) \qquad \text{eqn. (4)}$$

where $Q_t^{qf}$ is the quickflow discharge at time t; $q_0^{qf}$ is the max quickflow discharge, which is the difference between peak discharge $(Q_0)$ and $q_0^{bf}$ at time t equals 0 as shown in Figure 3; n is the reciprocal of $t_i$; and $\varepsilon$ is the slope of the trend line that best fits the curve created by the function Z (t):

$$Z(t) = \frac{q_0^{qf}}{Q_t^{bf}} (1 - nt) \qquad \text{eqn. (5)}$$

Equation 5 is only applied to the baseflow discharge between times $t_0$ and $t_i$.

For a recession with a convex form, the parameter $\Psi_t$ is expressed using the following formula:

$$\Psi_t = Q_t^{qf} = q_0^{qf} (1 - nt) \qquad \text{eqn. (6)}$$

The coefficient $\varepsilon$ defines the degree of concavity exhibited in the quick-flow component in terms of inverse time $(1 - nt)$. For the case of no concavity, ε equals 0, and the term (1 + εt) equals 1 and can be eliminated from equation 4. The elimination of the ε coefficient will produce a linear curve that has a convex shape in semi-log space.

Total discharge at any time t ($Q_t$) is the summation of the quickflow and baseflow discharges for the time increment of interest.

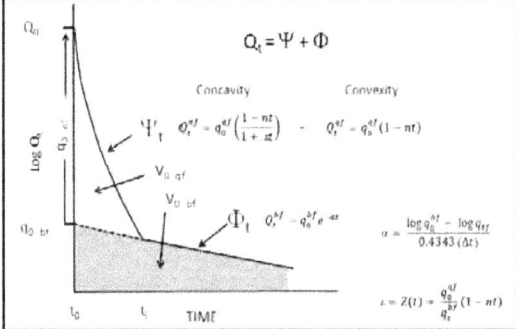

Figure 3. Karst-spring recession limb analysis using Mangin's (1975) approach. The shaded area represents baseflow and the unshaded area quickflow runoff (after Padilla et al., 1994).

Integration of the areas under the recession curves with respect to t provides the volume of water in the spring basin available for discharge (Padilla et al., 1994). The initial volume of stored water at t equals 0 available for discharge during the quickflow component ($V_0^{qf}$) of a recession limb (between t = 0 and $t_i$) with a concave quickflow depletion is calculated by the formula:

$$V_0^{qf} = \int_0^{t_i} q_0^{qf} \left( \frac{1 - nt}{1 + \varepsilon t} \right) dt$$

$$V_0^{qf} = \frac{q_0^{qf}}{\varepsilon} \left( \left( \ln(1 + \varepsilon t_i)(1 + \frac{n}{\varepsilon}) \right) - 1 \right) \quad \text{eqn. (7)}$$

The initial volume of water stored at t equals 0 available for discharge during the quickflow component of a recession limb (between t = 0 and $t_i$) with a convex quickflow depletion curve is calculated by the formula.

$$V_0^{qf} = \int_0^{t_i} q_0^{qf} (1 - nt) dt$$

$$V_0^{qf} = q_0^{qf} \left( t_i - \frac{nt_i^2}{2} \right) \quad \text{eqn. (8)}$$

The initial volume of stored water available for discharge for the baseflow component ($V_D$) from the onset of the recession (between t = 0 and ∞) is calculated from the formula:

$$V_D = \int_0^{\infty} q_0^{bf} \, e^{-\alpha t} \, dt$$

$$V_D = \frac{q_0^{bf}}{\alpha} \quad \text{eqn. (9)}$$

Equation 9 is known as the dynamic volume, and it represents the volume of water that would drain from the spring if the hydrograph was allowed to go from peak discharge to zero. To determine that actual volume of water that is discharged during the recession period ($V_0^{bf}$), equation 9 must be evaluated between t equals 0 and time final ($t_f$):

$$V_0^{bf} = \int_0^{t_f} q_0^{bf} \, e^{-\alpha t} \, dt$$

$$V_0^{bf} = \frac{q_0^{bf}}{\alpha} (1 - e^{-\alpha t_f}) \quad \text{eqn. (10)}$$

Summation of the $V_0^{qf}$ and $V_0^{bf}$ provides the total volume of water discharged during the recession to be determined, and summation of $V_0^{qf}$ and $V_D$ provides the thoretical storage volume.

## RESULTS

The functions presented above were applied to 42 recession hydrographs of Copperhead (14) and Langle (28) Springs. Both springs displayed recessions that were characterized by quickflow curves that exhibited concavity (8 for Copperhead and 16 for Langle), by quickflow curves that exhibited convexity (3 for Copperhead and 4 for Langle, or by recessions that exhibited no quickflow period (3 for Copperhead and 1 for Langle). In instances of no observed quickflow, only Maillet's exponential was used to simulate the recession.

For simulations that produced more than 20 percent average error (7 for Langle), extrapolating the curves beyond the influence of small deviations in flow to model the observed discharge was necessary, and results of these recessions are not discussed in this paper. A common reason that the functions failed to adapt to observed discharge was attributed to recession limbs characterized by more than two inflexion points, which separated the recession into more than three straight-line exponential segments in semi-log space. Despite the complexity of recession forms observed at Langle Spring, this spring was more sensitive to small-magnitude recharge events than Copperhead Spring producing a greater number of storm-event hydrographs to analyze.

Graphical representations of four recession analyses are shown in Figure 4. A summary of the main parameters that characterized the recessions are presented in Table 1, and a summary of volume determinations are presented in Table 2.

## DISCUSSION

The various recession forms observed at the SEW provide insight into the antecedent water levels/saturation prior to rainfall of, transmissive and storage properties of, and hydraulic control of and interaction between the three recognized flow components.

Recession limbs that can be characterized by a single exponent imply that there is a linear re-lation between hydraulic head and flow rate that is a function of the volume of water held in storage (Ford and Williams, 2007). The lack of a quickflow component indicates that the vadose zone (soil and epikarst) is effective in mod-erating infiltration and that the saturated zone (phreatic) is responsible for observed discharge response (Padilla et al., 1994). Exponential re-cessions tend to occur under low precipitation amounts (average of 38 mm), over long durations (average of 1.5 d), and during the dry season (October through February for this study). Under these circumstances, recharge is primarily autogenic in that rainfall is diffusely distributed through soil macropores and down many fissures across the epikarst zone. Low average total and dynamic volume determ-inations further support this

interpretation. The high average percent drained volumes (52% in 3.6 days for both springs) indicate that the phreatic zone is characterized by highly trans-missive elements.

Recession limbs that exhibit a quickflow curve indicate that recharge reaches the phreatic zone with less moderating affect from the vadose zone than exponential declines (Padilla et al., 1994). During heavy rains (average of 81 mm over 2 days) and/or when soil moisture and epikarst saturation is high, the infiltration capacity of the unsaturated zone is exceeded and inflow to the saturated zone occurs via preferential, vertical pathways. The porosity contrast between the epikarstic (typically 20%) and unweathered portions of the endokarstic (average of 1% at SEW) zones creates a permeability contrast that promotes a lateral component of flow (Ford and Williams, 2007). At the SEW, this interflow discharges from epikarstic springs into a sinking stream, providing an additional input of concentrated flow to the phreatic conduit network that occurs as allogenic recharge.

For quickflow components that exhibit con- cavity, the $\varepsilon$ term is a measure of the infiltration rate through the unsaturated zone (Padilla et al., 1994). Low $\varepsilon$ values, such as those of Langle Spring (0.70) suggest that infiltration through the vadose zone is slow. This observation is further supported by the average quickflow period ($t_i$) of 37 h. Contrastingly, Copperhead Spring has a fairly high $\varepsilon$ value (1.5) and an average $t_i$ of 53 h. In this timeframe, Copper-head Spring drains an average total volume that is one order of magnitude higher than Langle Spring. These parameters suggest that Copper-head Spring receives a greater percentage of flow from concentrated and highly transmissive zones with a direct hydraulic connection to the vadose zone. Despite the influence of the un-saturated zone on Copperhead, most of the total volume of water supplied to Copperhead and Langle Springs over the course of the recession is derived from the baseflow component (86% and 93%, respectively). The dominance of the baseflow recession suggests that the phreatic zone provides the primary control on discharge from these two springs. Whereas the average dynamic volume determinations are similar for each spring, Copperhead releases 66% of the available water from storage in about nine days compared to 31% for Langle over the same time

those of convex quickflow curves (an average of 81 mm over 2 days). Additionally, concave and

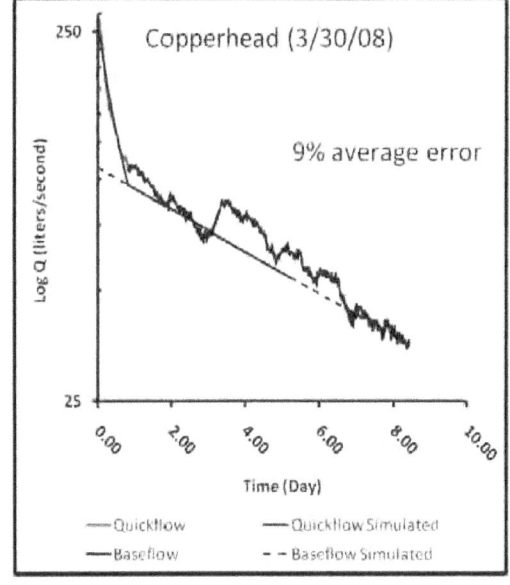

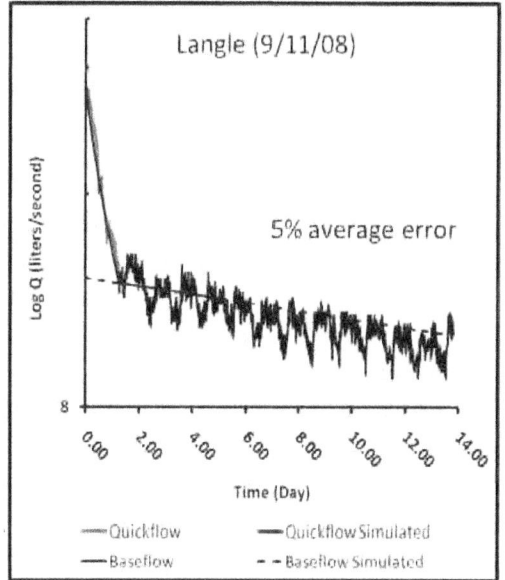

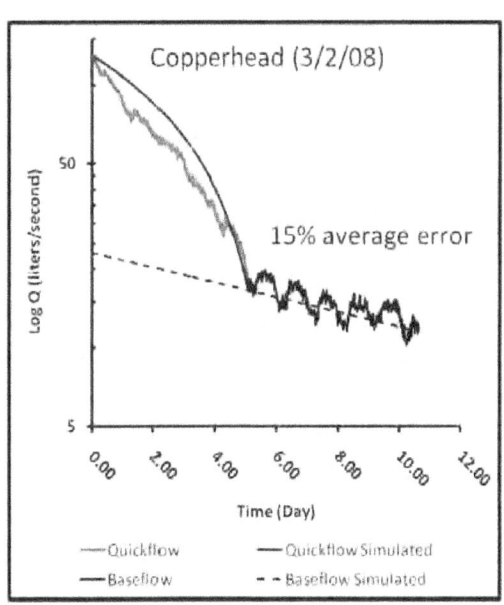

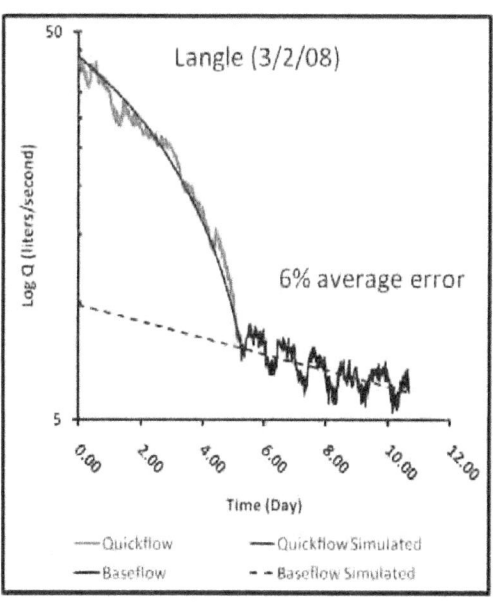

Figure 4. Hydrographs of two recessions (1 concave on top row, 1 convex on bottom row) from Copperhead and Langle Springs selected for analyses.

period. The high drainage rage from Copperhead suggests that the spring system is characterized by a better developed and more transmissive conduit network with shallower flowpaths.

Quickflow components that exhibit convexity occur under similar input conditions as

and convex quickflow curves develop under similar flow regimes. Parameters that were not monitored as a part of this study that could have assisted in understanding the provenance of the convex quickflow curve include rainfall intensity and antecedent saturation in the inter-face and interflow zones.

Table 1. Average Values of Recession Metrics by Recession Type

| Recession Type | Spring | $Q_0$ (l/s) | $q_0^{bf}$ (l/s) | $\alpha$ (day$^{-1}$) | $q_0^{qf}$ (l/s) | $t_i$ (day) | $\varepsilon$ (day$^{-1}$) | Avg. Error (%) |
|---|---|---|---|---|---|---|---|---|
| Concave | CH | 150 | 56.6 | 0.10 | 95.7 | 2.21 | 1.5 | 10.51 |
|  | L | 23.8 | 14.2 | 0.06 | 9.91 | 1.56 | 0.70 | 7.17 |
| Convex | CH | 76.5 | 14.2 | 0.15 | 62.3 | 3.19 | - | 14.17 |
|  | L | 31.7 | 10.8 | 0.07 | 20.7 | 4.85 | - | 7.09 |
| Exponential | CH | 40.2 | - | 0.24 | - | - | - | 5.74 |
|  | L | 6.23 | - | 0.24 | - | - | - | 4.34 |

Table 2. Average Values of Volume Determinations by Recession Type.

| Recession Type | Spring | $V_{qf}$ (l · 10$^5$) | %$V_T$ | $V_{bf}$ (l · 10$^5$) | %$V_T$ | $V_T$ (l · 10$^5$) | $V_D$ (l · 10$^5$) | Percent Drained |
|---|---|---|---|---|---|---|---|---|
| Concave | CH | 35 | 14 | 218 | 86 | 254 | 384 | 66 |
|  | L | 5 | 7 | 65 | 93 | 70 | 287 | 24 |
| Convex | CH | 119 | 62 | 72 | 38 | 151 | 251 | 60 |
|  | L | 46 | 33 | 95 | 67 | 140 | 244 | 57 |
| Exponential | CH | - | - | - | - | 117 | 185 | 63 |
|  | L | - | - | - | - | 9 | 22 | 41 |

Discharge through a conduit depends on the amount of available recharge (catchment control) and on the hydraulic capacity of the conduit (hydraulic control) (Palmer 1991). If instant-aneous inflow exceeds the hydraulic capacity of the conduits draining the epikarst zone, the ex-cess water must be stored in the epikarst or dis-charged from the system via lateral flowpaths.

Evidence of exceeding hydraulic capacity in the recessions is shown in the longer $t_i$ periods where the average percent volume drained over the course of the recession increases to 62% for Copperhead and 33% for Langle, indicating greater availability of water in the epikarst zone for draining conduits. The quickflow of Langle is 40 hours longer than Copperhead but drains a much lesser volume (7.3 x 10$^6$ l/s), indicating a less developed and transmissive conduit network with deeper flowpaths. Despite similar input rainfall amounts, Langle and Copperhead have an average dynamic volume that is one order of magnitude less than recessions with concave quickflow curves, which suggests that excess water is drained through other routes, such as overland flow or additional spring outlets observed along the escarpment in basin-1 that only operate under extreme flood events.

## CONCLUSION

A hydrograph decomposition method is presented in this paper that permits quantitative analysis of karst-spring recession hydrographs that exhibit multiple modes of decay. This methodology relies on an adaptation of the recession functions published by Mangin (1975) as described in Padilla et al. (1994). It is appropriate for springflow recession curves displaying convexity and requires no assumed variables, unlike the method of Bonacci (1993). Two karst springs located at the SEW that exhibit multiple recession forms were evaluated using this methodology. Comparisons of results show that despite a similar hydrogeologic setting, these two springs have very different transmission behaviors that are attributed to differences in flowpath properties and hydraulic connection with the epikarstic zone. Inter-pretations from analyses are consistent with observations from previous investigations of these two spring systems. Based on available monitored parameters, the recession form appears to be a function of antecedent saturation in aquifer components prior to rainfall, available recharge, and hydraulic control and interaction among the reservoir components. The proposed model incorporates and refines parts of previously published models. Whereas Bonacci (1987 and 1993) attributed the hydrograph form to conduit capacity along a reservoir-constricted flowpath, interpretations from this study suggest that the form could also be generated from instantaneous inflow from the epikarstic zone exceeding the endokarstic conduit capacity. This scenario is more likely to occur during the wet season when vadose saturation is high and the infiltration capacity can easily be exceeded producing focused flow.

# REFERENCES

Al-Rashidy, S.M., 1999, Hydrogeologic controls of groundwater in the shallow mantled karst aquifer, Copperhead Spring, Savoy Experimental Watershed, northwest Arkansas: Unpublished M.S. thesis, University of Arkansas, Fayetteville, 124 p.

Atkinson, T.C., 1977, Diffuse flow and conduit flow in limestone terrain in the Mendip Hills, Somerset (Great Britain): Journal of Hydrology, vol. 35, no. 1-2, p. 93-110.

Baedke, S.J. and Krothe, N.C., 2001, Derivation of effective hydraulic parameters of a karst aquifer from discharge hydrograph analysis: Water Resources Research, vol. 37, no. 1, p. 13-20.

Bailly-Comte, V., Martin, J.B., Jourde, H., Screaton, E.J., Pistre, S., and Langston, A., 2010, Water exchange and pressure transfer between conduits and matrix and their influence on hydrodynamics of two karst aquifers with sinking streams: Journal of Hydrology, vol. 386, p. 55-66.

Bonacci, O., 1987, Karst Hydrology: Springer Verlag, Heidelberg, Germany.

Bonacci, O., 1993, Karst spring hydrographs as indicators of karst aquifers: Journal des Sciences Hydrologiques, vol. 38, no. 1, p. 51-62.

Brahana, J.V., Hays, P.D., Kresse, T.M., Sauer, T.J., and Stanton, G.P., 1999, The Savoy experimental watershed; early lessons for hydrogeologic modeling from a well-characterized karst research site; Proceedings of the symposium: Karst Modeling: Karst Waters Institute Special Publication, vol. 5, p. 247-254.

Brodie, R.S. and Hostetler, S., 2005. A review of techniques for analyzing baseflow from stream hydrographs: Proceedings of the NZHS-International Association of Hydrogeologists-NZSSS 2005 Conference, 13 p.

Covington, M.D., Wicks, C.M., and Saar, M.O., 2009, A dimensionless number describing the effects of recharge and geometry on discharge from simple karstic aquifers: Water Resources Research, vol. 45, 16 p.

Ford, D. and Williams, P., 2007, Karst Hydrogeology and Geomorphology: John Wiley & Sons, Ltd, West Sussex, England, 562 p.

Halihan, T., Wicks, C.M, and Engeln, J.F., 1998, Physical response of a karst of a karst drainage basin to flood pulses: Example of Devil's Icebox cave system (Missouri, USA): Journal of Hydrology, vol. 204, p. 24-36.

Király, L., 2003, Karstification and groundwater flow: Speleogenesis and Evolution of Karst Aquifers, vol. 1, no. 3, p. 1-26.

Kovács, A. and Perrochet, P., 2008, A quantitative approach to spring hydrograph decomposition: Journal of Hydrology, vol. 352, no. 1-2, p. 16-29.

Mangin, A., 1975, Contrtibution a l'etude hydrodynamique des aquiferes karstiques: Annals of Speleology, vol. 26, no. 2, p. 283-339.

Maillet, E., 1905, Essais d'Hydraulique Souterraine et Fluviale: Hermann Paris, 218 p.

Padilla, A., Pulido-Bosch, A. and Mangin, A., 1994, Relative importance of baseflow and quickflow from hydrographs of karst springs: Ground Water, vol. 32, no. 2, p. 267-277.

Palmer, A.N., 1991, Origin and morphology of limestone caves: Geological Society of America Bulletin, vol. 103, p. 1-21.

Shevenell, L., 1996, Analysis of well hydrographs in a karst aquifer: estimates of specific yields and continuum transmissivities: Journal of Hydrology, vol. 174, no. 3-4, p. 331-355.

Soulios, G., 1991, Contribution a l'etude des courbes de recession des sources karstiques: Exemples du pays Hellenique: Journal of Hydrology, vol. 127, p. 29-42.

White, W.B., 1988, Geomorphology and Hydrology of Karst Terrains: Oxford University Press, New York, 464 p.

# Numerical Evaluations of Alternative Spring Discharge Conditions for Barton Springs, Texas, USA

By William R. Hutchison and Melissa E. Hill
Texas Water Development Board, 1700 North Congress Avenue, Austin, Texas 78711-3231

## Abstract

Numerical model simulations were performed to evaluate spring discharge conditions over a wide range of seven-year climatic conditions at Barton Springs, Texas, USA. The purpose for these predictive scenarios was to evaluate the effects of starting heads and alternative pumping conditions. Emphasis was placed on quantifying the relative frequency of discharges at or below 0.31 cubic meters per second ($m^3$/s), which is equivalent to the estimated historic minimum monthly discharges that occurred during the seven-year drought-of-record from the early 1950s. Each of the model scenarios included 342 seven-year simulations based on tree-ring data extending from 1648 through 1995. Results from the predictive scenarios were evaluated in terms of: i) average simulated discharges under intermediate starting head conditions and ii) low simulated discharges under various starting head and pumping conditions. Results from the predictive scenarios for the average simulated discharges indicate that under average precipitation conditions, the seven-year average simulated discharge can range from 0.90 to 2.57 $m^3$/s. Based on these results, 59 percent of the seven-year simulations had an average discharge above the long-term historical average discharge (1.5 $m^3$/s). Discharges at or below 0.31 $m^3$/s occur at a relative frequency of 5 percent under the low starting head condition, using an annual average pumping quantity of 8.4 million $m^3$/y, which is comparable to current groundwater withdrawal estimates, but decreases to 0 percent using this same pumping dataset under intermediate and high starting head conditions. Conversely, the relative frequency for discharges at or below 0.31 $m^3$/s under a high pumping scenario (20.1 million $m^3$/y) ranged from 4 to 17 percent regardless of starting head conditions. These results indicate that the relative frequency of low discharge events (at or below 0.31 $m^3$/s) using pumping quantities of 8.4 million $m^3$/y, or below, and with the pumping distribution applied in this analysis, will be highly dependent on the starting head conditions going into a seven-year drought. However, significant increases from current annual average pumping quantities would likely increase the relative frequency of low discharge events during drought conditions regardless of starting head conditions due to capture of natural discharge.

## INTRODUCTION

Barton Springs is located in central Austin within Travis County, Texas, USA, fig. 1 and is within the jurisdiction of the Barton Springs/Edwards Aquifer Conservation District. It consists of several vents emanating from the underlying Edwards (Balcones Fault Zone) Aquifer, which is comprised of the soluble Cretaceous-age Edwards Group and the Georgetown Formation (Scanlon and others, 2001). Recharge to the aquifer occurs from rain falling over the outcrop area and from streams that cross the outcrop area of the aquifer (Scanlon and others, 2001). Long-term average monthly discharge for Barton Springs from January 1917 to January 2007 was 1.5 $m^3$/s (Slade and others, 1986; Hunt, 2009). The minimum average monthly discharge of 0.31 $m^3$/s occurred in both July and August of 1956 during the seven-year historical drought-of-record (Slade and others, 1986). Discharges from the springs provide habitat for the federally listed endangered Barton Springs Salamander *(Eurycea sosorum)* (U.S. Fish and Wildlife Service, 2010).

The Texas Water Development Board is mandated by the Texas Legislature to develop numerical groundwater availability models for the major and minor aquifers of the state. The models typically utilize the MODFLOW groundwater flow simulator developed by the United States Geological Survey, and are typically large regional models intended to simulate regional flow conditions. The model for the Barton Springs segment of the Edwards (Balcones Fault Zone) Aquifer is a sub-regional model, with relatively smaller cell dimensions, that was developed to estimate groundwater availability and simulate spring discharges under future pumping scenarios and drought conditions.

Figure 1. Location of Barton Springs.

As part of an evaluation of spring flow conditions, the Barton Springs/Edwards Aquifer Conservation District requested that the Texas Water Development Board estimate groundwater pumping for drought conditions under average monthly spring discharges of 0.31, 0.25, 0.20, 0.14, and 0.08 m³/s.

## APPROACH

### Model Recalibration

The existing groundwater availability model for the Barton Springs segment of the Edwards (Balcones Fault Zone) Aquifer (Scanlon and others, 2001) was calibrated using data from 1989 through 1998, which did not include the drought-of-record period in the 1950s. Therefore, in order to increase the confidence in the ability of the model to simulate spring flow during historic drought conditions, the model

calibration period was extended and the model was recalibrated. Details of the recalibration effort are provided in Hutchison and Hill (2011). The calibration period for the recalibrated model was January 1943 through December 2004 (744 monthly stress periods), with a steady-state stress period (stress period 1) preceding the transient simulation for a total of 745 stress periods. The steady-state stress period did not represent a true pre-development condition, but it provided a stable initial head solution to initialize the transient simulation.

Although groundwater elevation measurements, and estimates for discharge and pumping were available for the historic drought-of record, stream gage data that would be useful for estimating recharge rates were not available. Therefore, recharge estimates for the extended calibration period were extrapolated from recharge estimates in Scanlon and others (2001) using 84 regression relationships developed for each month (12 total) and recharge zone (7 total) in the model. The recharge zones of Scanlon and others (2001) coincide with the outcrop area and stream locations. In addition, a relationship to estimate recharge from precipitation based on current month rainfall and rainfall in the preceding two months was developed in an effort to account for antecedent conditions.

In the recalibrated model, hydraulic conductivity was estimated within the same nine zones as used by Scanlon and others (2001). However, Scanlon and others (2001) assumed that the hydraulic conductivity is isotropic, whereas in the recalibrated model, hydraulic conductivity is assumed to be anisotropic. Significant anisotropy was expected given the geologic setting, and the estimated hydraulic conductivity values were consistent with that conceptualization.

The model of Scanlon and others (2001) was developed with MODFLOW-96 (Harbaugh and McDonald, 1996) and the recalibrated model was developed with MODFLOW-2000 (Harbaugh and others, 2000). Statistics from the recalibrated model for estimated/measured discharges and measured groundwater elevations are summarized in Table 1. A hydrograph showing the match between estimated/measured discharges versus simulated discharges for Barton Springs is provided in fig. 2.

| Calibration Statistic | Calibrated Model Simulated Discharges (m³/s) | Calibrated Model Simulated Groundwater Elevations (m) |
|---|---|---|
| Minimum Residual | -1.45 | -58.44 |
| Maximum Residual | 1.84 | 79.00 |
| Absolute Residual Mean | 0.38 | 9.59 |
| Standard Deviation of Residuals | 0.48 | 13.62 |
| Range of Measured Values | 3.51 | 141.48 |
| Standard Deviation/Range | 0.14 | 0.096 |
| Absolute Residual Mean/Range*100 | 11 | 7 |
| Sum of Squared Residuals | $6.31 \times 10^3$ | $1.37 \times 10^6$ |

Table 1. Calibration statistics for discharges and groundwater elevations.

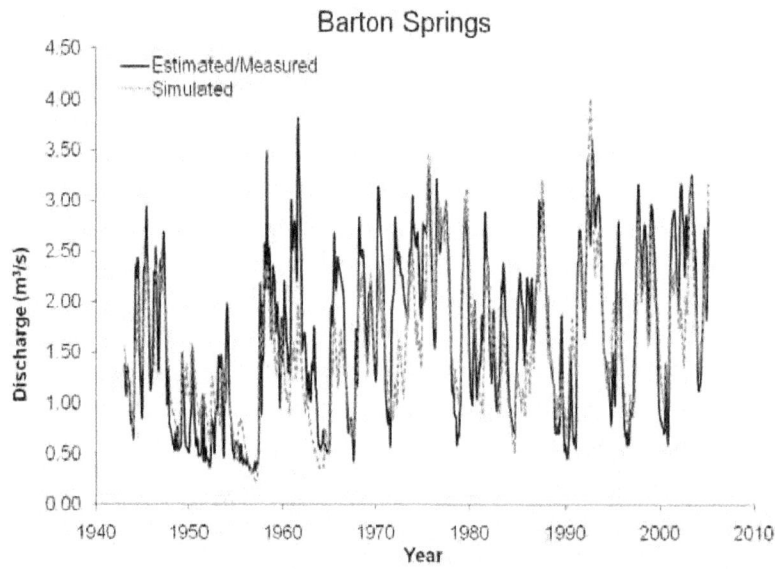

Figure 2. Hydrograph of estimated/measured discharges versus simulated discharges for Barton Springs using the recalibrated model.

## Predictive Simulations Using Recalibrated Model

Predictive simulations using the recalibrated model consisted of 15 scenarios that involved 3 different starting head conditions (low, intermediate, and high) and 5 pumping scenarios with annual pumping quantities of 4.7; 5.5; 6.7; 8.4; and 20.1 million m³/y. Current pumping is estimated to be approximately 8.4 million m³/y.

Starting head conditions for the predictive simulations were extracted from the recalibrated model by evaluating groundwater storage at the end of each stress period of the calibration period. Based on this review, February 1957 was selected as the low starting head condition, June 1992 was selected as the high starting head condition, and January 2004 simulated heads were selected as the intermediate starting head condition.

Groundwater pumping quantities and distribution for the 5 pumping scenarios were based on pumping estimates from the recalibrated model for 1982, 1987, and 2002. A factor of 1.25 and 3 was applied to the 2002 dataset to achieve 2 additional well datasets. The 1982 and 1987 amounts and distributions from the model's well package were used because the annual pumping quantities for these years are lower than the current pumping quantities. The 1.25 factor was applied to the 2002 dataset because groundwater withdrawals in the model for that year are lower than the estimates of current pumping provided by the Barton Springs/Edwards Aquifer Conservation District. The larger factor (3) was applied to the 2002 dataset to account for potential increases of groundwater withdrawals. The resulting annual average pumping quantities for the 5 well datasets were 4.7; 5.5; 6.7; 8.4; and 20.1 million $m^3/y$.

Each of these 15 scenarios included 342 seven-year simulations extending from 1648 through 1995 that were based on reconstructed rainfall based on the composite of six post-oak tree-ring chronologies for South Central Texas (Cleaveland, 2006). Rainfall data from the tree-ring dataset were used to estimate recharge from rainfall and streamflow. The first seven-year simulations were run using rainfall data from 1648 to 1654, followed by rainfall data from 1649 to 1655, and so forth with the last seven-year period ranging from 1989 to 1995.

Precipitation from the tree-ring dataset was estimated on an annual basis. Monthly rainfall estimates from these annual estimates were developed based on a lookup table that utilized the monthly distribution of rainfall during the calibration period. For example, if the annual rainfall for a given year in the tree-ring dataset expressed as a percent of long-term average (for example 84 percent) matched the annual rainfall of a year in the calibration period the monthly precipitation for that year were used to develop monthly precipitation estimates for that year of the simulation. If an exact match was not identified, the next closest match was selected and scaled to match the percentage based on the reconstructed rainfall values.

Results from the predictive simulations included spring discharge estimates under a wide variety of starting head conditions and pumping scenarios. These results can be used to quantify the relative frequency and magnitude of low discharges under a range of pumping scenarios. A more complete description of the predictive simulations is presented in Hutchison and Hill (2011).

## RESULTS

Results from the predictive simulations were evaluated in terms of: i) average simulated discharges under intermediate starting head conditions and ii) low simulated discharges under various starting head and pumping conditions. Results for the average simulated discharges were useful to evaluate the sensitivity of simulated discharges to increases in pumping, and to quantify discharge variation over a wide range of possible precipitation conditions over a seven-year period. Results related to low simulated discharges under various starting heads and pumping scenarios were used to quantify the relative frequency of low discharge events, and evaluate how starting head conditions and increased pumping could affect the occurrence of undesirably low discharges.

### Average Simulated Discharges Under Intermediate Starting Head Conditions

As expected, increases in pumping resulted in lower simulated discharges. As described in Bredehoeft and Durbin (2009), one effect of groundwater pumping is the capture of natural discharge from the groundwater flow system. In this case, under pre-development conditions, groundwater is recharged from precipitation and streamflow, and discharged at Barton Springs. Pumping increases result in groundwater storage depletion (which can be observed by declining groundwater elevations), and decreased discharges. The effects of increased pumping on the simulated seven-year average discharge using intermediate starting head conditions are summarized in fig. 3. Also, note that simulated spring discharge is equal to or above the long-term average of 1.5 $m^3/s$ when pumping is below about 10 million $m^3/y$.

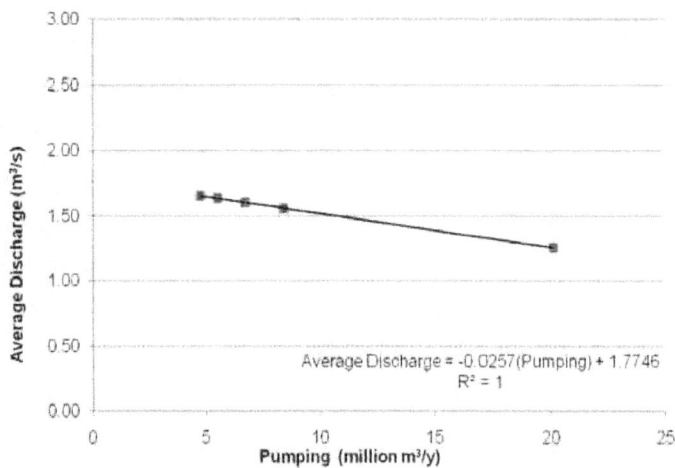

Figure 3. Effects of increased pumping on the simulated seven-year average discharge using intermediate starting head conditions.

An example for the variation of spring flow under the range of seven-year precipitation conditions for one of the five pumping scenarios (8.4 million $m^3$/y) under intermediate starting head conditions is shown in fig. 4. Note that each data point represents a seven-year average simulated discharge and under a particular seven-year period based on the tree-ring dataset. It can be seen that under average precipitation conditions, the seven-year average simulated discharge can range from 0.90 to 2.57 $m^3$/s. Moreover, 59 percent of those average seven-year simulated discharges (342 total) are at or above the long-term historical average discharge (1.5 $m^3$/s).

## Low Simulated Discharges Under Various Starting Head and Pumping Conditions

Results from the predictive scenarios were evaluated in terms of low simulated discharges to be responsive to the request of the Barton Springs/Edwards Aquifer Conservation District. The relative frequency for discharges at or below each of those listed in the request (0.31; 0.25; 0.20; 0.14; and 0.08 $m^3$/s) were calculated for each of the 15 scenarios, fig. 5. Simulated discharges for Barton Springs at or below 0.31 $m^3$/s, which are equivalent to the estimated minimum monthly discharges during the 1950 to 1956 drought-of-record, occurred at a relative frequency of 5 percent under the low starting head condition (essentially a seven-year drought that follows a seven-year drought), and an annual average pumping quantity of 8.4 million

$m^3$/y. The significance of starting head conditions is apparent as the relative frequency decreases to 0 percent using the same pumping scenario under intermediate and high starting head conditions, fig. 5.

The relative frequency of simulated discharges for Barton Springs at or below 0.31 $m^3$/s under low starting head conditions varies from 4 to 17 percent using the 4.7; 5.5; 6.7; 8.4 and 20.1 million $m^3$/y pumping scenarios. However, the relative frequency of simulated discharges at or below 0.31 $m^3$/s decreases to 0 percent under intermediate and high starting head conditions using the 4.7; 5.5; 6.7; and 8.4 million $m^3$/y scenarios, whereas the relative frequency of simulated discharges at or below 0.31 $m^3$/s using the largest pumping dataset (20.1 million $m^3$/y) varies from 4 to 6 percent under intermediate and high starting head conditions. This suggests that, under the highest pumping scenario, simulated discharges become more sensitive to pumping relative to starting head conditions as more of the natural discharge is captured (Hutchison and Hill, 2009).

Throughout this analysis, the pumping distributions did not vary from those originally assigned in the datasets extracted from the recalibrated model. Moreover, we did not evaluate the effects on simulated discharges that may result due to alternative pumping distributions. Results presented here will likely differ under alternative pumping distributions, particularly if pumping were to increase in the

186

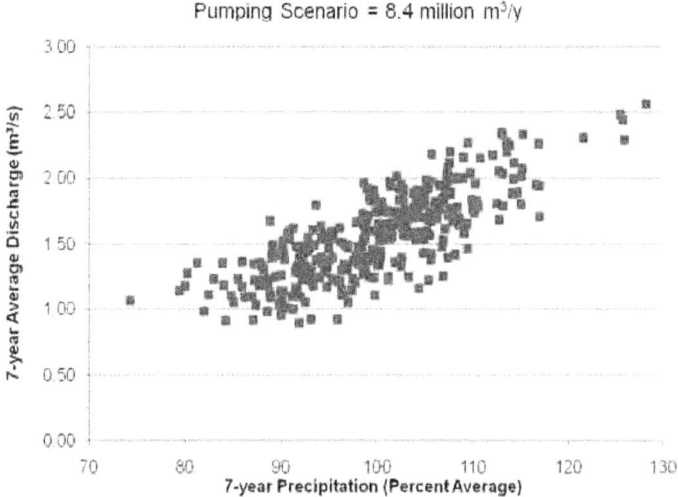

Figure 4. Plot showing the variation of precipitation versus simulated discharges for the 8.4 million m³/y pumping scenario. Fifty-nine percent of the seven-year average simulated discharges are at or above the long-term historical average discharge (1.5 m³/s).

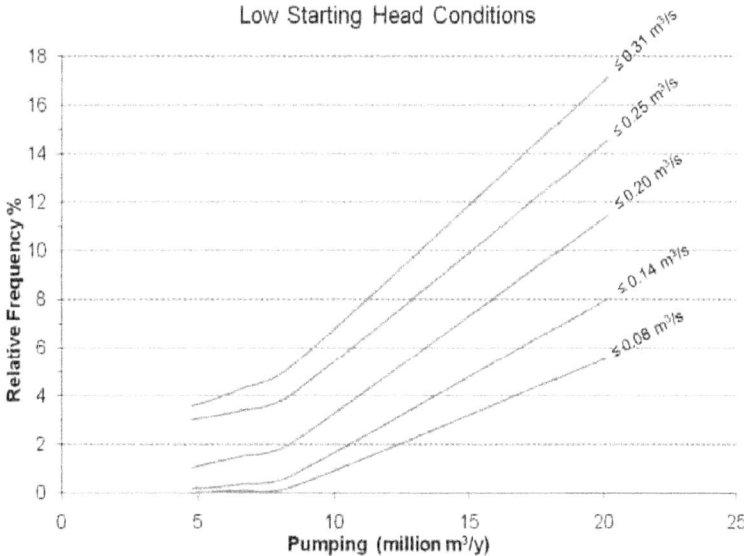

Figure 5. Curves for annual average pumping quantities (million cubic-meters per year) versus the relative frequency (percent) for 0.31, 0.25, 0.20, 0.14, and 0.08 cubic-meters per second spring discharge with low starting head conditions.

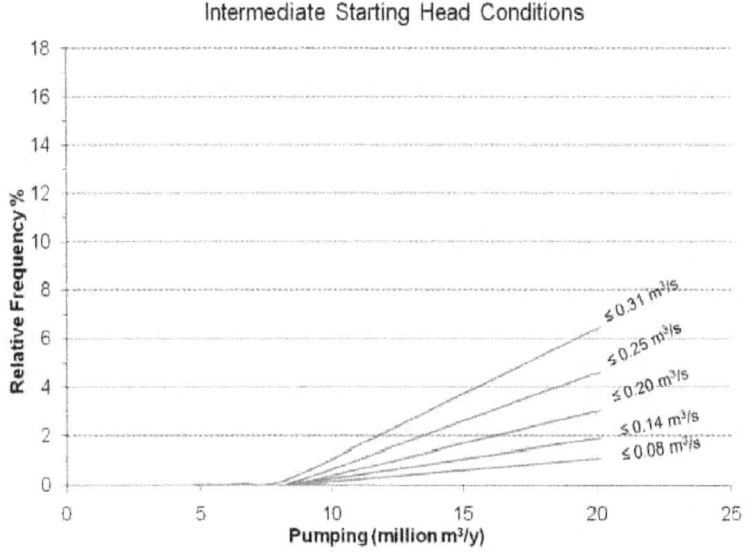

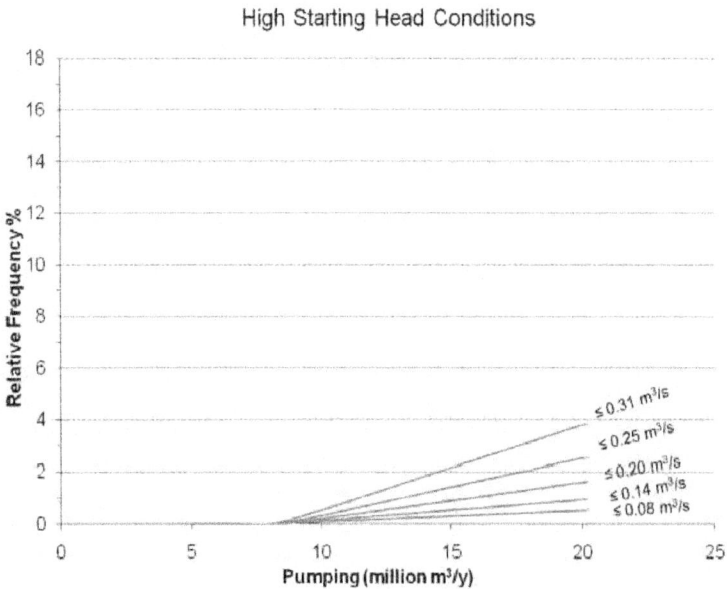

Figure 5 continued. Curves for annual average pumping quantities (million cubic-meters per year) versus the relative frequency (percent) for 0.31, 0.25, 0.20, 0.14, and 0.08 cubic-meters per second spring discharge with intermediate starting head conditions (top) and high starting head conditions (bottom).

vicinity of the springs which would result in more efficient capture of the natural discharge.

## SUMMARY AND CONCLUSIONS

Results from the predictive simulations, evaluated in terms of average simulated discharges, under intermediate starting head conditions demonstrate that the seven-year average discharge decreases with increases in pumping, but only the largest pumping dataset (20.1 million $m^3$/y) is below the long-term historical average discharge of 1.5 $m^3$/s. A graphical representation, of the seven-year percent average precipitation versus the seven-year average discharges, using intermediate starting head conditions, show that simulated discharges range from a low of 0.90 $m^3$/s to a high of 2.7 $m^3$/s. Additionally, 59 percent of those average seven-year simulated discharges (342 total) are at or above the long-term historical average discharge (1.5 $m^3$/s).

The predictive simulations for low discharges under low starting head conditions and various pumping scenarios resulted in relatively higher frequencies (4 to17 percent) for simulated discharges at or below 0.31 $m^3$/s compared to those using intermediate (0 to 6 percent) or high starting head conditions (0 to 4 percent). Conversely, the relative frequency for discharges at or below 0.31 $m^3$/s under the high pumping scenario (20.1 million $m^3$/y) ranged from 4 to 17 percent regardless of starting head conditions. In conclusion, these results indicate that the relative frequency of low discharge events (at or below 0.31 $m^3$/s) using pumping quantities of 8.4 million $m^3$/y or below, and with the pumping distribution applied in this analysis, will be highly dependent on the starting head conditions at the beginning of a seven year drought. However, significant increases from current annual average pumping quantities, would likely increase the relative frequency of low discharge events during drought conditions regardless of starting head conditions due to capture of natural discharge.

## ACKNOWLEDGMENTS

The authors thank Mr. Kirk Holland, Dr. Brian Smith, and Mr. Brian Hunt from the Barton Springs/Edwards Aquifer Conservation District for providing data and for fruitful discussions pertaining to the conceptualization of groundwater flow in the Barton Springs springshed. The authors also graciously thank Dr. Robert Mace and Ms. Cindy Ridgeway from the Texas Water Development Board for providing review comments that improved the overall quality of this manuscript.

## REFERENCES

Bredehoeft, J. and Durbin, T., 2009, Ground water development-the time to full capture problem: *Ground Water*, vol. 47, no. 4, p. 506-514.

Cleaveland, M.K., 2006, Extended chronology of drought in the San Antonio area, revised report: University of Arkansas, 29 p. http://www.gbra.org/documents/studies/treering/TreeRingStudy.pdf

Harbaugh, A.W., Banta, E.R., Hill, M.C., and McDonald, M.G., 2000, MODFLOW-2000, The U.S. Geological Survey modular ground-water model – user guide to modularization concepts and the ground-water flow process, U.S. Geological Survey Open-File Report 00-92, 121 p.

Harbaugh, A.W. and McDonald, M.G., 1996, User's documentation for MODFLOW-96, an update to the U.S. Geological Survey modular finite-difference ground-water flow model, U.S. Geological Survey Open-File Report 96-485, 56 p.

Hunt, B., 2009, *Written communication, June 19.* Austin, Texas: Barton Springs/Edwards Aquifer Conservation District.

Hutchison, W.R. and Hill, M.E., 2011, Recalibration of the Edwards Balcones Fault Zone (Barton Springs segment) Aquifer Groundwater Flow Model Draft Report: Texas Water Development Board, variously p.

Hutchison, W.R. and Hill, M.E., 2009, GAM Run 09-019 Draft Report: Texas Water Development Board, 28 p.

Scanlon, B.R., Mace, R.E., Smith, B., Hovorka, S., Dutton, A.R., and Reedy, R., 2001, Groundwater availability of the Barton Springs segment of the Edwards aquifer, Texas: Numerical simulations through 2050: Bureau of Economic Geology, 36 p.

Slade, R.M. Jr., Dorsey, M.E., and Stewart, S.L., 1986, Hydrology and water quality of the Edwards Aquifer associated with Barton Springs in the Austin area, Texas, U.S. Geological Survey Water-Resources Investigations Report 86-4036, 117 p.

U.S. Fish and Wildlife Service, 2010, Species Profile, http://ecos.fws.gov/speciesProfile/profile/species Profile.action?spcode=D010

# FIELD TRIP GUIDE
# Geology and Karst Landscapes of the Buffalo National River Area, Northern Arkansas

By Mark R. Hudson,[1] Kenzie J. Turner,[1] and Chuck Bitting[2]

[1]U.S. Geological Survey, Box 25046, Mail Stop 980, Denver, CO 80225
[2]National Park Service, 402 N. Walnut, Suite 136, Harrison, AR 72601

## Abstract

The Buffalo River watershed of northern Arkansas contains abundant karst features and associated karst hydrologic systems developed within Mississippian and Ordovician stratigraphic sequences. Cherty limestone of the Lower to Middle Mississippian Boone Formation is the predominant host of karst features and corresponds to the Springfield Plateau karst aquifer. Sinkholes are best developed at the top of the Boone Formation where it is overlain by Batesville Sandstone. Springs are concentrated at the base of the Boone Formation, with the largest karst systems localized within structural lows whose recharge may include interbasin transfer from the adjacent Crooked Creek topographic basin. A lower karst aquifer, the Ozark aquifer, is recognized where structural highs expose lower parts of the Ordovician Everton Formation within the Buffalo River watershed. Both modern karst systems and late Paleozoic lead-zinc mineralizing fluids were influenced by lithologic and structural characteristics of the bedrock, and modern karst systems may locally reoccupy ancient fluid pathways.

## INTRODUCTION

Extensive karst landscapes have developed within the Ozark Plateaus region due to their favorable geology and ample precipitation in a temperate climate. These landscapes host a variety of important and sometimes unique natural resources and ecosystems. Karst landscapes are well expressed at Buffalo National River, a park administered by the National Park Service (NPS) in northern Arkansas (fig. 1) that was established in 1972 to preserve this 214-km-long free-flowing river. The watershed for the Buffalo River and adjacent areas contains abundant caves, sinkholes, losing streams, and springs that are linked in karst hydrologic systems. During the last 15 years, geologic mapping at 1:24,000 scale by the U.S. Geological Survey and the Arkansas Geological Survey has advanced knowledge of the lithologic and structural characteristics that have potentially influenced karst development in this region (figs. 1 and 2). Stops on this field trip (fig. 3) will examine the geologic setting of karst features in the west-central part of the Buffalo River watershed.

## GEOLOGIC, PHYSIOGRAPHIC, AND HYDROLOGIC SETTINGS

Geologically, the Buffalo River watershed lies within the southern flank of the Ozark dome (fig. 1), a late Paleozoic uplift developed in the foreland to the Ouachita orogenic belt. The watershed includes part of the Boston Mountains and Springfield Plateau physiographic provinces that are held up by Pennsylvanian sandstone and shale and by Mississippian cherty limestone, respectively. The river incises Ordovician rocks that hold up the Salem Plateau farther to the northeast (Purdue and Miser, 1916).

Erosion by the Buffalo River and its tributaries has enhanced bedrock exposures of an approximately 1,600 ft (490 m) thick stratigraphic sequence (fig. 4; Appendix) of carbonate and clastic sedimentary rocks (Purdue and Miser, 1916; McKnight, 1935) that were deposited predominantly in shelf-marine, near-shore-marine, or fluvial environments. The distribution of carbonate lithologies determines the potential for karst development. Ordovician rocks are alternating dolostones, limestones, and carbonate-cemented sandstones, as typified by the heterogeneous Middle Ordovician Everton Formation (Suhm, 1974) that crops out widely in lower elevations of the Buffalo River valley.

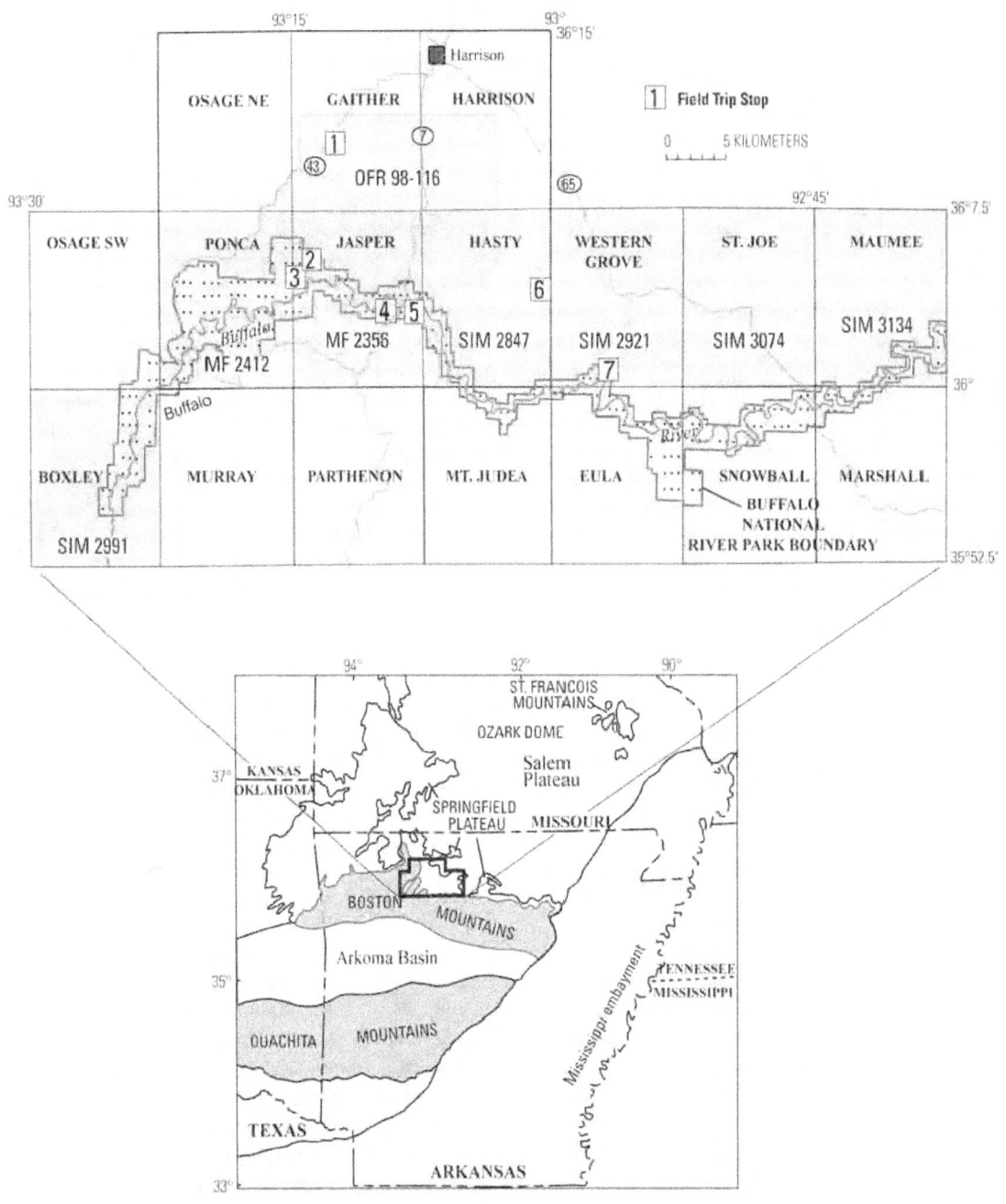

Figure 1. Location of study area within northern Arkansas, overlapping the western part of Buffalo National River. Field trip stops are approximately located. U.S. Geological Survey geologic maps in area (Hudson, 1998; Hudson and others, 2001, 2006; Hudson and Murray, 2003, 2004; Hudson and Turner, 2007, 2009; Turner and Hudson, 2010) are listed by publication number. Lower regional map illustrates geological and selected physiographic provinces of Arkansas and adjacent areas. The Ozark Plateaus region includes Salem and Springfield Plateaus and Boston Mountains physiographic provinces.

Buffalo National River Geology Compilation, North Central Arkansas

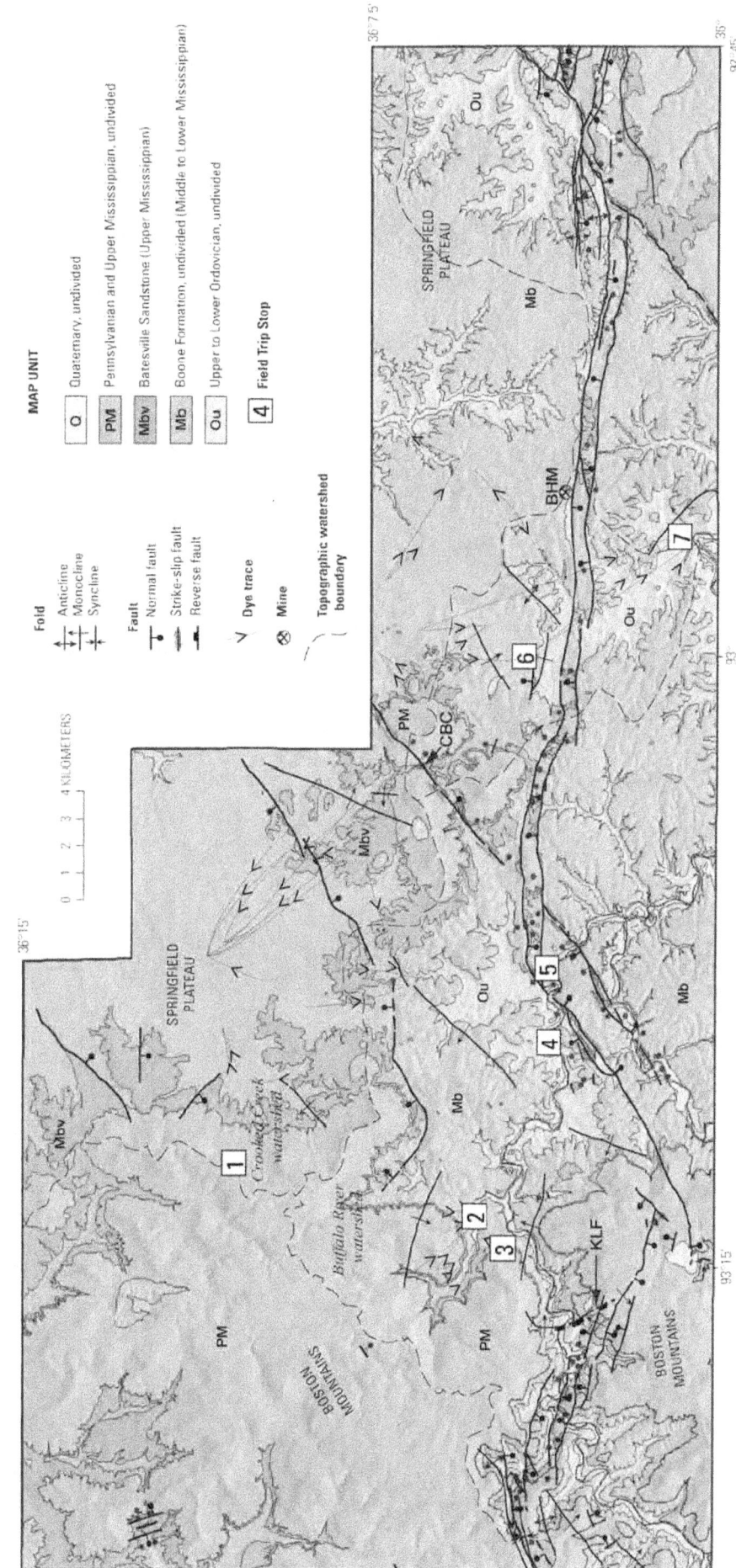

Figure 2. Geologic map for west-central part of Buffalo River watershed and adjacent areas compiled from published maps listed in fig. 1 as well as McMoran (1968) and Lucas (1971). Locations for field trip stops and dye tracer tests are also shown. Other abbreviations include: KLF, Kyles Landing fault; BHM, Big Hurricane Mine; CBC, Chilly Bowl Cave.

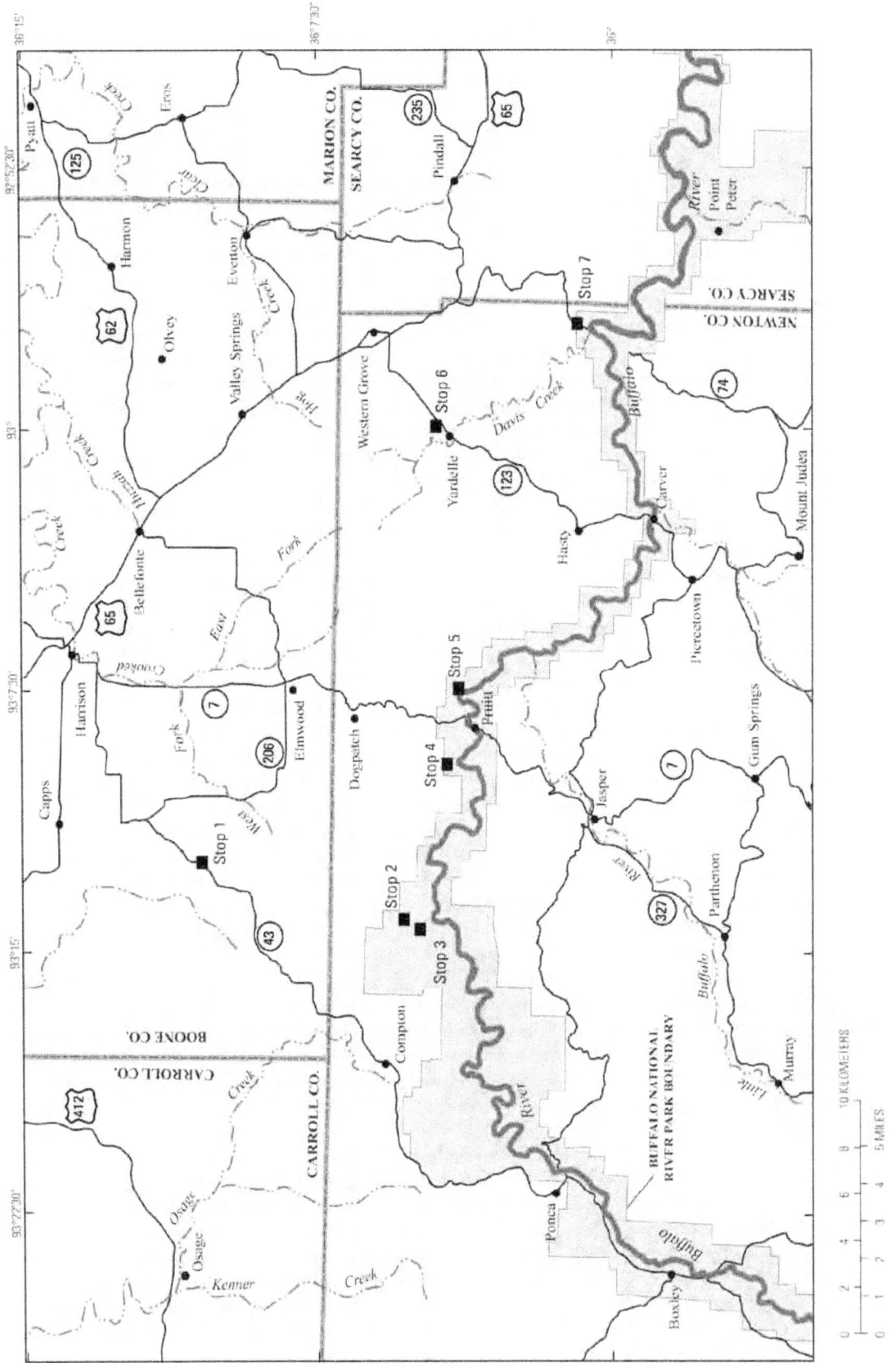

Figure 3. Regional road map of northern Arkansas with Buffalo National River (shaded area) and locations of field-trip stops.

194

Thin, but hydrologically significant, shale beds are present within both the Ordovician Powell Dolomite and St. Peter Sandstone. The exposed Ordovician rocks comprise the upper part of the Ozark aquifer of Adamski and others (1995). Mississippian rocks are predominantly limestone but also have significant thickness of shale and sandstone in their upper part. The Lower to Middle Mississippian Boone Formation is divided into the basal St. Joe Limestone Member (Kinderhookian to Osagean) and the main body of the Boone (Osagean to Meramecian); together these form the Springfield Plateau aquifer of Adamski and others (1995). Upper Mississippian Pitkin Limestone is present in parts of the watershed. Pennsylvanian rocks are dominantly sandstone and shale, with carbonate-cemented sandstone and limestone restricted to the Prairie Grove Member of the Hale Formation and lower Bloyd Formation, respectively. Due to their significant thickness of low-permeability shale, the Upper Mississippian and Pennsylvanian strata were included in the Western Interior Plains confining hydrogeologic unit of Adamski and others (1995). Regional unconformities (fig. 4) separate the Ordovician, Mississippian, and Pennsylvanian sequences; uppermost strata of both the Ordovician and Mississippian sequences are truncated northward beneath these two unconformities. Quaternary surficial deposits are present as colluvial wedges, landslides, and as a series of river-terrace deposits that are linked to the history of modern landscape development.

Paleozoic strata were deformed by a series of faults and open folds (fig. 2) that developed in three late Paleozoic deformation phases (Hudson, 2000; Hudson and Cox, 2003). The oldest deformation, **D1**, is aerially restricted to the >4-km-long, north-northwest-trending Kyles Landing fault and an associated fold that were active in latest Mississippian-earliest Pennsylvanian time (Hudson and Murray, 2003). A widespread deformation phase, **D2**, formed most of the west- to west-northwest-striking normal faults, northeast-striking dextral strike-slip faults, and variably oriented monoclinal folds throughout the area. These structures acted in a coordinated fashion to accommodate north-south extension during flexure of the continental margin as it was being loaded beneath the Ouachita orogenic belt to the south, probably during Middle-Late Pennsylvanian time (Hudson, 2000). In the final deformation phase, **D3**, preexisting structures were locally reactivated in north-south shortening related to final docking of the Ouachita orogenic belt. Regionally in northern Arkansas, north-south shortening of **D3** is recorded in calcite-twin deformation in limestones (Chinn and Konig, 1973) and by a pervasive north-trending joint set (Hudson and others, 2010).

Late Paleozoic lead-zinc mineralization was widespread in the Ozarks region and has been linked to a northward flow of mineralizing brines out of the adjacent Arkoma Basin during the Ouachita orogeny (Leach and Rowan, 1986). Leach and Rowan (1986) suggest that regional flow of ore-bearing fluids was focused along basal Cambrian clastic strata, but in northern Arkansas lead-zinc deposits are hosted by Middle Ordovician and Lower Mississippian strata near faults (McKnight, 1935), demonstrating that faults acted to divert mineralizing fluids up section (Erickson and others, 1988). Late Pennsylvanian to early Permian ages for the mineralization are indicated by isotopic dating (Brannon and others, 1996) and paleomagnetic studies (for example, Pan and others, 1990). Commonly associated with lead-zinc deposits (McKnight, 1935), collapse breccias demonstrate that ancient episodes of carbonate dissolution predate modern karst development.

## Karst Features

In the modern landscape, carbonate rocks in the Buffalo River region were subjected to dissolution as capping Pennsylvanian and Mississippian sandstones and shales were eroded from the area. In particular, the thick limestone interval of the Mississippian Boone Formation developed karst systems from the interchange of surface and groundwater via recharge from sinkholes and losing-stream segments and discharge at springs. Dye tracer tests (fig. 2) indicate subsurface transport of groundwater occurs rapidly (1000's of ft/day) within the Springfield Plateau aquifer (Mott and others, 2000), consistent with transport via solution conduits. Many cavern systems are known within the Boone Formation. An unpublished

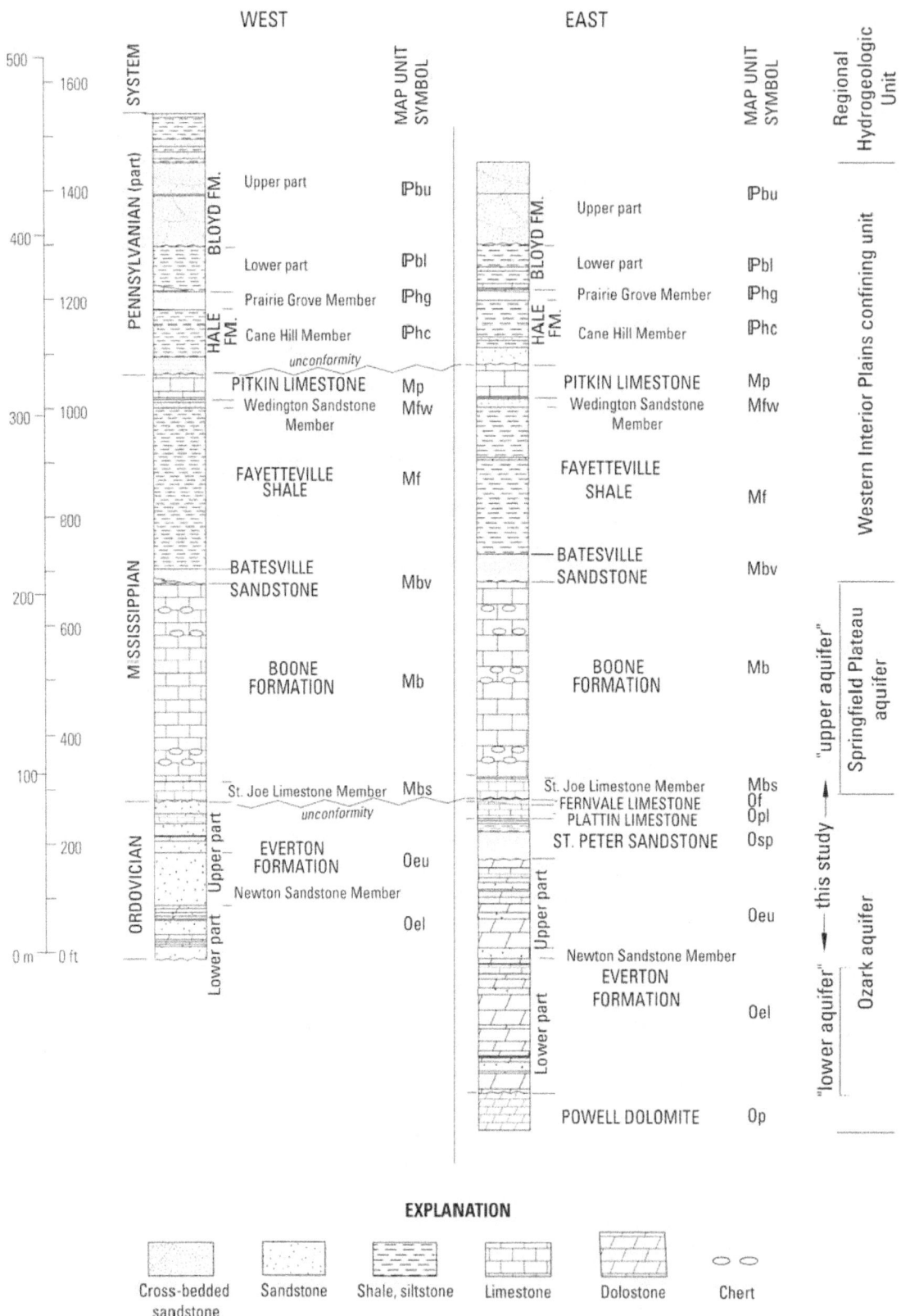

Figure 4. Representative stratigraphic columns for Paleozoic rocks for the eastern and western parts of the study area. Regional hydrogeologic units follow Adamski and others (1995) and local karst aquifers recognized in this study are indicated. FM.–Formation

NPS inventory of caves within the park demonstrates that Boone Formation is the dominant host of caves within the area (fig. 5). Discharge occurs at springs in both the Buffalo River and adjacent Crooked Creek watersheds and recharge areas for springs in the Buffalo River watershed locally extend beyond topographic divides of surface drainages (Mott and others, 2000). An inventory of springs within the Mill Creek subbasin (Hudson, 1998) of the Buffalo River watershed demonstrates a strong frequency maximum at or near the basal contact of the St. Joe Limestone Member of the Boone Formation (fig. 5). Although these characteristics indicate that the Boone Formation is the main karst aquifer in the watershed, the largest spring in the Park (Mitch Hill Spring) discharges from the lower part of the Ordovician Everton Formation, another common host of caves (fig. 5). Lower parts of the Ordovician stratigraphic sequence are exposed by several structural highs within the watershed. Comparison of springs and connecting dye traces to the underlying geology in these areas indicates that a limestone-bearing interval in the lower part of the Everton Formation forms a second karst aquifer discharging into the Buffalo River watershed.

## Field Trip Stops

This field trip will provide an overview of the west-central Buffalo National River region and will include key stops to illustrate probable stratigraphic and structural controls on karst systems. It will address the effects of both a spatially variable Paleozoic stratigraphy and late Paleozoic faulting and folding on the location of major springs, caves, and recharge basins within two stacked karst aquifers (fig. 4). Most stops will be near roads (fig. 3), although other interesting features will be noted that may be reached on foot with more time. All location coordinates for trip stops are given relative to the North American Datum of 1927 to match 1:24,000-scale topographic maps. This trip is designed to be easily completed in a single day.

### Stop 1 – Gaither Mountain (36°10.464'N, 93°12.474'W)

This overlook serves as an introduction to the physiographic setting of the southern Ozark

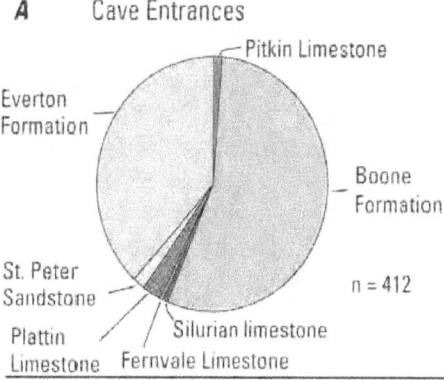

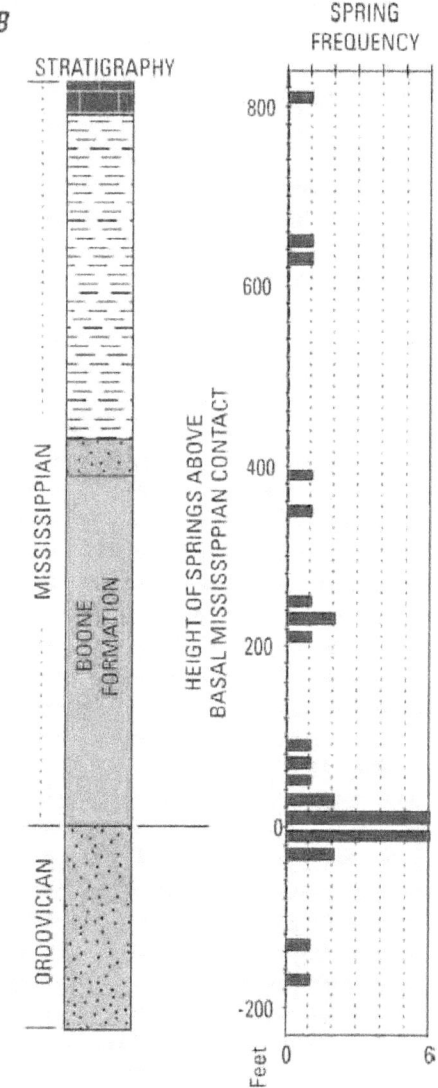

Figure 5. Frequency of caves and springs within stratigraphic units of part of the study area. (A) Pie diagram of cave entrances within park boundaries of Buffalo National River from unpublished NPS inventory. Description of Silurian limestone can be found in Turner and Hudson (2010). (B) Relative frequency of springs versus stratigraphic position from parts of Ponca, Jasper, Hasty, Gaither, and Harrison quadrangles (from Hudson, 1998).

Plateau region. The southern Ozark Plateau region contains three regional geomorphic surfaces. The highest, Boston Mountains, is held up by Pennsylvanian strata, mostly sandstone and shale. The intermediate Springfield Plateau is held up by Lower to Middle Mississippian limestone. The Salem Plateau is held up by Cambrian and Ordovician strata, predominantly dolostone with lesser limestone and sandstone.

The overlook stop is sited within Lower Pennsylvanian (Morrowan) strata of the Bloyd Formation (figs. 2, 3, and 4) with a notable thick cross-bedded sandstone member (middle Bloyd sandstone) that forms prominent bluffs throughout the region (fig. 6). The view to the east gives an overview of the Springfield Plateau that is developed on top of the Mississippian Boone Formation. A cluster of mountains (Sulphur, Boat, and Pinnacle Mountains) are capped by Pennsylvanian strata and represent remnants of the Boston Mountains plateau that have not yet been lost to erosion. Looking east, the surface drainage drains northward into Crooked Creek. A topographic divide to the southeast separates the Crooked Creek watershed from the deep valley of the Buffalo River to the south that cuts through Pennsylvanian and Mississippian strata into Ordovician strata. Dye tracer studies have documented interbasin transfer of groundwater from the southern Crooked Creek topographic watershed into the Mill Creek subbasin of the Buffalo River watershed (Mott and others, 2000).

Figure 6. Photograph of thick cross-bedded sandstone of middle part of the Pennsylvanian Bloyd Formation present at Stop 1. This unit holds up a significant part of the high Boston Mountains geomorphic surface in the region.

## Stop 2 – Fitton Spring (36°5.273'N, 93°14.001'W)

Springs within karst systems in the Buffalo River watershed reflect the discharge of groundwater, and larger springs may be associated with cavern systems. Fitton Spring (figs. 7, and 8) represents the discharge point for Fitton Cave, the longest known cave system in Arkansas; cave mapping has identified more than 12.8 km (8 mi) of mapped passages with multiple vertical levels spanning a 120 m (393 ft) elevation range. The deepest cave passages have been mapped to within about 150 m (500 ft) distance of Fitton Spring, and dye traces have confirmed the intervening hydrologic connection.

A series of dye tracer tests conducted by T. Aley for the NPS in the late 1990's demonstrated that Fitton Spring and nearby Van Dyke Spring gather their recharge from upstream segments of Cecil Creek. Traces detected at Fitton Spring received input from the Bartlett Cove branch and from slopes northwest of the spring. Van Dyke Spring received input from traces farther west in the Cecil Creek basin.

Geologically, the Fitton Spring and Cave system are fully hosted within Mississippian Boone Formation and the spring emerges from the upper part of the basal St. Joe Limestone Member. Three geologic factors may have contributed to development of this large karst system. (1) Facies of the main body of the Boone Formation limestone in this part of the watershed contain less insoluble chert than typical for northern Arkansas, facilitating connectivity of caverns. (2) An uncommonly thick section of Batesville Sandstone and Fayetteville Shale overlies Boone Formation in the Cecil Creek area. These units contain disseminated pyrite whose oxidation may have provided chemically aggressive waters that aided limestone dissolution. Such pyrite oxidation also provides a source of sulfate for gypsum cave formations that are common in highest passages within Fitton Cave, closest to the contact with the overlying formations. (3) The Fitton Cave karst system is localized at the base of a structural trough bounded by two west-northwest-trending monoclines. The association

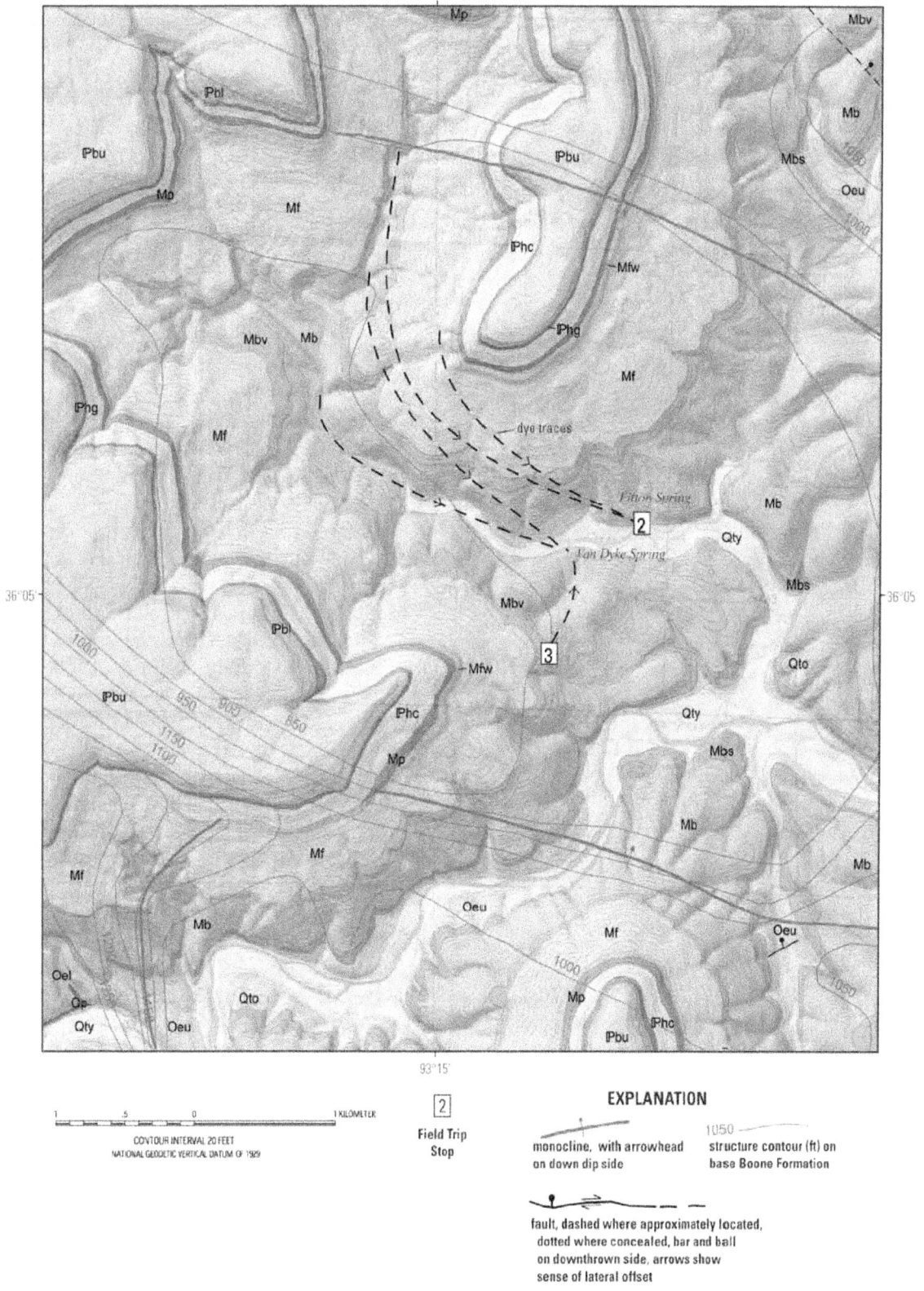

Figure 7. Geologic map of the northeastern Ponca quadrangle (Hudson and Murray, 2003) and northwestern Jasper quadrangle (Hudson and others, 2001) with location of stops 2 and 3. Map unit symbols are keyed to Fig. 4 and Appendix. Dye tracer tests were conducted by T. Aley in the late 1990's for National Park Service.

199

points in the Springfield Plateau aquifer gather groundwater where they are perched above less-permeable Ordovician rocks at the surface.

## Stop 3 – Batesville sinkhole (36°4.782'N, 93°14.430'W)

Field observations indicate that all carbonate lithologies in the Buffalo River region undergo dissolution and thus have potential to develop shallow surface depressions. However, sinkholes of sufficient size and depth to have expression on 1:24,000-scale topographic maps are strongly concentrated at or near the contact of the Batesville Sandstone with the Boone Formation. This stop (fig. 7) examines a sinkhole (fig. 9A) developed at the edge of a local Batesville Sandstone topographic flat where the overlying Fayetteville Shale was removed by erosion. Batesville Sandstone is interpreted to have sufficient strength to hold up topography until caverns in the underlying Boone Formation enlarge to sufficient size to cause collapse and form large sinkholes. In contrast, solution cavities preserved on lower slopes of Boone Formation are mostly smaller, in part because they are infilled by residual chert blocks weathered from higher in the formation. An additional factor in creation of large sinkholes in the stratigraphic interval of the upper Boone Formation-Batesville Sandstone may be the presence of pyrite in the Batesville Sandstone and overlying Fayetteville Shale whose oxidation may have increased acidity of descending waters. Oxidized pyrite framboids are common (fig. 9B) and form red spheres distributed through the tan, parallel laminated, very fine-grained sandstone blocks of Batesville Sandstone.

Within the sinkhole, vertical conduits developed within limestone of the underlying Boone Formation are elongated along northeast-trending joints, one of several dominant regional joint sets (fig. 10) (Hudson and others, 2010). Such apparent local control of dissolution features by joints is prevalent at outcrop scale throughout the Buffalo River region, although at regional scales the intersection of multiple joint trends probably leads to near-isotropic fracture permeability.

Figure 8. For Stop 2, (A) Map view of survey lines for Fitton Cave passages, from unpublished mapping data gathered by NPS. (B) View to north of Fitton Spring, the main discharge for Fitton Cave that emits from the uppermost part of the St. Joe Limestone Member of the Mississippian Boone Formation. Within Fitton Cave, gypsum cave formation types include (C) polycrystalline flowers, (D) needles, and (E) crusts and pendants.

of large karst systems with structural troughs in the Boone Formation has been documented elsewhere in the Buffalo River watershed (Hudson, 1998) and demonstrates that these low

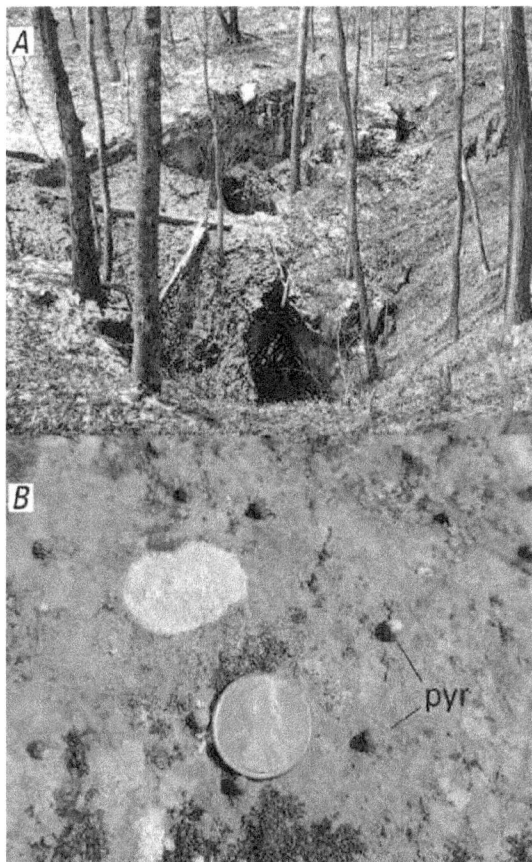

Figure 9. For Stop 3, (*A*) photograph of sinkhole developed in Batesville Sandstone and (*B*) oxidized pyrite (pyr) framboids standing in relief on weathered surface of Batesville Sandstone block (1.9-cm diameter coin for scale).

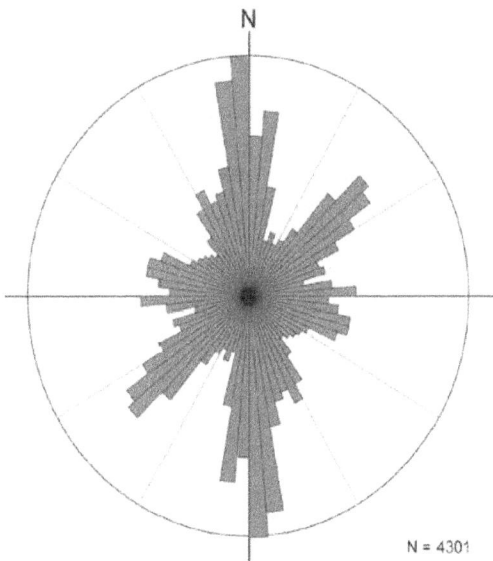

Figure 10. Rose diagram of joints measured within Buffalo River mapping areas.

## Stop 4 - Ozark (36°3.876'N, 93°9.535'W)

This stop illustrates both Ordovician Everton Formation and multiple Quaternary terrace deposits of the Buffalo River (fig. 11). The Middle Ordovician Everton Formation is a heterogeneous mix of dolostone, sandstone, and limestone that were deposited in tidal flat and barrier island environments, and have been divided into several members (Suhm, 1974). Although grouped within the Ozark aquifer of Adamski and others (1995), the Everton Formation contains significant sandstone intervals that act as local aquitards. Across the river at this stop, Briar Bluff exposes the thick Newton Sandstone Member overlying an interval containing thin beds of stromatolitic limestone (figs. 12*A* and *B*). Local dissolution is evident in the limestone interval below the Newton Sandstone Member. This boundary is used to separate informal upper and lower map units of the Everton Formation and represents the top of a proposed lower carbonate aquifer within the watershed.

Sandy deposits form a terrace level about 20 ft (6 m) above the river base level that is extensive along the Buffalo River. An older terrace deposit is preserved as an interval of cobbles and capping sand 80 ft to 110 ft (24 m to 33.5 m) above the river base level that is inset into Ordovician bedrock to the south. The cobbles are subrounded to well rounded and include clasts of sandstone and chert derived from a mixture of Ordovician, Mississippian, and Pennsylvanian bedrock sources. Remnants of similar cobble deposits at similar heights above the current river base level have been mapped for many kilometers both upstream and downstream from this site (fig. 11). Terrace deposits provide a record of valley incision and thus the lowering of the ultimate base level for karst springs discharging into the watershed. Studies are underway to better characterize the style and rate of Buffalo River valley incision and how this may have influenced karst development.

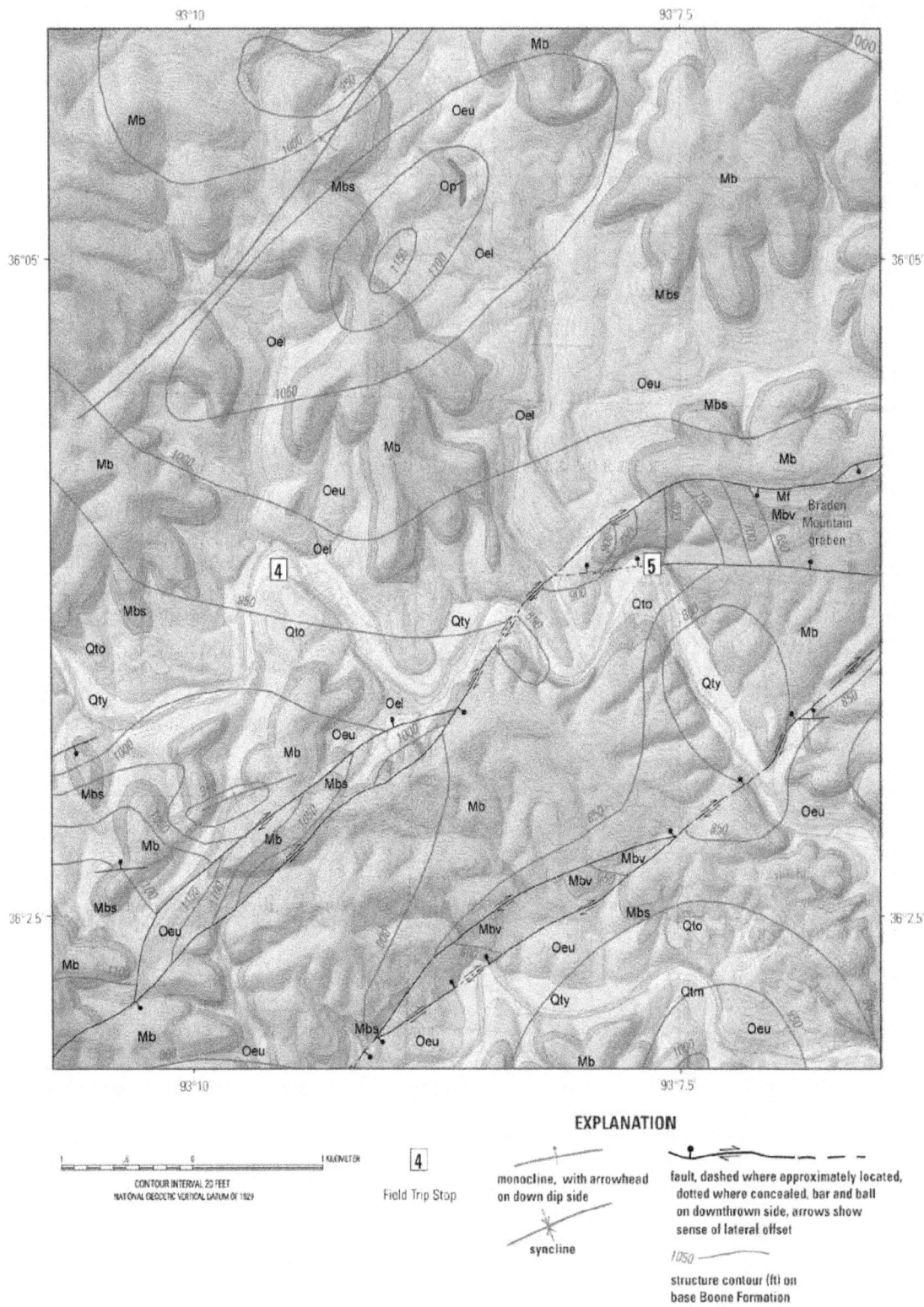

**EXPLANATION**

CONTOUR INTERVAL 20 FEET
NATIONAL GEODETIC VERTICAL DATUM OF 1929

4

Field Trip Stop

monocline, with arrowhead
on down dip side

syncline

fault, dashed where approximately located,
dotted where concealed, bar and ball
on downthrown side, arrows show
sense of lateral offset

1050

structure contour (ft) on
base Boone Formation

Figure 11. Geologic map modified from the eastern Jasper quadrangle (Hudson and others, 2001) and western Hasty quadrangle (Hudson and Murray, 2004) with location of stops 4 and 5. Map unit symbols are keyed to fig. 4 and Appendix.

Figure 12. Photographs of (A) Briar Bluff exposing Ordovician Everton Formation, with designation of upper and lower parts. Top of lower part placed below Newton Sandstone Member at uppermost stromatolitic limestone beds. See lower right box for location of (B) close view of stromatolitic limestone beds with local dissolution conduits. Note hemispherical casts of algal heads in lower bed.

## Stop 5 - Welch Bluff (36°3.841'N, 93°7.617'W)

This stop (fig. 11) displays a well-exposed normal fault zone (fig. 13A) in a bluff on the opposite side of the Buffalo River. The fault zone forms the southern margin of the Braden Mountain graben, part of an east-striking graben system that continues eastward for about 29 km (18.3 mi). This graben system accommodated north-south **D2** extension that developed during late Paleozoic flexure of the southern Ozarks region in the foreland of the Ouachita orogenic belt. At Welch Bluff, the contact between the Ordovician Everton Formation and the basal, reddish St. Joe Limestone Member of the Mississippian Boone Formation rises 20 m (66 ft) above the river across two fault splays to form the southern footwall of the fault. On the hanging wall, this Boone-Everton contact is

dropped below river level adjacent to the fault but it rises above the surface farther to the northwest. The southeast-dipping strata in the hanging wall (fig. 13B) are an example of an extensional relay ramp caused by a decrease of the vertical displacement of faults toward the western termination of the graben. Farther east, throw on the southern fault of the Braden Mountain graben is as much as 120 m (393 ft). In the bluff, the fault zone is composed of two fault planes that dip 74°N with a 3-m- (10-ft-) wide intervening fault block in which thin limestone beds are folded about an axis parallel to the fault strike. This fold geometry and the preservation of several high-rake slickensides on these fault planes indicate a dip-slip, normal displacement for the main movement on this fault zone.

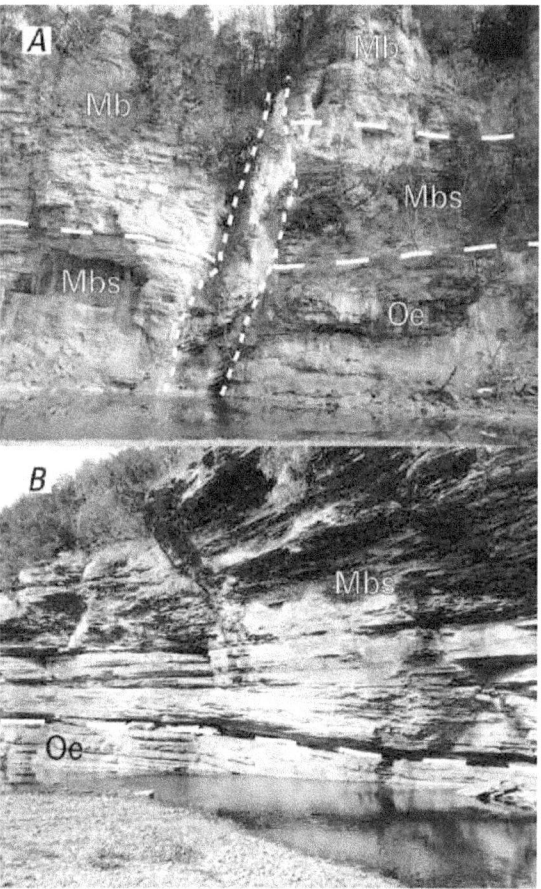

Figure 13. For Stop 5, (A) View looking east of South Braden Mountain fault zone with two fault splays that drop contact between the St. Joe Limestone Member (Mbs) of the Boone Formation and the Everton Formation (Oe) below river level on the northern hanging wall block. Mb—main body Boone Formation. (B) View to west of gently southeast-dipping strata in hanging wall of fault zone.

Fault zones commonly coincide with topographic lineaments in the Buffalo River area. In several instances, parallel alignments of small sinkholes are coincident with faults, particularly where both affect the stratigraphic interval near the Boone Formation-Batesville Sandstone contact. These observations suggest that karst dissolution is enhanced near fault zones due to an increased fracture density and associated fracture permeability. Fault zones also localized late Paleozoic fluid flow associated with lead-zinc mineralization in northern Arkansas, as illustrated by ore extracted from Big Hurricane Mine (McKnight, 1935) located about 17 km (10.7 mi) to the east along the same graben system (fig. 2). These relations demonstrate the potential for modern cavern systems located near fault zones to have had hypogene, late Paleozoic origins, as has been documented at Chilly Bowl Cave (Tennyson and others, 2008) located about 9 km (5.7 mi) northeast of the stop (fig. 2).

## Stop 6 – Yardelle Spring (36°4.115'N, 93°0.156'W)

Yardelle Spring discharges from the St. Joe Limestone Member of the Mississippian Boone Formation (figs. 14 and 15) at the bottom of a structural trough framed by a northeast-trending down-to-southeast monocline on the west, the northeast-trending Shaddock Branch anticline to the east, and the west-northwest-trending down-to-north Yardelle monocline to the south (figs. 14 and 16). The spring emerges just above the basal contact with sandstone of Ordovician Everton Formation. Dye tracer studies conducted by the National Park Service demonstrate that Yardelle and the nearby Yardelle Branch Spring receive water from sites across the topographic divide within the Clear Creek watershed to the north. Similar interbasin transfer of groundwater discharging from springs in structural lows within the Springfield Plateau aquifer (Boone Formation) has also been documented in areas to the east and west (fig. 2). Due to their location within a perched, unconfined karst aquifer, we interpret that the groundwater gradient linked to these low-elevation discharge points allows these associated springs to gather recharge from long distances that can extend into the adjacent watershed. In the Mill Creek subbasin to the west, such interbasin recharge has practical

resource management concerns because increased nutrient loads are introduced into the springs from agricultural lands within the adjacent Crooked Creek watershed (Mott and others, 2000).

## Stop 7 – Mitch Hill Spring (36°0.895'N, 92°57.044'W)

Mitch Hill Spring (fig. 14), located near Mt. Hersey, is the largest spring within the park with an average flow of about 2 cubic feet per second. It is sourced from the lower part of the Ordovician Everton Formation and thus represents discharge from a lower karst aquifer within the Buffalo River watershed. In contrast to springs within the Boone Formation that respond quickly to precipitation, Mitch Hill Spring has more constant flow with a modest artesian head (fig. 17A) that suggests the source aquifer is partially confined. Geologically, the lower part of the Everton Formation and underlying Powell Dolomite are brought to the surface in a structural high by the northwest-trending down-to-northeast Cane Branch monocline. Powell Dolomite is an argillaceous dolostone (McKnight, 1935) that is interpreted to act as a lower confining unit forcing groundwater discharge at Mitch Hill Spring (Hudson and others, 2005; Turner and others, 2007). Nearby the spring to the west, McKnight (1935) reported that zinc carbonate ore was extracted at Old Granby mine from the lowermost Everton Formation as early as 1875. Sandstone ledges directly above the Mitch Hill Spring horizon commonly contain silicified collapse breccias (fig. 17B), giving indirect evidence for karst dissolution in underlying strata. However, the strong silicification and the observation that breccias are cut by joints that probably formed during late Paleozoic orogeny (Hudson and others, 2010), indicates that the breccias are probably ancient and likely formed due to dissolution associated with flow of lead-zinc mineralizing fluids. Thus, the modern karst system of Mitch Hill Spring is in part coincident with, and may have reused, ancient fluid-flow pathways.

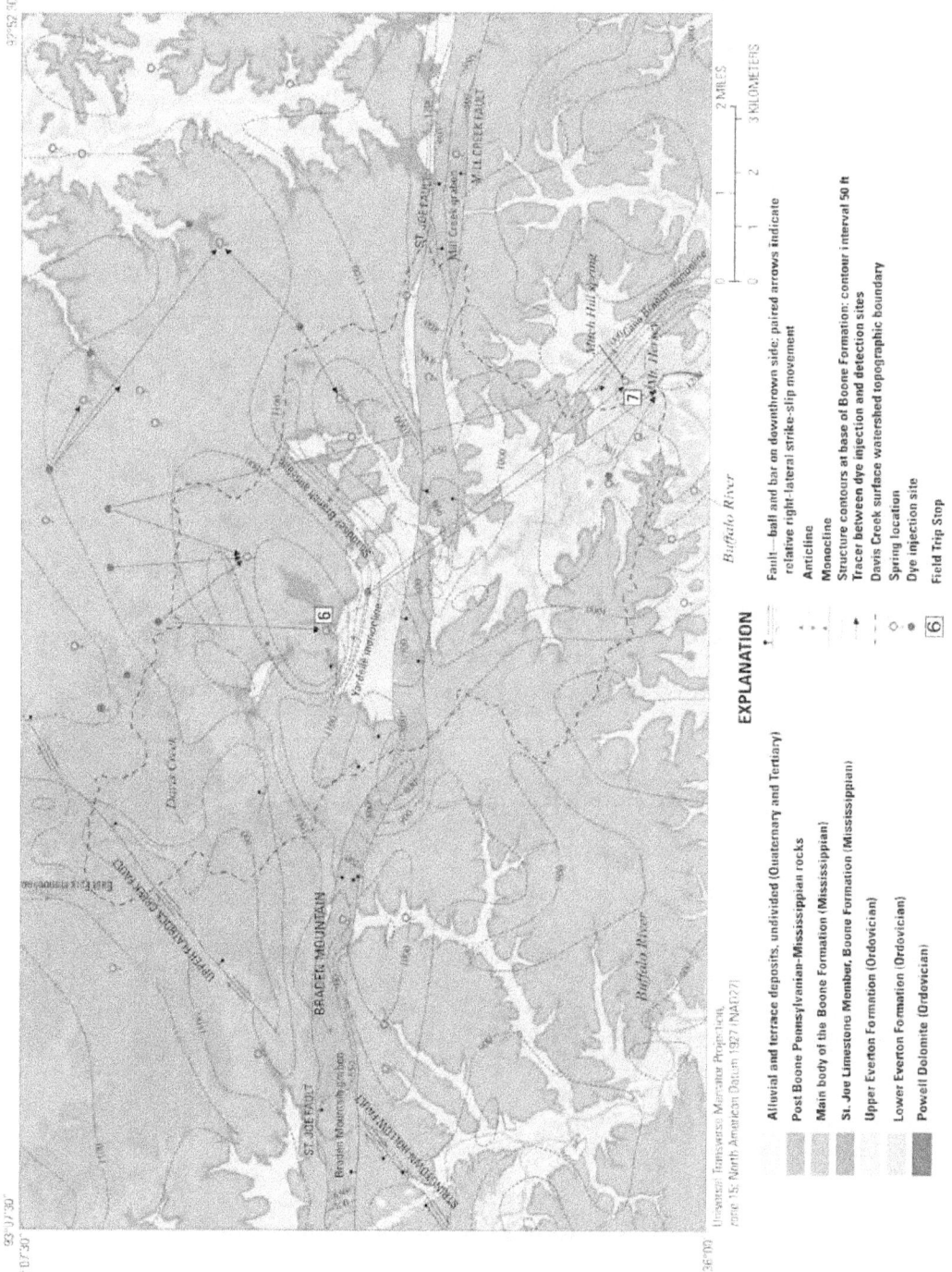

Figure 14. Generalized geologic map from Turner and others (2007) based on geologic data from Hudson and Murray (2004) and Hudson and others (2006). Also shown are Davis Creek subbasin of Buffalo River watershed, springs, and dye traces conducted by National Park Service.

205

Figure 15 (left). View looking west of Yardelle Spring, discharging from St. Joe Limestone Member of Boone Formation.

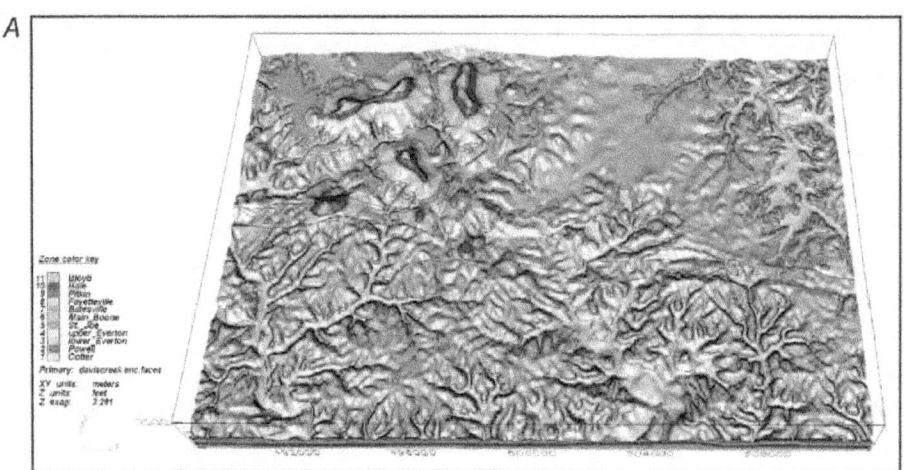

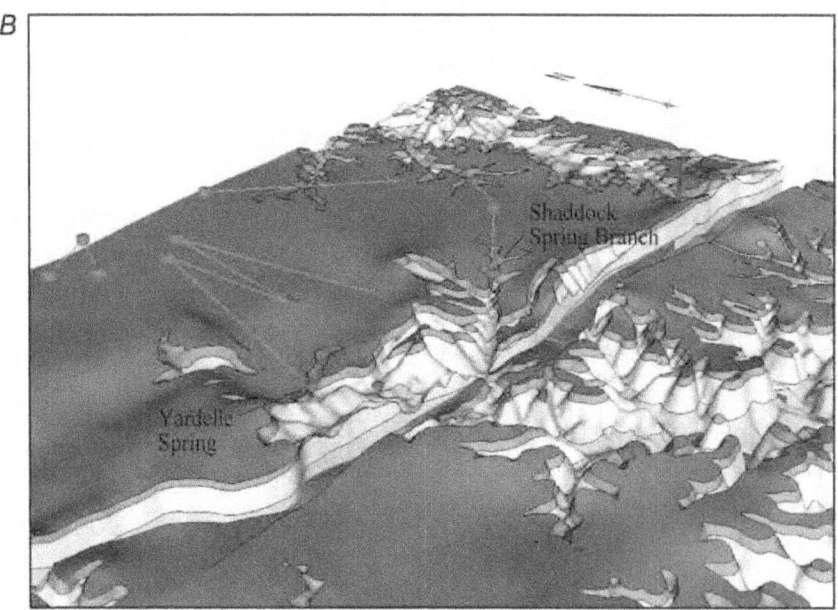

Figure 16 (above and right). Images from three-dimensional geologic model (Turner and others, 2007) of the Hasty and Western Grove quadrangles (area of fig. 14). (A) Overview of model, looking to north. (B) View to northeast with layers above the St. Joe Limestone Member of the Boone Formation removed. Boxes represent dye trace input and detection

206

points with connecting lines. (*C*) View to northwest showing potential dye trace path from sink in lower part of the Everton Formation, in Davis Creek below Yardelle Spring through graben toward Mitch Hill Spring. (*D*) View to west of potential groundwater flow path across graben between lower Everton Formation on graben flanks via St. Joe Limestone Member of Boone Formation within graben. (*E*) View to northwest showing end of potential dye trace at Mitch Hill Spring at upwarp of underlying Powell Dolomite at Cane Branch monocline.

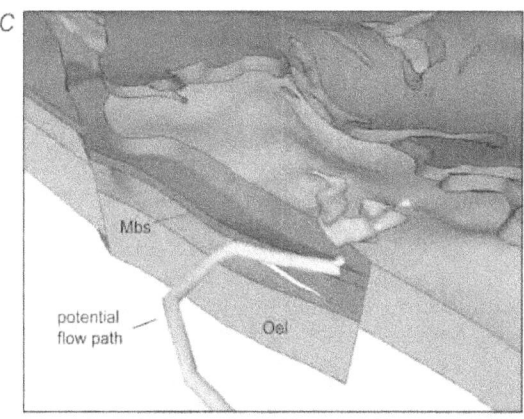

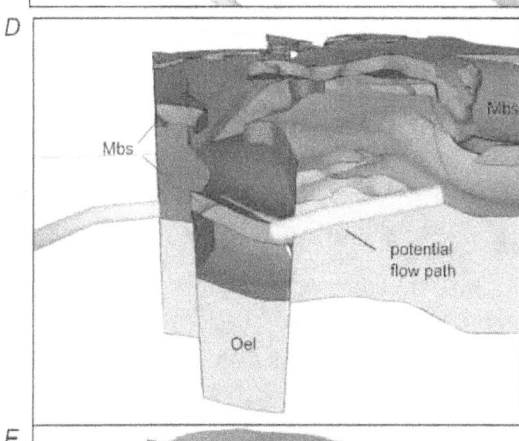

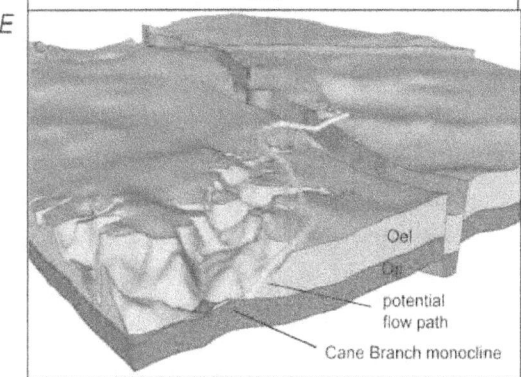

losing stream segment emerge from Mitch Hill Spring, but the flow path must cross an intervening graben (fig. 14). A three-dimensional geology model (Turner and others, 2007) allows visualization of likely pathways for groundwater flow across the graben (fig. 16). This model demonstrates likely paths of groundwater flow from the lower Everton Formation through down-faulted Boone Formation within the graben, and back into the lower Everton Formation.

Figure 17. Photographs of (*A*) Mitch Hill Spring, and (*B*) Breccias within silicified sandstone layers that lie above Mitch Hill Spring horizon. North-trending joints that cut breccia are probably late Paleozoic in age.

## REFERENCES CITED

Adamski, J.C., Petersen, J.C., Freiwald, D.A., and Davis, J.V., 1995, Environmental and hydrologic

Dye tracer studies from the upper part of the immediately adjacent Davis Creek drainage demonstrate a transition of groundwaters and surface waters from the upper Springfield Plateau aquifer (Boone Formation) to the lower Everton aquifer, but this transfer is complicated by intervening faults and folds. Downstream from Yardelle Spring, Davis Creek can go dry where the lower part of the Everton Formation is brought to the surface. Dye traces from this setting of the Ozark Plateaus study unit, Arkansas, Kansas, Missouri, and Oklahoma: U.S. Geological Survey Water-Resources Investigation Report 94–4022, 69 p.

Brannon, J.C., Podosek, F.A., and Cole, S.C., 1996, Radiometric dating of Mississippi Valley-type ore deposits, in D.F. Sangster, ed., Carbonate-hosted lead-zinc deposits: Society of Economic Geologists Special Publication 4, p. 536–545.

Chinn, A.A., and Konig, R.H., 1973, Stress inferred from calcite twin lamellae in relation to regional structure of northwest Arkansas: Geological Society of America Bulletin, v. 84, p. 3731–3736.

Erickson, R.L., Chazin, B., Erickson, M.S., Mosier, E.L., and Whitney, H., 1988, Tectonic and stratigraphic control of regional subsurface geochemical patterns, midcontinent, U.S.A., in Kisvarsanyi, G., and Grant, S.K., eds., Proceedings of tectonic control of ore deposits and the vertical and horizontal extent of ore systems: Rolla, Missouri, University of Missouri, Rolla Press, p. 435–446.

Glick, E.E., and Frezon, S.E., 1953, Lithologic character of the St. Peter Sandstone and Everton Formation in the Buffalo River Valley, Newton County, Arkansas: U.S. Geological Survey Circular 249, 39 p.

Hudson, M.R., 1998, Geologic map of parts of the Gaither, Hasty, Harrison, Jasper, and Ponca quadrangles, Boone and Newton Counties, northern Arkansas: U.S. Geological Survey Open-File Report 98–116, scale 1:24,000.

Hudson, M.R., 2000, Coordinated strike-slip and normal faulting in the southern Ozark dome of northern Arkansas: Deformation in a late Paleozoic foreland: Geology, v. 28, p. 511–514.

Hudson, M.R., and Cox, R.T., 2003, Late Paleozoic tectonics of the southern Ozark dome, in Cox, R.T., compiler, Field trip guidebook for Joint South-Central and Southeastern Sections, Geological Society of America: Tennessee Division of Geology Report of Investigations 51, p. 15–32.

Hudson, M.R., Mott, D.N., Turner, K.J., and Murray, K.E., 2005, Geologic controls on a transition between karst aquifers at Buffalo National River, northern Arkansas, in Kuniansky, E.L., 2005, U.S. Geological Survey Karst Interest Group proceedings, Rapid City, South Dakota, September12–15, 2005: U.S. Geological Survey Scientific Investigations Report 2005–5160, 296 p., http://pubs.usgs.gov/sir/2005/5160/.

Hudson, M.R., and Murray, K.E., 2003, Geologic map of the Ponca Quadrangle, Newton, Boone, and Carroll Counties, Arkansas: U.S. Geological

Survey Miscellaneous Field Studies Map MF-2412, scale 1:24,000, http://pubs.usgs.gov/mf/2003/mf-2412.

Hudson, M.R., and Murray, K.E., 2004, Geologic map of the Hasty Quadrangle, Boone, and Newton Counties, Arkansas: U.S. Geological Survey Scientific Investigations Map 2847, scale 1:24,000, http://pubs.usgs.gov/sim/2004/2847/.

Hudson, M.R., Murray, K.E., and Pezzutti, Deborah, 2001, Geologic map of the Jasper Quadrangle, Newton and Boone Counties, Arkansas: U.S. Geological Survey Miscellaneous Field Studies Map MF–2356, scale 1:24,000, http://pubs.usgs.gov/mf/2001/mf-2356.

Hudson, M.R., and Turner, K.J., 2007, Geologic map of the Boxley Quadrangle, Newton and Madison Counties, Arkansas: U.S. Geological Survey Scientific Investigations Map 2991, scale 1:24,000, http://pubs.usgs.gov/sim/2991/.

Hudson, M.R., and Turner, K.J., 2009, Geologic map of the St. Joe Quadrangle, Searcy and Marion Counties, Arkansas: U.S. Geological Survey Scientific Investigations Map 3074, scale 1:24,000, http://pubs.usgs.gov/sim/3074/.

Hudson, M.R., Turner, K.J., and Repetski, J.E., 2006, Geologic map of the Western Grove Quadrangle, northwestern Arkansas: U.S. Geological Survey Scientific Investigations Map 2921, scale 1:24,000, http://pubs.usgs.gov/sim/2006/2921/.

Hudson, M.R., Turner, K.J., and Trexler, C.C., 2010, Geometry, relative ages, and tectonic implications of joints from a regional assessment of the Buffalo National River area, northern Arkansas: Geological Society of America Abstracts with Programs, v. 42, no. 2, p. 71.

Leach, D.L., and Rowan, E.L., 1986, Genetic link between Ouachita foldbelt tectonism and Mississippi Valley-type lead-zinc deposits of the Ozarks: Geology, v. 14, p. 931–935.

Lucas, P.E., 1971, Geology of the Osage NE quadrangle, Arkansas (Master's thesis): Fayetteville, University of Arkansas, 77 p.

McKnight, E.T., 1935, Zinc and lead deposits of northern Arkansas: U.S. Geological Survey Bulletin 853, 311 p.

McMoran, W.D., 1968, Geology of the Gaither Quadrangle, Boone County, Arkansas: (Master's thesis): Fayetteville, University of Arkansas, 86 p.

Mott, D.N., Hudson, M.R., and Aley, T., 2000, Hydrologic investigations reveal interbasin recharge contributes significantly to detrimental nutrient loads at Buffalo National River, Arkansas: Proceedings of Arkansas Water

Resources Center Annual Conference MSC–284, Fayetteville, Ark., p. 13–20

Pan, H., Symons, D.T.A., and Sangster, D.F., 1990, Paleomagnetism of the Mississippi Valley-type ores and host rocks in the northern Arkansas and Tri-State districts: Canadian Journal of Earth Sciences, v. 27, p. 923–931.

Purdue, A.H., and Miser, H.D., 1916, Descriptions of the Eureka Springs and Harrison quadrangles: U.S. Geological Survey Atlas, Folio 202, scale 1:125,000.

Suhm, R.A., 1974, Stratigraphy of Everton Formation (early medial Ordovician), northern Arkansas: American Association of Petroleum Geologists Bulletin, v. 58, p. 685–707.

Tennyson, R., Terry, J., Brahana, V., Hays, P., and Pollack, E., 2008, Tectonic control of hypogene speleogenesis in the southern Ozarks— Implications for NAWQA and beyond, *in* Kuniansky, E.L., ed., U.S. Geological Survey Karst Interest Group proceedings, Bowling Green, Kentucky, May 27–29, 2008: U.S. Geological Survey Scientific Investigations Report 2008–5023, p. 37–46.

Turner, K.J., and Hudson, M.R., 2010, Geologic map of the Maumee quadrangle, Searcy and Marion Counties, Arkansas: U.S. Geological Survey Scientific Investigations Map 3134, scale 1:24,000, http://pubs.usgs.gov/sim/3134/.

Turner, K.J., Hudson, M.R., Murray, K.E., and Mott, D.N., 2007, Three-dimensional geologic framework model for a karst aquifer system, Hasty and Western Grove quadrangles, northern Arkansas: U.S. Geological Survey Scientific Investigations Report 2007–5095, 12 p., http://pubs.usgs.gov/sir/2007/5095/

Zachry, D.L., 1977, Stratigraphy of middle and upper Bloyd strata (Pennsylvanian, Morrowan), northwestern Arkansas, *in* Sutherland, P.K., and Manger, W.L., eds., Upper Chesterian-Morrowan stratigraphy and the Mississippian-Pennsylvanian boundary in northeastern Oklahoma and northwestern Arkansas: Oklahoma Geological Survey Guidebook 18, p. 61–66.

# APPENDIX

Geologic Map Unit Descriptions (modified from Hudson and others, 2001; Hudson and Murray, 2003, 2004; Hudson and others, 2006; Hudson and Turner, 2009).

**Qal Alluvial deposits (Quaternary)—** Unconsolidated sand and gravel. Active-channel gravel deposits are composed of subangular to rounded Paleozoic rock clasts of mixed lithology along drainages, and interspersed with bedrock exposures (not mapped). Low terraces of light-brown, fine sand are locally present adjacent to creeks. As thick as 10 ft

**Qty Younger terrace and channel alluvial deposits (Quaternary)—**Unconsolidated sand and gravel of Buffalo River. Terrace deposits are principally light-brown fine sand and have smooth upper surfaces that are about 20 ft above the base-flow level of river. Channel gravel is subangular to rounded and composed of Paleozoic sandstone, chert, dolostone, and limestone; unmapped bedrock exposures are interspersed along channel. Deposits are as thick as 20 ft

**Qto Older terrace and alluvial deposits (Quaternary)—**Unconsolidated gravel, sand, and silt deposits adjacent to Buffalo River. Deposits are principally a lag of brown-weathered, subrounded to rounded cobbles of Paleozoic sandstone and chert in a reddish-brown sandy to silty matrix. Fine sand, silt, and clay locally overlie cobble deposits. Deposits are 80–120 ft above base-flow level of river. Deposits are as thick as 20 ft

**Qc Colluvial deposits (Quaternary)—** Unconsolidated deposits of subrounded to angular blocks as large as 20 ft in diameter, commonly in an orange-brown silty clay matrix. Blocks are mostly derived from the basal sandstone of the upper part of the Bloyd Formation (ℙbu) or the Cane Hill Member (ℙhc) of the Hale Formation. Deposits have fan-like morphology and were mapped where sufficiently thick to mask typical ledge-flat topography of underlying bedrock. Smaller, thinner colluvial deposits elsewhere were not mapped. Thickness uncertain but probably more than 10 ft

**Bloyd Formation (Lower Pennsylvanian, Morrowan)—**Interbedded sequence of sandstone, siltstone, shale, and thin limestone beds separated into lower and upper intervals. As much as 400-ft thick

**ℙbu Upper part—**Dominantly sandstone with interbedded siltstone and shale. Upper part of sequence contains dark-gray to black shale and siltstone beds that form topographic flats interbedded with sandstone beds that form ledges. Upper sandstone intervals are 5–20 ft thick and vary from orange-brown, fine to coarse grained with local quartz pebbles, and medium to thick planar bedded to crossbedded to tan or olive, fine to very fine grained, ripple cross-laminated to planar bedded. Base of unit is crossbedded sandstone as thick as 80 ft that forms prominent cliffs in area. This basal sandstone is white to light-brown, fine- to medium-grained quartz arenite that has a sharp erosional base and is commonly a composite of several tabular and trough-crossbed sets. Sandstone contains local concentrations of white quartz pebbles and casts of wood fragments. Rocks of the upper part of the Bloyd Formation were originally assigned to the Winslow Formation by Purdue and Miser (1916). Zachry (1977) concluded that the basal sandstone was a time-equivalent unit with the Woolsey Member of the Bloyd Formation farther west and designated it informally as the "middle Bloyd sandstone." Thickness is 200–300 ft

**Pbl Lower part**—Sequence of predominantly dark-gray to black shale and siltstone with thin beds of sandstone and limestone; poorly exposed under moderate to steep hillslopes. Siltstone is medium to dark gray, fissile, and moderately bioturbated. Medium beds of rippled, very fine grained sandstone are locally exposed as ledges. Coarse bioclastic limestone is reddish gray and weathers dark brown but is poorly exposed and mostly observed in loose blocks in lower part of sequence. Thickness 90–120 ft

**Hale Formation (Lower Pennsylvanian, Morrowan)**—Interbedded sequence of sandstone, siltstone, shale, and thin limestone. Thickness 100–180 ft

**Phg Prairie Grove Member**—Brown to reddish-brown, fine- to medium-grained, thick-bedded, calcite-cemented sandstone. Locally contains quartz pebbles in its base and thin interbeds of fossiliferous limestone. Beds are planar or crossbedded, and crossbeds may have bi-directional dips. Weathered sandstone forms rounded surfaces. Sandstone forms steep hillslopes or ledges that are commonly covered by slope debris from overlying units. Thickness 10–60 ft

**Phc Cane Hill Member**—Interbedded sequence of shale, siltstone, and sandstone. Most of unit is poorly exposed dark-gray fissile shale and thin-bedded siltstone that form gentle to moderately steep slopes. Upper part locally contains rippled, thin-bedded, very fine and fine-grained sandstone intervals as thick as 5 ft. Lower part includes 10-ft-thick sandstone interval that changes downward from olive-brown, very fine-grained, thin-bedded sandstone with ripple cross-lamination to reddish-brown, medium- to thick-bedded, very fine-grained to medium-grained sandstone with trough crossbeds. Lower sandstone contains sparse casts of wood fragments and conglomerate lenses with quartz pebbles and angular to subrounded clasts of shale, siltstone, and sandstone. Unit unconformably overlies the Pitkin Limestone. Thickness 80–160 ft

**Mp Pitkin Limestone (Upper Mississippian, Chesterian)**—Generally medium- to dark-gray, fetid limestone. Limestone in medium to thick wavy beds varies from micrite at base to coarse grained and locally oolitic near top. Limestone beds may contain abundant crinoids, brachiopods, corals, and bryozoan *Archimedes*. Basal contact with the Fayetteville Shale is conformable, although rarely exposed. The Pitkin generally forms a prominent ledge or steep slope. Thickness 40–100 ft

**Fayetteville Shale (Upper Mississippian, Chesterian)**—Fine-grained sandstone and siltstone of the Wedington Sandstone Member that grades downward into main body of black, slope-forming shale. Thickness 250–390 ft

**Mfw Wedington Sandstone Member**—Brown to light-gray, well-indurated, calcite-cemented sandstone and siltstone. The Wedington caps a steep slope and is commonly separated from the overlying Pitkin Limestone by a topographic bench. Sandstone is very fine grained and is present in thick to thin planar beds with internal parallel laminations and locally developed low-angle crossbeds. Sandstone grades downward into siltstone beds that are ripple cross-laminated and bioturbated. The Wedington grades downward into main body. Thickness 20–40 ft

**Mf Main body**—Below the Wedington Sandstone Member, the Fayetteville Shale is poorly exposed black shale that locally contains thin, rippled beds of olive-brown siltstone near its top. The lower part of the Fayetteville crops out along stream gullies where it consists of black fissile shale that may contain medium- to light-gray, fetid septarian concretions as large as 2 ft in diameter. The Fayetteville Shale is susceptible to landslides. Thickness 230–350 ft

**Mbv Batesville Sandstone (Upper Mississippian, Chesterian)**—Fine-grained to very fine grained, light- to medium-brown, calcite-cemented sandstone with sparse interbedded limestone. Thin to medium beds are typically parallel laminated; low-angle crossbeds common in upper part of unit. Sandstone commonly contains burrows on bedding plane surfaces. Sandstone breaks into thin flat blocks. One or more discontinuous, 1- to 3-ft-thick, medium- to dark-gray, fetid, fossiliferous limestone beds are locally interbedded with sandstone; limestone beds contain crinoids and brachiopods. Both sandstone and limestone beds may contain 2- to 10-mm-diameter oxidized pyrite framboids that weather to reddish-brown spheres. Where stripped of the overlying Fayetteville Shale, the top of the Batesville forms topographic flats. Unit hosts sinkholes formed by collapse into dissolution cavities in the underlying Boone Formation. Thickness 40–60 ft

**Boone Formation (Upper to Lower Mississippian)**—Limestone and cherty limestone of main body that grade into the basal St. Joe Limestone Member. The Boone Formation is a common host of caves and sinkholes. Total thickness is 380–400 ft

**Mb Main body (Upper to Lower Mississippian, Meramecian to Osagean)**—Medium- to thick-bedded, chert-bearing bioclastic limestone. Limestone is light to medium gray on fresh surfaces and generally coarsely crystalline with interspersed crinoid ossicles. A 1- to 3-ft-thick bed of oolitic limestone is common in upper part of unit. Beds of dense, fine-grained limestone are present in upper one-third of unit. Beds are typically parallel planar to wavy, but channel fills are locally present in lower part of unit. Chert content varies vertically and laterally within the Boone and forms lenticular to anastomosing lenses. Chert-rich horizons are generally poorly exposed but produce abundant float of white-weathered chert on hillslopes. Chert in uppermost part of unit contains prominent brachiopod molds. Thickness 310–350 ft

**Mbs St. Joe Limestone Member (Lower Mississippian, Osagean to Kinderhookian)**—Thin- and wavy-bedded bioclastic limestone containing ubiquitous 3- to 6-mm-wide crinoid fragments in a finely crystalline matrix. Limestone is commonly pink to red on fresh surfaces due to hematite in matrix, but color varies depending on hematite concentration. Lenticular red to pink chert nodules are locally present in upper part. Contact with overlying main body of Boone Formation is gradational and marked by increase in chert and thicker beds upward. Middle to lower part of St. Joe Limestone Member contains greenish-gray shale interbeds. Base of St. Joe is a 0.3- to 8-ft-thick, very fine to fine-grained, moderately sorted, tan sandstone containing phosphate pebbles. Total thickness of member is 30–50 ft

**Of Fernvale Limestone (Upper Ordovician)**—Thick- to massive-bedded, coarsely crystalline bioclastic limestone. Limestone is light gray to pink on fresh surfaces and contains abundant 3- to 10-mm-wide cylindrical to barrel-shaped crinoid ossicles in a coarsely crystalline matrix. Thickness 10–20 ft

**Opl Plattin Limestone (Middle Ordovician)**—Thin- to medium-planar-bedded, fine-grained, dense limestone locally interbedded with calcarenite. Characteristically, dense, very fine grained, medium-gray limestone that breaks with conchoidal fracture and weathers to very light gray, thin, tabular blocks. Locally in upper parts there are interbeds of laminated, tan to light-gray, fine-grained to very fine grained, limey to dolomitic calcarenite as thick as 10

ft; calcarenite has fetid odor when freshly broken. Thickness 10–100 ft

**Osp St. Peter Sandstone (Middle Ordovician)**—Very fine grained, grayish-yellow sandstone interbedded with blue-green siltstone, shale, and sparse dolostone. Sandstone is calcite-cemented arenite containing well-rounded quartz grains. Upper part of unit, as thick as 25 ft, is sandstone interbedded with blue-green fissile shale, thin-bedded siltstone, and sparse 0.5- to 3-ft-thick beds of brown-gray, fine crystalline dolostone; shale, siltstone, and dolostone; this interval is commonly marked by zone of slumped sandstone blocks on hill slopes. Main part of unit contains a sequence of 1- to 2-ft-thick sandstone beds overlying a massive basal sandstone ledge as thick as 20 ft. Bioturbation is common in most sandstone beds and includes cylindrical burrows (*Skolithos*) perpendicular to bedding that weather to distinctive straw-like forms. Basal sandstone is crossbedded, contains inclined fractures, and discordantly overlies Everton Formation. Thickness ranges from 0–70 ft

**Everton Formation (Middle Ordovician)**—Interbedded dolostone, limestone, and sandstone sequence, divided into upper and lower parts. Unit is 250–450 ft thick

**Oeu Upper part**—Interbedded dolostone, sandstone, and limestone. Limestone is finely crystalline, light gray, thick bedded, and locally fossiliferous, and is present in intervals as thick as 20 ft. Dolostone is light to dark gray, finely to medium crystalline, laminated, and medium to thick bedded commonly containing sandstone stringers. Sandstone is arenite composed of well-sorted, well-rounded, fine to medium quartz grains. Sandstone is light tan to white in planar, medium to thick beds and is variably cemented by dolomite or calcite; sandstone locally grades into sandy dolostone or limestone. Limestone interbedded with limy sandstone and lesser dolostone at top of unit comprises Jasper Member of Glick and Frezon (1953). Dolostone and dolomitic sandstone underlying Jasper Member make up most of upper part of Everton. Base of upper part of Everton is marked by base of Newton Sandstone Member (McKnight, 1935), a planar, medium-bedded quartz arenite composed of well-sorted, well-rounded, fine to medium quartz grains. Newton Sandstone Member is 10–90 ft thick. Upper part of Everton Formation is 180–200 ft thick

**Oel Lower part**—Interbedded dolostone, sandstone, and limestone. Limestone is restricted to upper part of unit and is interbedded with dolostone and sandstone. Limestone is light gray, finely crystalline, thin to medium bedded, commonly laminated, and contains sparse stromatolite hemispheres as large as 1 ft in diameter. Dolostone, present throughout unit, is light gray to grayish brown, finely to medium crystalline, thin to thick bedded, and commonly laminated, and it commonly contains stromatolites. Sandstone is light-tan to white quartz arenite, composed of well-sorted, well-rounded, fine to medium grains, variably cemented by dolomite or calcite. Sandstone intervals, composed of planar, thin to medium beds as thick as 6 ft, are present throughout unit. Sandstone also forms lenses and stringers within limestone and dolostone intervals. Light- to dark-brownish-gray dolostone and sandy dolostone interspersed with sandstone are predominant in lower part of unit. Collapse breccia with clasts of planar laminated dolostone and dolomitic sandstone is present. Unit is 140–250 ft thick

**Op Powell Dolomite (Lower Ordovician)**—Fine-grained, thin-bedded, argillaceous dolostone. Yellowish gray on fresh surfaces. Weathers to thin, platy blocks. As much as 65 ft of Powell is exposed along the Buffalo River where it is exposed in the eroded upthrown side of Cane Branch monocline near Mt. Hersey (Glick and Frezon, 1953). Regionally, formation thickness ranges from 40–200 ft